AF594995

Microwave Remote Sensing of Soil Moisture

Microwave Remote Sensing of Soil Moisture

Editors

Jiangyuan Zeng
Jian Peng
Wei Zhao
Chunfeng Ma
Hongliang Ma

Basel • Beijing • Wuhan • Barcelona • Belgrade • Novi Sad • Cluj • Manchester

Editors
Jiangyuan Zeng
Aerospace Information Research Institute, Chinese Academy of Sciences
Beijing, China

Jian Peng
Helmholtz Centre for Environmental Research—UFZ
Leipzig, Germany

Wei Zhao
Institute of Mountain Hazards and Environment, Chinese Academy of Sciences
Chengdu, China

Chunfeng Ma
Northwest Institute of Eco-Environment and Resources, Chinese Academy of Sciences
Lanzhou, China

Hongliang Ma
INRAE, UMR 1114 EMMAH, UMT CAPTE
Avignon, France

Editorial Office
MDPI
St. Alban-Anlage 66
4052 Basel, Switzerland

This is a reprint of articles from the Special Issue published online in the open access journal *Remote Sensing* (ISSN 2072-4292) (available at: https://www.mdpi.com/journal/remotesensing/special_issues/microwave_hydrology).

For citation purposes, cite each article independently as indicated on the article page online and as indicated below:

Lastname, A.A.; Lastname, B.B. Article Title. *Journal Name* **Year**, *Volume Number*, Page Range.

ISBN 978-3-0365-9094-3 (Hbk)
ISBN 978-3-0365-9095-0 (PDF)
doi.org/10.3390/books978-3-0365-9095-0

Contents

About the Editors

Jiangyuan Zeng

Jiangyuan Zeng, Ph.D., is currently an Associate Professor with the State Key Laboratory of Remote Sensing Science, Aerospace Information Research Institute, Chinese Academy of Sciences (CAS), Beijing, China. His research interests include microwave remote sensing of geophysical parameters (particularly soil moisture), hydrological applications of satellite remote sensing, and bistatic scattering of land surfaces. Dr. Zeng received several well-recognized international awards, including the International Society for Photogrammetry and Remote Sensing (ISPRS) Best Young Author Award in 2020, the Young Scientist Award (including Cash Award) from the Progress in Electromagnetics Research Symposium (PIERS) in 2018, and the Young Scientist Award from the International Union of Radio Science (URSI) in 2017. He has been an Editorial Board Member of Remote Sensing of Environment (RSE) since 2020, an Associate Editor of IEEE Journal of Selected Topics in Applied Earth Observations and Remote Sensing (IEEE JSTARS) since 2021, and a Youth Editor of The Innovation since 2022. He also served as an Executive Guest Editor for the RSE Special Issue of "Emerging remote sensing techniques for hydrological applications", the leading Guest Editor for the Remote Sensing Special Issue of "Microwave remote sensing of soil moisture", and a Guest Editor for the GIScience & Remote Sensing Special Issue of "Soil moisture for earth ecosystems: methods, products, and applications". He is an excellent member of the Youth Innovation Promotion Association, CAS, and a senior member of IEEE.

Jian Peng

Jian Peng (Professor) is an Earth system scientist and Head of the Department of Remote Sensing at the UFZ in Leipzig. He is also a full professor for Hydrology and Remote Sensing at the University of Leipzig. His research interests are the quantitative retrieval of land surface parameters from remote sensing data, the assimilation of remote sensing data into climate and land surface process models, understanding land-atmosphere interactions using earth system models and observational data, and quantification of climate change impact on water resources. His research in particular focuses on estimation of high-resolution land surface water and energy fluxes from satellite observations, and the investigation of hydrological and climatic extremes as well as their impacts on ecosystems. He has been involved in various national and international research projects funded by, e.g., ESA, EU, UK space agency, NERC, and DFG. He has received numerous international awards, most recently in 2019 the Remote Sensing Young Investigator Award of the Swiss scientific publisher MDPI. He is Co-Editor-in-Chief of the journal "Geoscience Data Journal".

Wei Zhao

Dr. Wei Zhao received his B.S. degree in Geographic Information System from Beijing Normal University, Beijing, China, in 2006 and his Ph.D. degree in Cartography and Geographic Information System from Institute of Geographic Sciences and Natural Resources Research, Chinese Academy of Sciences (CAS), Beijing, China, in 2012. During 2009 and 2011, He has been funded by China Scholarship Council and conducted two years visit in University of Strasbourg, France. Currently, he is a professor with the Institute of Mountain Hazards and Environment (IMHE), CAS, Chengdu, China. He devotes himself to mountain remote sensing, with special focuses on quantitative estimation methods for mountain surface soil moisture, mountain thermal environment monitoring, and mountain ecosystem and hazards monitoring and assessment. Currently, he has published more than 100 papers in peer-reviewed journals such as Remote Sensing of Environment, ISPRS Journal of Photogrammetry and Remote Sensing, IEEE Transactions on Geoscience and Remote Sensing, Journal of Geophysical Research-Atmospheres, and Journal of Hydrology. He serves as an editorial board member of "Mountain Research and Development", "Journal of Mountain Science" and a Youth Editor of "The Innovation". Based his scientific contribution, He was elected as the Excellent Young Scientists Fund awarded by the National Natural Science Foundation of China (NSFC).

Chunfeng Ma

Chunfeng Ma, Ph.D., is currently an Associate Professor with the Northwest Institute of Eco-Environment and Resources, Chinese Academy of Sciences, Lanzhou, China. His research interests include remote sensing of agriculture and ecohydrology, water and food security. He is a senior member of IEEE and a secretary of the remote sensing panel in the China Society of Cryospheric Science.

Hongliang Ma

Hongliang Ma received the B.S. degree in GIS from Huazhong Agriculture University in 2016, and the Ph.D. degree in cartography and geographic information engineering in LIESMARS, Wuhan University in 2022, respectively, in which joint as the visiting PhD in INRAE-Bordeaux, France for one year (supervisor: Jean-Pierre Wigneron). He is currently working as Post-Doc in INRAE-Avignon, France (supervisor: Frederic Baret and Marie Weiss). His study interests include remote sensing (both microwave and optical) of soil moisture and vegetation water/biomass, and global water-carbon couplings through vegetation. Dr. Ma served as the guest editor for Guest Editor for the GIScience & Remote Sensing Special Issue of "Soil moisture for earth ecosystems: methods, products, and applications", Remote Sensing Special Issues on "Microwave Remote Sensing of Soil Moisture" and "Spatial or Temporal Analysis of Soil Moisture from Space", respectively.

Preface

Soil moisture is well recognized as a pivotal parameter linking the water, energy, and carbon cycles. Active and passive microwave remote sensing has been well-recognized as the most promising means to infer soil moisture spatially and temporally. Active microwave remote sensing, particularly the synthetic aperture radar (SAR), has a much finer spatial resolution than passive sensors but suffers more from geometrical features of the scene (e.g., surface roughness, vegetation, and topography). Passive microwave remote sensing has higher sensitivity to soil moisture than active radar but is limited by its coarse spatial resolution. Moreover, active and passive microwave signals respond differently to soil and vegetation parameters and thus can provide complementary information for each other.

Over the past several decades, great progress has been made in microwave remote sensing of soil moisture. Several field or aircraft experiments (e.g., SGP, SMEX, HiWATER, SMAPEx1-5, and SMAPVEX) have been organized to support the assessment and refinement of active and passive microwave soil moisture retrieval algorithms. At the same time, a number of microwave spaceborne satellites/sensors have been successfully launched to provide valuable opportunities to obtain soil moisture data at various spatial scales from meters to tens of kilometers. These include passive microwave instruments, such as the multi-frequency AMSR-E/2 (2002-), FY-3 MWRI (2008-), L-band SMOS (2009-), and SMAP (2015-), as well as active microwave instruments, such as the scatterometer-based Metop/ASCAT series (2006-), monostatic ALOS-2 (2014-), Sentinel-1 (2014-), and Gaofen-3 (2016-), bistatic CYGNSS (2016-), and the P-band Biomass (planned launch in the next few years). All of these open a wide range of possibilities to estimate soil moisture at regional and global scales.

In this context, this book aims to present the most advanced theories, models, algorithms, and products related to microwave remote sensing of soil moisture. The book is aimed at a wide range of readers, from graduate students, university faculty members, and scientists, to policy makers and managers.

We acknowledge the funding from the National Natural Science Foundation of China (Grant No. 41971317, 42271402, 42222109) and the Youth Innovation Promotion Association CAS (Grant No. Y2022050).

Jiangyuan Zeng, Jian Peng, Wei Zhao, Chunfeng Ma, and Hongliang Ma
Editors

remote sensing

Editorial

Microwave Remote Sensing of Soil Moisture

Jiangyuan Zeng [1,*], Jian Peng [2,3], Wei Zhao [4], Chunfeng Ma [5] and Hongliang Ma [6]

1 State Key Laboratory of Remote Sensing Science, Aerospace Information Research Institute, Chinese Academy of Sciences, Beijing 100101, China
2 Department of Remote Sensing, Helmholtz Centre for Environmental Research—UFZ, 04318 Leipzig, Germany; jian.peng@ufz.de
3 Remote Sensing Centre for Earth System Research—RSC4Earth, Leipzig University, 04103 Leipzig, Germany
4 Institute of Mountain Hazards and Environment, Chinese Academy of Sciences, Chengdu 610299, China; zhaow@imde.ac.cn
5 Heihe Remote Sensing Experimental Research Station, Key Laboratory of Remote Sensing of Gansu Province, Northwest Institute of Eco-Environment and Resources, Chinese Academy of Sciences, Lanzhou 730000, China; machf@lzb.ac.cn
6 INRAE, UMR 1114 EMMAH, UMT CAPTE, Provence-Alpes-Côte d'Azur, F-84000 Avignon, France; hongliang.ma@inrae.fr
* Correspondence: zengjy@radi.ac.cn

1. Introduction

Soil moisture is an important component of the global terrestrial ecosystem and has been recognized as an Essential Climate Variable (ECV) by the Global Climate Observing System (GCOS) [1]. The change in soil moisture content is a critical representation and driving factor of the terrestrial water cycle which has a significant impact on the spatial distribution and intensity of land evapotranspiration, rainfall, and runoff processes, and thus affects a series of important issues related to sustainable development, such as water resources and food security, drought and flood disasters, soil erosion, and ecological degradation [2–4]. Therefore, obtaining accurate spatiotemporal distribution of soil moisture is both necessary and highly interesting.

Microwave remote sensing, in both active and passive forms, is one of the most effective ways to detect soil moisture content on a large scale. Over the past few decades, significant efforts have been made to develop empirical/semi-empirical/theoretical models, retrieval algorithms, downscaling methods, and validation strategies related to the microwave remote sensing of soil moisture [5–12]. Following the turn of the century, a series of microwave-based satellites/sensors have been successfully launched (Figure 1), such as the passive Soil Moisture and Ocean Salinity (SMOS), Advanced Microwave Scanning Radiometer-Earth Observing System (AMSR-E), AMSR2, Fengyun (FY)-3B/C/D, the active Advanced Scatterometer (ASCAT), Sentinel-1, Advanced Land Observing Satellite-2 (ALOS-2), Gaofen-3 (GF-3), and the active-passive Soil Moisture Active Passive (SMAP), and Aquarius. Therefore, satellite soil moisture products have become increasingly abundant, greatly promoting the various application of satellite soil moisture datasets [13–15]. Despite numerous studies and achievements in this field, great challenges remain, such as the spatial resolution, retrieval accuracy, and validation strategies related to satellite soil moisture datasets.

This Special Issue aims to present the most recent scientific advances in the theories, models, algorithms, and products associated with the microwave remote sensing of soil moisture. Ten articles are published in this Special Issue, covering research progress on the following topics: (1) downscaling passive microwave-based soil moisture products, (2) estimating soil moisture from active microwave observations, (3) presenting some new algorithms (freeze–thaw state detection algorithm) and models (soil dielectric models) that are closely related to the microwave remote sensing of soil moisture, (4) evaluating

Citation: Zeng, J.; Peng, J.; Zhao, W.; Ma, C.; Ma, H. Microwave Remote Sensing of Soil Moisture. *Remote Sens.* **2023**, *15*, 4243. https://doi.org/10.3390/rs15174243

Received: 3 August 2023
Revised: 18 August 2023
Accepted: 24 August 2023
Published: 29 August 2023

microwave-based soil moisture products, (5) reviewing the state-of-the-art techniques and algorithms used to estimate and improve the quality of soil moisture estimations.

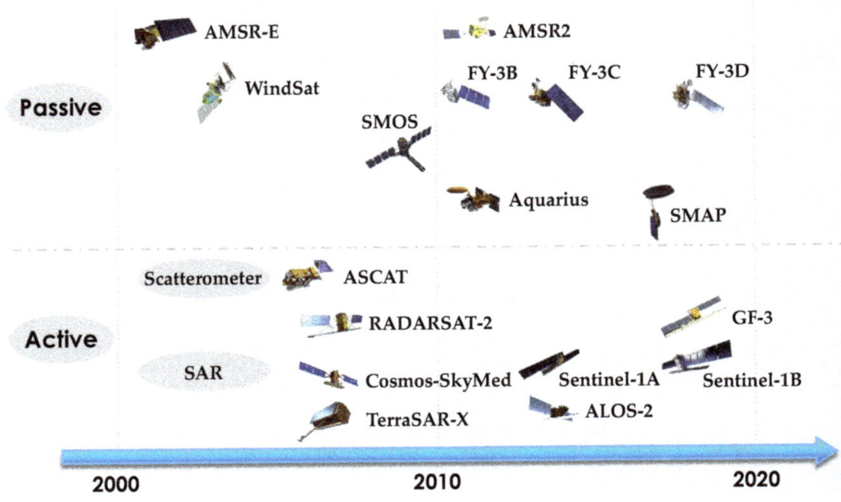

Figure 1. The primary microwave satellites/sensors that have been used to estimate soil moisture since the 2000. Note that both active and passive microwave sensors are mounted on the SMAP and Aquarius missions.

2. Highlights of the Research Articles

Brightness temperature has strong sensitivity to soil moisture [16], making passive microwave remote sensing a valuable tool to estimate soil moisture globally [17]. A number of passive microwave-based soil moisture products, such as SMAP, SMOS, AMSR2, FY-3, are available to the public. However, the coarse spatial resolution of such products (often dozens of kilometers) limits their various applications in the field and at a local scale. Three papers published in this Special Issue address this issue. Zhao et al. [18] evaluated four commonly used auxiliary variables, including NDVI (Normalized Difference Vegetation Index), LST (Land Surface Temperature), TVDI (Temperature Vegetation Dryness Index), and SEE (Soil Evaporative Efficiency), against in situ soil moisture in an arid region of China (Heihe River Basin). They found that SEE was an optimal auxiliary variable for the scaling and mapping of soil moisture, and the combination of multiple auxiliary variables (LST, NDVI, and SEE) was recommended for improving the scaling and mapping accuracy of soil moisture. Llamas et al. [19] proposed a modular spatial inference framework, which was the foundation of a cyberinfrastructure tool named SOil MOisture SPatial Inference Engine (SOMOSPIE), to downscale ESA CCI soil moisture products to 1 km using terrain parameters and examined the skill of two modeling methods, i.e., Kernel-Weighted K-Nearest Neighbor (KKNN) and Random Forest (RF). The results indicated that the SOMOSPIE framework provided a feasible approach to downscaling satellite soil moisture data, and RF performed better in the cross-validation compared to the reference ESA CCI data, but as part of independent validation, KKNN had a slightly higher consistency with ground soil moisture observations. In addition, a soil moisture retrieval and spatiotemporal fusion model (SMRFM) was proposed by Jiang et al. [20] to reduce the dependence of the method on the optical/thermal infrared data. They successfully downscaled the AMSR-E soil moisture from 25 km to 1 km using the MODIS-derived soil moisture and the SMRFM over the Central Tibetan Plateau.

Compared to passive microwave remote sensing, active microwave remote sensing, e.g., the synthetic aperture radar (SAR), can provide soil moisture estimates with much finer spatial resolution but are negatively affected by the geometry of the land surface (e.g., surface roughness and vegetation structure). Therefore, it is still a challenge to obtain SM retrievals with a high accuracy via active microwave measurements. In Dong et al. [21], the response of radar signal to surface parameters was analyzed using the database simulated from the advanced integral equation model (AIEM), and soil moisture was retrieved from Sentinel-1 using empirical models and machine learning methods. It was found the machine learning algorithms performed much better than the empirical models, and the skill of the RF algorithm surpassed that of the other machine learning approaches. Two hybrid methodologies, namely improving a change detection approach with regard to vegetation, and combining a change detection approach with a neural network algorithm, were proposed and tested using Sentinel-1 and Sentinel-2 data in the study by Nativel et al. [22]. Their results indicated that using hybrid algorithms (particularly change detection via a neural network) could improve the accuracy of estimating soil moisture content.

Furthermore, previous studies generally focused on estimating soil moisture in mineral soils since the soil dielectric models used in soil moisture retrieval algorithms were usually mineral-soil-based models. Zhang et al. [23] compared the performance of nine soil dielectric models, four of which incorporate soil organic matter (SOM) in organic soil in Alaska within the framework of the SMAP single-channel algorithm at vertical polarization (SCA-V). Using the SMAP SCA-V algorithm, they reported that the Mironov 2009 and Mironov 2019 models were the best choices for mineral soils (SOM $< 15\%$) and organic soils (SOM $\geq 15\%$), respectively. Meanwhile, there are large uncertainties in soil moisture retrievals when the soil becomes frozen. Thus, soil moisture values are often masked in satellite soil moisture products such as SMAP, SMOS, and AMSR2. In Lv et al. [24], a new freeze–thaw state detection algorithm was developed based on the daily variation of the SMAP H-pol brightness temperature. The physical foundation of the algorithm lied in the fact that the difference in the microwave brightness temperature between 6 a.m. (descending overpass) and 6 p.m. (ascending overpass) was relatively small over frozen soil owing to the large penetration depth, resulting in a higher temperature stability in deeper soils.

Moreover, microwave-based soil moisture products have been extensively evaluated in previous studies using in situ observations. However, most research has ignored the possible vertical mismatch between in situ data and satellite retrievals. Yang et al. [25] investigated the stratification characteristics of in situ soil moisture and assessed SMOS L2, SMOS-IC SMAP L2, SMAP L4 soil moisture products using multilayer in situ data (5, 10, 20, 5.08, 10.16, 20.32 cm) collected from the International Soil Moisture Network (ISMN). They discovered that (1) the differences in soil moisture content between layers were close to or even beyond the 0.04 $m^3 m^{-3}$ nominal retrieval accuracy of SMOS and SMAP; (2) satellite products showed the highest correlation and the smallest bias with 5/5.08 cm in situ data, and the SMAP L4 product was closest to in situ measurements compared to the other datasets.

In addition, a good summary of the state-of-the-art progress in the microwave remote sensing of soil moisture is of great interest to the soil moisture research community. Two review papers were published in this Special Issue. In Wu and Wen [26], the research progress in observing and simulating L-band microwave emissions, ground soil moisture measurements, and soil moisture retrieval from L-band passive microwave observations over the Third Pole, i.e., the Tibetan Plateau, was summarized. Moreover, Liu and Yang [27] presented a systematic review of the primary methodologies for detecting soil moisture content and the current approaches used to enhance the quality of soil moisture products.

3. Conclusions and Outlook

This Special Issue entitled "Microwave Remote Sensing of Soil Moisture" covers a wide range of research on the satellite detection of soil moisture, including developing retrieval algorithms and downscaling methods, comparing soil dielectric models, freeze–thaw state detection approaches, and satellite soil moisture products. The theories, methods, validations, and applications of satellite soil moisture datasets are reviewed in detail. Notably, there is much room for improvement regarding algorithms and datasets related to the microwave remote sensing of soil moisture and their applications in various disciplines. The selected papers should help the soil moisture research community to better understand the current development status and future trends of microwave remote sensing of soil moisture.

The following aspects could be considered in future research: (1) developing new methods (e.g., upscaling method) for validating satellite soil moisture products, particularly in regions with high spatial heterogeneity; (2) developing new technologies to identify and suppress the influence of radio frequency interference and open water to further improve the quality of microwave signals used for estimating soil moisture; (3) combining active and passive microwave, multi-polarization, and multi-frequency observations to alleviate ill-posed problems, and improve the spatial resolution of soil moisture; (4) developing P-band related theoretical technologies to obtain deeper soil moisture and soil moisture profile information; (5) using bistatic radar (e.g., upcoming Tandem-L) to decouple the effects of soil moisture and other perturbing parameters (e.g., surface roughness) to obtain more reliable soil moisture data with a high spatial resolution.

Author Contributions: Conceptualization, J.Z., J.P., W.Z., C.M. and H.M.; formal analysis, J.Z., J.P., W.Z., C.M. and H.M.; investigation, JZ., J.P., W.Z., C.M. and H.M.; writing—original draft preparation, J.Z.; writing—review and editing, J.P., W.Z., C.M. and H.M.; funding acquisition, J.Z. and C.M. All authors have read and agreed to the published version of the manuscript.

Funding: This research was funded by the National Natural Science Foundation of China (Grant No. 41971317, 42271402) and the Youth Innovation Promotion Association CAS (Grant No. Y2022050).

Conflicts of Interest: The authors declare no conflict of interest.

References

1. Mason, P.J.; Zillman, J.W.; Simmons, A.; Lindstrom, E.J.; Harrison, D.E.; Dolman, H.; Bojinski, S.; Fischer, A.; Latham, J.; Rasmussen, J.; et al. *Implementation Plan for the Global Observing System for Climate in Support of the UNFCCC (2010 Update)*; Word Meteorological Organization (WMO): Geneva, Switzerland, 2010.
2. Brocca, L.; Zhao, W.; Lu, H. High-resolution observations from space to address new applications in hydrology. *Innovation* **2023**, *4*, 100437. [CrossRef] [PubMed]
3. Yang, H.; Wang, Q.; Zhao, W.; Tong, X.; Atkinson, P.M. Reconstruction of a global 9 km, 8-day SMAP surface soil moisture dataset during 2015–2020 by spatiotemporal fusion. *J. Remote Sens.* **2022**, *2022*, 9871246. [CrossRef]
4. Wang, F.; Harindintwali, J.D.; Wei, K.; Shan, Y.; Mi, Z.; Costello, M.J.; Grunwald, S.; Feng, Z.; Wang, F.; Guo, Y.; et al. Climate change: Strategies for mitigation and adaptation. *Innov. Geosci.* **2023**, *1*, 100015. [CrossRef]
5. Chaubell, M.J.; Yueh, S.H.; Dunbar, R.S.; Colliander, A.; Chen, F.; Chan, S.K.; Entekhabi, D.; Bindlish, R.; O'Neill, P.E.; Asanuma, J.; et al. Improved SMAP dual-channel algorithm for the retrieval of soil moisture. *IEEE Trans. Geosci. Remote Sens.* **2020**, *58*, 3894–3905. [CrossRef]
6. Zeng, J.; Chen, K.S.; Cui, C.; Bai, X. A physically based soil moisture index from passive microwave brightness temperatures for soil moisture variation monitoring. *IEEE Trans. Geosci. Remote Sens.* **2020**, *58*, 2782–2795. [CrossRef]
7. Bauer-Marschallinger, B.; Freeman, V.; Cao, S.; Paulik, C.; Schaufler, S.; Stachl, T.; Modanesi, S.; Massari, C.; Ciabatta, L.; Brocca, L.; et al. Toward global soil moisture monitoring with Sentinel-1: Harnessing assets and overcoming obstacles. *IEEE Trans. Geosci. Remote Sens.* **2019**, *57*, 520–539. [CrossRef]
8. Kim, S.B.; Van Zyl, J.J.; Johnson, J.T.; Moghaddam, M.; Tsang, L.; Colliander, A.; Dunbar, R.S.; Jackson, T.J.; Jaruwatanadilok, S.; West, R.; et al. Surface soil moisture retrieval using the L-band synthetic aperture radar onboard the soil moisture active–passive satellite and evaluation at core validation sites. *IEEE Trans. Geosci. Remote Sens.* **2017**, *55*, 1897–1914. [CrossRef]
9. Ma, C.; Li, X.; Chen, K.S. The discrepancy between backscattering model simulations and radar observations caused by scaling issues: An uncertainty analysis. *IEEE Trans. Geosci. Remote Sens.* **2019**, *57*, 5356–5372. [CrossRef]
10. Peng, J.; Loew, A.; Zhang, S.; Wang, J.; Niesel, J. Spatial downscaling of satellite soil moisture data using a vegetation temperature condition index. *IEEE Trans. Geosci. Remote Sens.* **2015**, *54*, 558–566. [CrossRef]
11. Zhao, W.; Sánchez, N.; Lu, H.; Li, A. A spatial downscaling approach for the SMAP passive surface soil moisture product using random forest regression. *J. Hydrol.* **2018**, *563*, 1009–1024. [CrossRef]
12. Ma, H.; Li, X.; Zeng, J.; Zhang, X.; Dong, J.; Chen, N.; Fan, L.; Sadeghi, M.; Frappart, F.; Liu, X.; et al. An assessment of L-band surface soil moisture products from SMOS and SMAP in the tropical areas. *Remote Sens. Environ.* **2023**, *284*, 113344. [CrossRef]

13. Peng, C.; Zeng, J.; Chen, K.S.; Li, Z.; Ma, H.; Zhang, X.; Shi, P.; Wang, T.; Yi, L.; Bi, H. Global spatiotemporal trend of satellite-based soil moisture and its influencing factors in the early 21st century. *Remote Sens. Environ.* **2023**, *291*, 113569. [CrossRef]
14. Jung, M.; Reichstein, M.; Ciais, P.; Seneviratne, S.I.; Sheffield, J.; Goulden, M.L.; Bonan, G.; Cescatti, A.; Chen, J.; De Jeu, R.; et al. Recent decline in the global land evapotranspiration trend due to limited moisture supply. *Nature* **2010**, *467*, 951–954. [CrossRef]
15. Rigden, A.J.; Mueller, N.D.; Holbrook, N.M.; Pillai, N.; Huybers, P. Combined influence of soil moisture and atmospheric evaporative demand is important for accurately predicting US maize yields. *Nat. Food* **2020**, *1*, 127–133. [CrossRef] [PubMed]
16. Ma, C.; Li, X.; Wang, J.; Wang, C.; Duan, Q.; Wang, W. A comprehensive evaluation of microwave emissivity and brightness temperature sensitivities to soil parameters using qualitative and quantitative sensitivity analyses. *IEEE Trans. Geosci. Remote Sens.* **2017**, *55*, 1025–1038. [CrossRef]
17. Ma, H.; Zeng, J.; Chen, N.; Zhang, X.; Cosh, M.H.; Wang, W. Satellite surface soil moisture from SMAP, SMOS, AMSR2 and ESA CCI: A comprehensive assessment using global ground-based observations. *Remote Sens. Environ.* **2019**, *231*, 111215. [CrossRef]
18. Zhao, Z.; Jin, R.; Kang, J.; Ma, C.; Wang, W. Using of Remote Sensing-Based Auxiliary Variables for Soil Moisture Scaling and Mapping. *Remote Sens.* **2022**, *14*, 3373. [CrossRef]
19. Llamas, R.M.; Valera, L.; Olaya, P.; Taufer, M.; Vargas, R. Downscaling Satellite Soil Moisture Using a Modular Spatial Inference Framework. *Remote Sens.* **2022**, *14*, 3137. [CrossRef]
20. Jiang, H.; Chen, S.; Li, X.; Wu, J.; Zhang, J.; Wu, L. A Novel Method for Long Time Series Passive Microwave Soil Moisture Downscaling over Central Tibet Plateau. *Remote Sens.* **2022**, *14*, 2902. [CrossRef]
21. Dong, L.; Wang, W.; Jin, R.; Xu, F.; Zhang, Y. Surface Soil Moisture Retrieval on Qinghai-Tibetan Plateau Using Sentinel-1 Synthetic Aperture Radar Data and Machine Learning Algorithms. *Remote Sens.* **2023**, *15*, 153. [CrossRef]
22. Nativel, S.; Ayari, E.; Rodriguez-Fernandez, N.; Baghdadi, N.; Madelon, R.; Albergel, C.; Zribi, M. Hybrid methodology using sentinel-1/sentinel-2 for soil moisture estimation. *Remote Sens.* **2022**, *14*, 2434. [CrossRef]
23. Zhang, R.; Chan, S.; Bindlish, R.; Lakshmi, V. A Performance Analysis of Soil Dielectric Models over Organic Soils in Alaska for Passive Microwave Remote Sensing of Soil Moisture. *Remote Sens.* **2023**, *15*, 1658. [CrossRef]
24. Lv, S.; Wen, J.; Simmer, C.; Zeng, Y.; Guo, Y.; Su, Z. A Novel Freeze-Thaw State Detection Algorithm Based on L-Band Passive Microwave Remote Sensing. *Remote Sens.* **2022**, *14*, 4747. [CrossRef]
25. Yang, N.; Xiang, F.; Zhang, H. The Characterization of the Vertical Distribution of Surface Soil Moisture Using ISMN Multilayer In Situ Data and Their Comparison with SMOS and SMAP Soil Moisture Products. *Remote Sens.* **2023**, *15*, 3930. [CrossRef]
26. Wu, X.; Wen, J. Recent Progress on Modeling Land Emission and Retrieving Soil Moisture on the Tibetan Plateau Based on L-Band Passive Microwave Remote Sensing. *Remote Sens.* **2022**, *14*, 4191. [CrossRef]
27. Liu, Y.; Yang, Y. Advances in the Quality of Global Soil Moisture Products: A Review. *Remote Sens.* **2022**, *14*, 3741. [CrossRef]

Article

The Characterization of the Vertical Distribution of Surface Soil Moisture Using ISMN Multilayer In Situ Data and Their Comparison with SMOS and SMAP Soil Moisture Products

Na Yang [1,*], Feng Xiang [2] and Hengjie Zhang [1]

[1] School of Surveying and Land Information Engineering, Henan Polytechnic University, Jiaozuo 454000, China; zhanghengjie123@foxmail.com
[2] College of Computer Science and Technology, Henan Polytechnic University, Jiaozuo 454000, China; xiangfeng821008@foxmail.com
* Correspondence: yangna@hpu.edu.cn; Tel.: +86-0391-3987661

Abstract: In this paper, we investigated the vertical distribution characteristics of surface soil moisture based on ISMN (International Soil Moisture Network) multilayer in situ data (5, 10, and 20 cm; 2, 4, and 8 in) and performed comparisons between the in situ data and four microwave satellite remote sensing products (SMOS L2, SMOS-IC, SMAP L2, and SMAP L4). The results showed that the mean soil moisture difference between layers can be −0.042~−0.024 (for the centimeter group)/−0.067~−0.044 (for the inch group) m^3/m^3 in negative terms and 0.020~0.028 (for the centimeter group)/0.036~0.040 (for the inch group) m^3/m^3 in positive terms. The surface soil moisture was found to have very significant stratification characteristics, and the interlayer difference was close to or beyond the SMOS and SMAP 0.04 m^3/m^3 nominal retrieval accuracy. Comparisons revealed that the satellite retrievals had a higher correlation with the field measurements of 5 cm/2 in, and SMAP L4 had the smallest difference with the in situ data. The mean difference caused by using 10 cm/4 in and 20 cm/8 in in situ data instead of the 5 cm/2 in data could be about −0.019~−0.018/−0.18~−0.015 m^3/m^3 and −0.026~−0.023/−0.043~−0.039 m^3/m^3, respectively, meaning that there would be a potential depth mismatch in the data validation.

Keywords: soil moisture; calibration and validation; Soil Moisture and Ocean Salinity (SMOS); Soil Moisture Active Passive (SMAP)

Citation: Yang, N.; Xiang, F.; Zhang, H. The Characterization of the Vertical Distribution of Surface Soil Moisture Using ISMN Multilayer In Situ Data and Their Comparison with SMOS and SMAP Soil Moisture Products. *Remote Sens.* **2023**, *15*, 3930. https://doi.org/10.3390/rs15163930

Academic Editor: Gabriel Senay

Received: 5 July 2023
Revised: 29 July 2023
Accepted: 3 August 2023
Published: 8 August 2023

1. Introduction

The SMOS (Soil Moisture and Ocean Salinity, ESA, November 2009) and SMAP (Soil Moisture Active Passive, NASA, January 2015) missions are dedicated to the acquisition of global soil moisture information. They both use the L band (1.4/1.41 GHz) in the mode of passive microwave remote sensing, as there would be a greater depth of penetration due to the longer wavelength [1,2]. The soil moisture products (retrievals and estimates) nominally released by SMOS and SMAP are the average soil moisture at the top of the surface, and they are conventionally compared with 5 cm in situ data [3,4]. However, the response depth of the L band is likely to vary from a very thin surface to a certain deep layer due to the variety and instability of the observing conditions in practice, which are difficult or impossible to measure accurately at present [5–7]. The depth mismatch would potentially be present in the comparisons of SMOS and SMAP soil moisture products and also in their comparisons with soil moisture field measurements, which is commonly thought to introduce uncertainties in the validation of multisource data [8–17].

From a data flow perspective, the soil moisture products from SMOS and SMAP can be considered the comprehensive results of three main processes, namely, brightness temperature (TB) observation, brightness temperature simulation, and soil moisture retrieval [18–23]. The numerical difference presented in validation and comparison can, in this context, consist of two parts. The first one would be collectively called the "retrieval error" and may be caused by upstream phases, including the instruments' performances, observing conditions, reconstruction methods, radiative transfer models and parameter settings, auxiliary information inputs, and iterative computational strategies [24–29]. The other one is generally referred to as the "verification uncertainty" and is mainly caused by the difference in scale, depth, and time between the multisource data [30–33]. To some extent, in order to accurately find out the source of the "retrieval error", further understand its propagation mechanism, and make corresponding improvements, one should first exclude the "verification uncertainty" due to the spatial and temporal mismatch of the multisource data; in other words, they must adopt a way of tracing back from the downstream stage to the upstream stage, which is exactly the opposite of the flow of data production.

Based on high-frequency in situ measurements, the soil moisture at 5 cm undergoes natural fading of a very small magnitude during the time intervals between SMOS and SMAP, with an average variation (0.003 m^3/m^3 minimum; 0.007 m^3/m^3 maximum), that is insufficient to be identified using satellites (nominal accuracy 0.04 m^3/m^3), and the temporal mismatch may not cause external uncertainty and is negligible in data validation [34]. Similarly, by using multilayer in situ data as a reference, the effect of depth mismatch on the validation of SMOS and SMAP soil moisture products can be assessed to some extent. This paper attempts to make comparisons between L band microwave remote sensing soil moisture products and in situ soil moisture measurements, and the main objectives are as follows:

- To investigate the vertical distribution characteristics of surface soil moisture, the numerical characteristics of each layer, and the similarities and differences between the layers;
- To quantify the numerical difference between satellite soil moisture retrievals and multilayer in situ measurements;
- To demonstrate the effect of the depth mismatch, the rationality of using in situ data at one depth as a reference, and the feasibility of using another depth as a substitute.

2. Materials and Methods

2.1. Data

Five datasets were selected, and the time span was set to 1 January 2015~31 December 2020. The ISMN (International Soil Moisture Network) provides multilayer in situ soil moisture measurements, which were used to study the stratification characteristics and as a reference for the comparison with satellite products. SMOS L2 and SMAP L2 (passive) products are soil moisture retrievals; SMOS-IC and SMAP L4 products can be considered independent retrievals and estimates, respectively.

2.1.1. ISMN In Situ Soil Moisture Data

The ISMN is an international collaboration to establish and maintain a global in situ soil moisture database. It brings together in situ soil moisture measurements collected and freely shared by a variety of organizations, harmonizes them in terms of units and sampling rates, applies advanced quality control, and stores them in a database [35,36]. In addition to single/multilayer soil moisture, static information (land cover, clay fraction, sand fraction, etc.) and other dynamic variables (soil temperature, air temperature, precipitation, etc.) are also included in the ISMN datasets. In general, soil moisture is quantified in terms of volumetric water content (m^3/m^3) and an hourly sampling rate.

2.1.2. SMOS L2 Soil Moisture Product

The SMOS L2 Soil Moisture User Data Product (MIR_SMUDP2) consists of swath-based retrieved information over land surfaces. The base product includes fields for soil moisture, vegetation water content, calculated brightness temperatures at 42.5 °C, and dielectric constant from pole to pole. The product is organized in the form of a Discrete Global Grid (DGG) in the ISEA 4H9 (Icosahedral Snyder Equal Area) grid projection, and the average distance between nodes is close to 15 km. The soil moisture retrievals (field: Soil_Moisture) are volumetric water content in m^3/m^3, and the accuracy requirement is set to 0.04 m^3/m^3 (i.e., 4% volumetric soil moisture) or better [37,38].

2.1.3. SMOS-IC Soil Moisture Product

The SMOS INRA-CESBIO (SMOS-IC) product provides global daily soil moisture and L band vegetation optical depth (L-VOD) from the ascending and descending orbits at a spatial resolution of 25 km (EASE grid 2.0). The SMOS-IC corresponds to the SMOS "original algorithm"; it is to be as independent as possible from auxiliary data, thus avoiding circular evaluation/validation [39]. The soil moisture retrievals (field: Soil_Moisture) are released in m^3/m^3 and with a dry bias of~−0.045 m^3/m^3 against ISMN in situ sites [40]. The SMOS-IC V2 soil moisture product is the latest release (January 2020), and comparisons with in situ measurements and other "official" satellite products may help to better understand its characteristics.

2.1.4. SMAP L2 Soil Moisture Product

The SMAP L2 Radiometer Half-Orbit 36 km EASE-Grid Soil Moisture (L2_SM_P) product contains gridded data of passive soil moisture retrievals (in the top 5 cm of the soil column), ancillary data, and quality assessment flags on the 36 km global cylindrical Equal-Area Scalable Earth (EASE) Grid 2.0 projection and is presented in half-orbit granules. The soil moisture retrievals (field: Soil_Moisture) are volumetric water content in m^3/m^3, with an accuracy requirement of ~±0.04 m^3/m^3 [41,42].

Attention needs to be paid to the SMOS L2 and SMAP L2 soil moisture products. They are the direct retrieval outputs with Level 1 (L1) instrument brightness temperature observations as the input, and also the inputs used to generate Level 3 (L3) global daily soil moisture composites. The L2 products inherit the location and time codes of the L1 products but do not undergo the spatiotemporal resampling of the L3 products, thus avoiding the uncertainties introduced by data processing and ensuring reverse traceability from data validation to error location. For this reason, SMOS and SMAP L2 soil moisture products were used in this paper.

2.1.5. SMAP L4 Soil Moisture Product

The SMAP L4 Global 3-hourly 9 km EASE-Grid Surface and Root Zone Soil Moisture Geophysical Data (SPL4SMGP) contains global estimates of surface soil moisture (0–5 cm vertical average), root zone soil moisture (0–100 cm vertical average), and additional research products (soil temperature, evapotranspiration, etc.), based on the assimilation of SMAP L band brightness temperatures. This product appears on the EASE-Grid 2.0 projection at 9 km grid resolution, the soil moisture estimates (field: SM_Surface) are 3-hourly time-averaged volumetric water content in m^3/m^3, and the accuracy requirement is 0.04 m^3/m^3 [43,44]. It should be noted that SMOS also provides the L4 soil moisture product, but the coverage is limited to European and Mediterranean countries and therefore could not be used in this research.

The SMAP L4 soil moisture product has greater temporal continuity and spatial integrity than the L2 soil moisture product and is more application-oriented. The L4 product is formally at a higher level in the data system because it has more added value, but it is equivalent to the L2 product in terms of the data collection process because they both use the L1 product as input. The L2 and L4 products represent the two main ways of obtaining soil moisture information from satellite remote sensing; they reflect different implementation concepts, calculation methods, and spatiotemporal visualization systems, but both need to be verified and evaluated. It is therefore worth including the SMAP L4 soil moisture product in this study.

2.2. Methods

There are four parts to this section: the quality control of ISMN multilayer in situ data; the spatial and temporal matching of SMOS, SMAP products, and ISMN data; methods for the analysis of the stratification properties of soil moisture; and methods for the verification of the depth mismatch.

2.2.1. Quality Control of the In Situ Data

The ISMN in situ data of 1871 sites from 42 networks met the download conditions (global, 1 January 2015~31 December 2020). Although discussions on the accuracy and reliability of the data are beyond the scope of this article, quality control is still required. Following the three-level hierarchy of ISMN data storage, from network to site to variable file, the quality requirements were set as follows: First, networks with more than 10 sites should be retained. Second, sites should be selected that can provide 5, 10, and 20 cm soil moisture as well as 5, 10, and 20 cm soil temperature, i.e., there were 6 variables (must but not limited to) and only one sensor per depth (no multiple observations). It should be noted that some sites set the observation depth at 2 in, 4 in, and 8 in, which after conversion are 5.08 cm, 10.16 cm, and 20.32 cm respectively; such sites are also reserved as long as they have the six variables. Third, for each record (once per hour) in the variable file, it is considered "valid" if the 6 variables are all marked with "G" (good, ISMN Quality Flag), the number of such records should exceed 50% per year and every year from 2015 to 2020. In the end, 83 sites from 3 networks passed the quality check. The 3 networks are USCRN (U.S. Climate Reference Network), SCAN (Soil Climate Analysis Network), and SNOTEL (Snow Telemetry), and all 83 sites are located within the continental U.S., as shown in Figure 1A. Information on land cover, sand fraction, and clay fraction was read from the static variables file, as shown in Figure 1B.

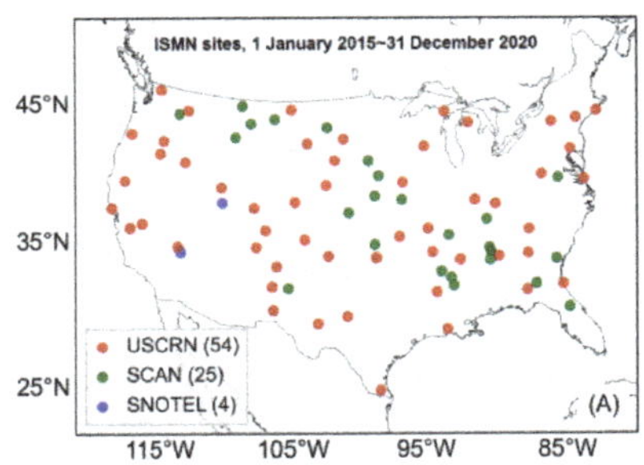

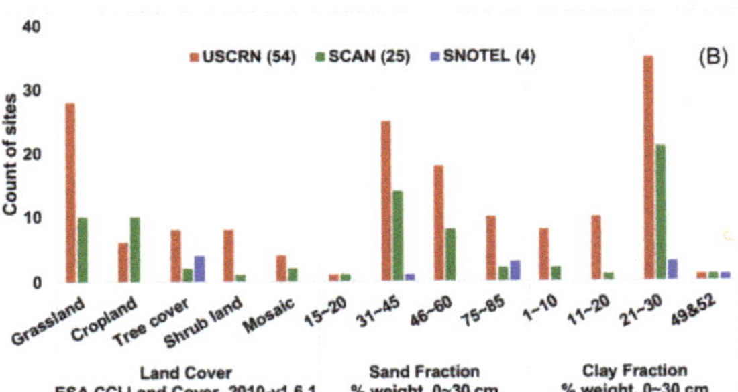

Figure 1. The 83 sites that passed the quality control: (**A**) spatial distribution of the sites; (**B**) information on land cover and soil properties of the sites.

2.2.2. Spatiotemporal Matching of In Situ Data and Satellite Products

Discussions on the retrieval and estimation accuracy of satellite products are beyond the scope of this article, and thus only the comparative differences between satellite soil moisture and multilayer in situ soil moisture were examined. As the data have different spatial and temporal characteristics, they had to be matched before performing any comparison.

The first step was spatial matching. The 83 sites from the 3 networks (USCRN, SCAN, and SNOTEL) provide hourly multilayer in situ soil moisture measurements; their locations are marked by longitude and latitude and are usually thought of as points in space. SMOS L2, SMOS-IC, SMAP L2, and SMAP L4 products are mapped in the ISEA 4H9 (~15 km), EASE-Grid 2.0–25 km, EASE-Grid 2.0–36 km, and EASE-Grid 2.0–9 km systems, respectively; the grids correspond to a specific area in space; and only the latitude and longitude of the grid center are given. The spatial matching of satellite products and in situ data can be performed according to the principle of closest distance. Taking the location of each site as a reference, a five-element matching group (ISMN site, SMOS L2 grid center, SMOS-IC grid center, SMAP L2 grid center, and SMAP L4 grid center) can be formed to search separately for the satellite grid center that is closest to the site.

It should be noted that no horizontal rescaling processing was applied to the ISMN sites and satellite grids, and neither their spatial difference nor their representativeness was considered in this article. The ISMN multilayer in situ soil moisture measurements were used as a reference for comparison with SMOS and SMAP soil moisture products [45,46]. Although potentially accompanied by the spatial mismatch, this type of absolute difference could turn into a relative difference similar to a "system bias" when all products were compared to the same reference object.

The second step was temporal matching. All five types of data have UTC timestamps but in different formats. Timing can be adjusted to the nearest time by rounding minutes and seconds to hours. No additional processing was required as the sampling rate of the in situ data is hourly. The timestamps of the SMOS L2, SMOS-IC, and SMAP L2 products include minutes and seconds, which were rounded to the nearest hour. The timestamp of the SMAP L4 product corresponds to the center of the 3 h averaging interval; therefore, it was mapped to this 3 h time set in a left-closed and right-open fashion. It can be assumed that the SMAP L4 product is complete on the hourly time axis as there was an estimate for each hour; the in situ data were nearly complete except for a small number of missing (invalid) records; and the SMOS L2, SMOS-IC, and SMAP L2 products were discrete due to their temporal resolution.

There were two temporal matching schemes. The first was matching the in situ data with the satellite products one at a time. This type of comparison was expected to independently reflect the numerical characteristics of the satellite soil moisture. The second was matching all data simultaneously, which can be considered as eliminating the influence of the temporal mismatch and therefore allowing a comparison between satellite products [47,48]. The sample size of each matching group is shown in Table 1. It should be noted that timestamp is only one of the auxiliary information and cannot be utilized to discuss the temporal representativeness and rationality of the products.

Table 1. Sample size of temporal matching groups.

		Temporal Matching Groups			**Counts**
ISMN	SMOS L2				128,619
ISMN		SMOS-IC			86,646
ISMN			SMAP L2		123,635
ISMN				SMAP L4	3,257,075
ISMN	SMOS L2	SMOS-IC	SMAP L2	SMAP L4	7848

2.2.3. Analysis of the Vertical Distribution Characteristics of Surface Soil Moisture

The overall distribution of soil moisture in each layer can be represented by its mean value. According to the maximum record of the ISMN data and the nominal retrieval accuracy of SMOS and SMAP products, the detailed distribution can be expressed by the segmented statistics of sample size in the total range of 0~0.52 m^3/m^3. The distribution analysis was based on all samples without distinguishing the site to which they belong.

The similarity of soil moisture between the layers can be quantified by the (Pearson) correlation coefficient (R) [49,50]. Three correlation sets were formed, namely, 5/5.08 and 10/10.16 cm (5/5.08 and 10/10.16); 10/10.16 and 20/20.32 cm (10/10.16 and 20/20.32); and 20/20.32 and 5/5.08 cm (20/20.32 and 5/5.08). The correlation coefficient was calculated separately for each site and was also presented in groups according to the static variables (land cover, sand fraction, and clay fraction), which were designed to reflect, to some extent, the potential influence of external environmental factors on the vertical distribution of soil moisture. It should be noted that the correlation coefficient only indicates the similarity between the two sets of samples from a numerical point of view and cannot explain the coupling mechanism of soil moisture between the layers.

The difference in soil moisture between the layers can be directly expressed by their actual numerical differences, and the detailed distribution can also be reflected by the segmented statistics of sample size. Three sets were formed, namely, 5/5.08 minus 10/10.16 cm (5/5.08 − 10/10.16); 10/10.16 minus 20/20.32 cm (10/10.16 − 20/20.32); and 5/5.08 minus 20/20.32 cm (5/5.08 − 20/20.32). The soil moisture difference was calculated for all samples without distinguishing the site to which they belonged. The positive and negative differences were counted separately, as well as the average and the total number of samples on both sides.

2.2.4. Comparison between the Satellite Products and the In Situ Data

The comparison between the SMOS/SMAP products and the ISMN data was carried out on the basis of temporal matching (Table 1), using the actual numerical difference as an indicator to present the difference between them. Similarly, segmented statistics of sample size were used to visualize the detailed distribution of the differences. The positive and negative differences were counted separately, as were the mean and total sample sizes on both sides. The mean difference (*MD*) and mean absolute difference (*MAD*) were used as quantification indices according to the following equations:

$$MD = \frac{\sum(satellite - in\_situ)}{sample\ size}. \tag{1}$$

$$MAD = \frac{\sum|satellite - in\_situ|}{sample\ size}. \tag{2}$$

3. Results

3.1. Stratification Characteristics of Surface Soil Moisture

3.1.1. Single-Layer Distribution

As shown in Figure 2, there seemed to be a turning point at 0.24~0.28 m^3/m^3. For the 5/10/20 cm group, when it was below this range, the distribution of 5 and 10 cm showed some similarity. With an increase in depth, the peaks of the three layers gradually moved to higher ranges (0.04~0.08, 0.08~0.12, 0.16~0.20 m^3/m^3), especially in the ranges of 0~0.04 m^3/m^3 and 0.16~0.20 m^3/m^3, and the low-value characteristics of 5 cm and the high-value characteristics of 20 cm were very significant. However, above this range, a strong similarity was found between 10 and 20 cm, and the distribution difference among the three layers was reduced. For the 5.08/10.16/20.32 cm group, the sample size ranking of the three layers showed opposite trends below and above the inflection point; the distribution of 5.08 and 10.16 cm was also found to be similar, and their peaks were both located around 0.20~0.24 m^3/m^3. The distribution of 20.32 cm was very different from the other two layers, as its peak appeared at 0.32~0.40 m^3/m^3 where the soil moisture was very high. Although the difference in depth was small, the soil moisture of the two groups behaved quite differently; their means indicated that the soil moisture of the 5.08/10.16/20.32 cm group was always slightly higher than that of the 5/10/20 cm group (0.178/0.196/0.200 vs. 0.200/0.223/0.244 m^3/m^3). However, both showed a pattern of increasing soil moisture with the increase in depth, which appeared to be a stable distribution state of soil moisture.

The mean values of soil moisture in each layer were compared in groups according to the static variables of land cover, sand fraction, and clay fraction, and the results are shown in Figure 3. The general trends of the two sets of curves appear to be similar at first sight. For the 5/10/20 cm group (Figure 3A), 5 cm soil moisture showed a significantly low-value characteristic; the mean values of 10 and 20 cm soil moisture were very close, but the latter was slightly higher. Regardless of the static variables, the order of the three soil moisture layers from low to high remained unchanged with the increase in depth. For the 5.08/10.16/20.32 cm group (Figure 3B), 5.08 cm soil moisture was still the lowest, the difference between 10.16 cm and 20.32 cm became larger, and in some cases, 10.16 cm soil moisture was higher.

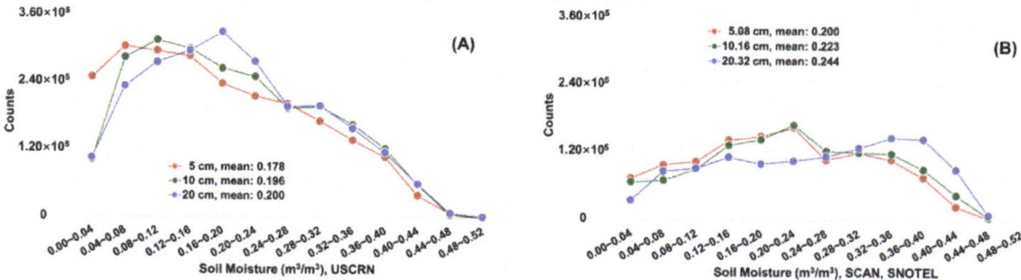

Figure 2. Soil moisture distribution in each layer: (**A**) the distribution at 5, 10, and 20 cm; (**B**) the distribution at 5.08, 10.16, and 20.32 cm (2, 4, and 8 in).

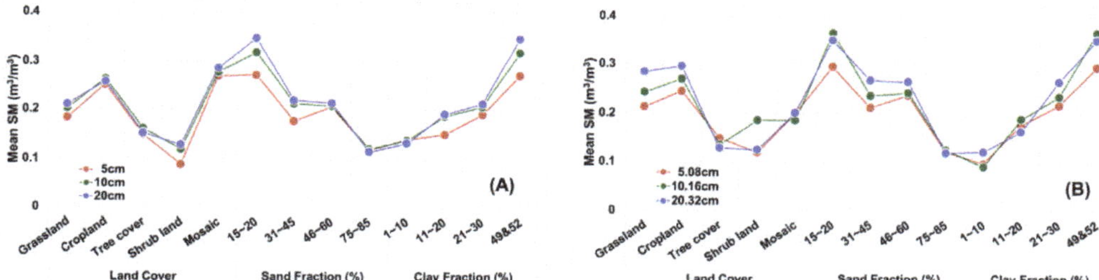

Figure 3. Mean soil moisture under static variables: (**A**) for the 5/10/20 cm group; (**B**) for the 5.08/10.16/20.32 cm group (2, 4, and 8 in).

The difference between the two groups was most obvious in terms of land cover. The 5.08/10.16/20.32 cm group showed stronger stratification characteristics than the 5/10/20 cm group in grassland, cropland, and shrubland conditions. Although the means of the three layers were close within each group under the conditions of tree cover and mosaic (mainly multiple vegetation types), there were large differences between the two groups.

As the sand and clay fractions increased, soil moisture tended to decrease and increase, respectively. The three layers differed significantly within and between the two groups, especially the pair of 10.16 and 20.32 cm. It appeared that the difference in the vertical distribution of soil moisture between the three layers became smaller with the increase in sand and larger with the increase in clay. For the 5/10/20 cm group, the influence of soil properties was slightly higher than that of land cover, but both types of static variables played a significant role for the 5.08/10.16/20.32 cm group.

3.1.2. Interlayer Correlation

The correlation coefficients of soil moisture between layers were calculated for each site, and the results are shown in Figure 4. The two groups showed a common pattern, i.e., the correlation coefficients decreased with increasing depth difference, although those of the 5/10/20 cm group were higher than those of the 5.08/10.16/20.32 cm group (0.899/0.884/0.813 vs. 0.809/0.767/0.690), and their distributions appeared to be very different.

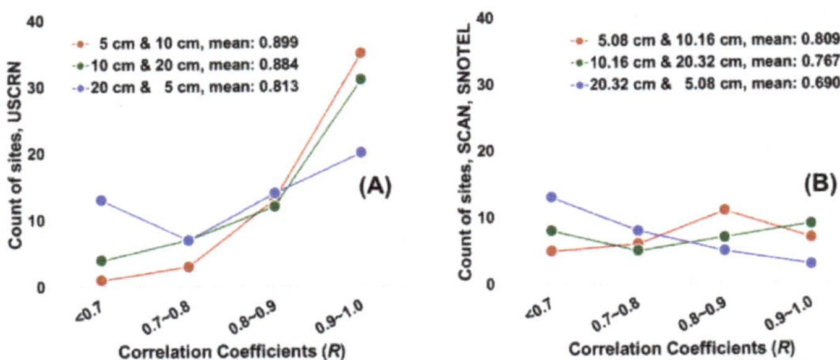

Figure 4. Correlation coefficients of soil moisture between layers: (**A**) interlayer correlation coefficients for the 5/10/20 cm group; (**B**) interlayer correlation coefficients for the 5.08/10.16/20.32 cm group (2, 4, and 8 in).

For the 5/10/20 cm group (Figure 4A), the distribution of the correlation coefficients of the three sets all showed an upward trend. Taking 0.8~0.9 as the turning point, in areas where the correlation coefficient was below 0.8, the order of the number of sites from small to large was "5&10", "10&20", and "20&5"; in areas where the correlation coefficient was above 0.9, the order was reversed. For the 5.08/10.16/20.32 cm group (Figure 4B), the three sets were distributed differently and no uniform trend was found. The downward trend of "20.32&5.08" looked very significant, while both "5.08&10.16" and "10.16&20.32" had an upward trend, although their peak and trough were located at 0.8~0.9 and 0.7~0.8, respectively. However, it can still be seen that the number of sites of "5.08&10.16" was highest in the intervals where the correlation coefficient was high, that of "20.32&5.08" was highest in the intervals where the correlation coefficient was low, and that of "10.16&20.32" always remained in the middle of the other two sets. This also reflected, to some extent, the tendency for the interlayer correlation coefficient to decrease as the depth difference increased.

The correlation coefficients were also grouped according to static variables, as shown in Figure 5. Firstly, in most cases, the order from lowest to highest was still "20/20.32&5/5.08", "10/10.16&20/20.32", and "5/5.08&10/10.16", with the difference between the three sets also increasing as the depth difference increased. Secondly, the correlation coefficients of the 5/10/20 cm group were always higher than those of the 5.08/10.16/20.32 cm group, except for the conditions of mosaic and the 75~85 sand fraction. Thirdly, for the 5/10/20 cm group (Figure 5A), the distribution of "20&5" appeared quite different from the other two sets, especially in shrubland, the 15~20 sand fraction, and the "49&52" clay fraction; there was not much difference between "5&10" and "10&20", and they were almost the same in some conditions. For the 5.08/10.16/20.32 cm group (Figure 5B), the three sets were quite different from the 5/10/20 cm group. They seemed to have a good synchronous trend, but the differences were very pronounced in the land cover condition.

It can be observed that the correlation coefficients of soil moisture between the layers decreased with the increase in depth difference, where the depth difference was 5/5.08 cm (10/10.16–5/5.08), 10/10.16 cm (20/20.32–10/10.16), and 15/15.24 cm (20/20.32–5/5.08). However, this could only indicate that the vertical similarity of soil moisture is related to the depth difference, but it was not possible to confirm where the depth difference lay. The correlation coefficients of "5/5.08&10/10.16" might not be the highest if the in situ measurements of 15/15.24 cm were provided, as there would be two more sets of depth differences also equal to 5/5.08 cm (15/15.24–10/10.08 and 20/20.32–15/15.24).

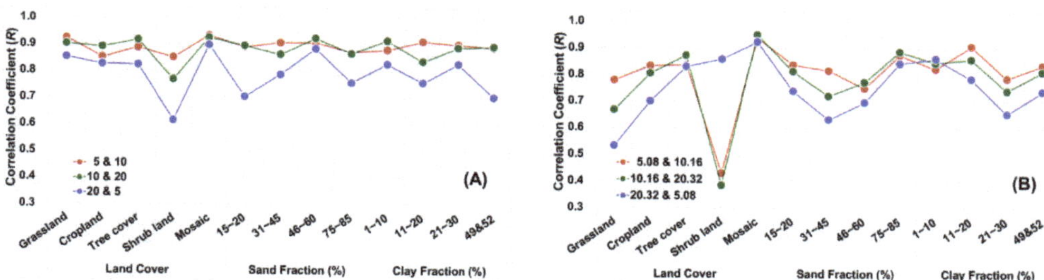

Figure 5. The correlation coefficients grouped according to the static variables: (**A**) the average correlation coefficients for the 5/10/20 cm group; (**B**) the average correlation coefficients for the 5.08/10.16/20.32 cm group (2, 4, and 8 in).

3.1.3. Interlayer Difference

As shown in Figure 6, the two groups reflected two types of vertical distribution in terms of mean and sample size for both the negative and positive values. For the 5/10/20 cm group, the order of the negative difference from small to large was "10–20", "5–10", and "5–20", indicating that the soil moisture of 10 cm was close to that of 20 cm, and the soil moisture difference between 5 cm and the other two lower layers (10 and 20 cm) increased with the increase in depth difference (−0.32, −0.42 m^3/m^3). The positive difference showed a consistent increasing trend (0.020, 0.024, and 0.028 m^3/m^3), but the sample size was much smaller than that of the negative difference; perhaps it can be assumed that this reverse increase with distance between the layers was random rather than conventional and was probably caused by precipitation. For the 5.08/10.16/20.32 cm group, the negative difference between the layers became more significant (−0.044, −0.048, and −0.067 m^3/m^3), with the sample size on both sides, showing a consistent trend of increase and decrease. The cases where the upper soil moisture was higher than the lower can also be explained by the influence of precipitation. The basic characteristics of soil moisture increasing with depth were more pronounced and showed a uniform variation in the vertical direction.

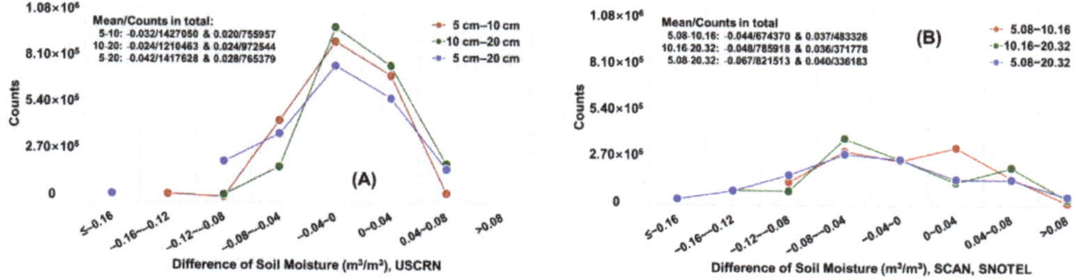

Figure 6. Soil moisture difference between layers: (**A**) the interlayer difference for the 5/10/20 cm group; (**B**) the interlayer difference for the 5.08/10.16/20.32 cm group (2, 4, and 8 in).

In terms of detailed distribution, for the 5/10/20 cm group, the peaks of the three sets of differences were all within −0.04~0 m^3/m^3; the upper limit of the positive differences was the same and did not exceed 0.04~0.08 m^3/m^3, while the lower limit of the negative differences was inconsistent, with the order from small to large being "10–20 (−0.12~−0.08 m^3/m^3)", "5–10 (−0.16~−0.12 m^3/m^3)", and "5–20 ($\leq$−0.16 m^3/m^3)". The 5.08/10.16/20.32 cm group seemed to be spread over a wider range than the other group, with the peaks moving backward to around −0.08~−0.04 m^3/m^3; the maximum positive difference was all above 0.08 m^3/m^3, and the descending order of the minimum negative difference became "5–10 (−0.12~−0.08 m^3/m^3)", "10–20 (−0.16~−0.12 m^3/m^3)", and "5–20 (≤−0.16 m^3/m^3)".

The mean positive and negative differences under static variables and the difference in sample size between the two sides were shown in Figure 7. The difference in soil moisture between the layers at the two sets of depth was very different (Figure 7A,B) and was more pronounced in cases of low vegetation (grassland, cropland, and shrubland), lower sand fraction (15~20 and 31~45), and higher clay fraction (21~30, 49, and 52). Consistent with the difference in soil moisture, the difference in sample size appeared to be greater among the static variables (Figure 7C,D). The difference in soil moisture between the layers can be more clearly distinguished not only within each group but also between the two groups, further demonstrating the influence of land cover and soil properties on the water-holding capacity. It is worth mentioning that in terms of soil property, although the magnitude of the two sets of soil moisture difference was very different, the overall trend was similar. The sand and clay fractions given by ISMN refer to the soil property of 0~30 cm, and if we focus only on the surface layer of 0~5 cm, the difference in the composition may not be large; in other words, the soil property may not be the main factor affecting the vertical distribution characteristics of shallow soil moisture. Based on years of "big data", it may be possible to model the behavior of soil moisture under normal and disturbed conditions to provide more straightforward optimization solutions for soil moisture retrieval algorithms (brightness temperature simulation, parameter modeling, ancillary information assimilation, etc.).

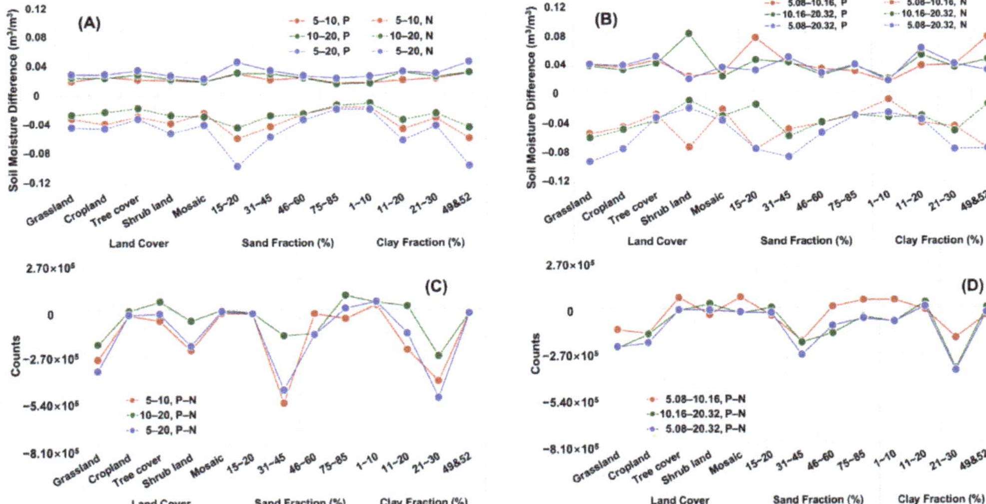

Figure 7. The interlayer difference of soil moisture and sample size under static variables; P and N refer to positive and negative, and P-N refers to positive minus negative: (**A**) the interlayer difference for the 5/10/20 cm group; (**B**) the interlayer difference for the 5.08/10.16/20.32 cm group (2, 4, and 8 in); (**C**) the difference of the sample size between the positive and negative sides for the 5/10/20 cm group; (**D**) the difference of the sample size between the positive and negative sides for the 5.08/10.16/20.32 cm group (2, 4, 8 and in).

3.2. Comparisons between the Satellite Products and the In Situ Data

The comparison was carried out in two ways based on temporal matching (Table 1). The first was comparing each type of satellite product separately with the three-layer in situ data (SMOS L2/SMOS-IC/SMAP L2/SMAP L4—in situ), and the second was comparing all four types of satellite products simultaneously with each single-layer in situ data (satellite products—5/5.08/10/10.16/20/20.32 cm).

3.2.1. Separate Comparison

The correlation coefficients are shown in Table 2. For within groups, it decreased with the increase in depth. For between groups, the correlation coefficients with the 5/10/20 cm group were slightly higher than those with the 5.08/10.16/20.32 cm group, and for the satellite products, the order from small to large was SMOS L2, SMOS-IC, SMAP L2, and SMAP L4. It can be seen that the satellite soil moisture products correlate better with the 5/5.08 cm in situ data than with the other two layers.

Table 2. Correlation coefficient of satellite soil moisture products and multilayer in situ measurements, separate comparison.

R	5 cm	10 cm	20 cm	5.08 cm	10.16 cm	20.32 cm
SMOS L2	0.461	0.510	0.397	0.462	0.334	0.404
SMOS IC	0.675	0.607	0.610	0.559	0.538	0.493
SMAP L2	0.648	0.629	0.586	0.580	0.524	0.500
SMAP L4	0.701	0.654	0.655	0.613	0.602	0.572

As shown in Figure 8, each satellite product had its own unique performance. For SMOS L2 (Figure 8A), the peaks of its difference with the in situ data were around -0.04~0 (5 cm), -0.1~-0.04 (5.08, 10, 20 cm), and -0.2~-0.1 m^3/m^3 (10.16, 20.32 cm), reflecting, to some extent, the dry bias referred to in the literature. For SMOS-IC (Figure 8B) the peaks of the difference shifted to -0.2~-0.1 m^3/m^3 for all but 5.08 cm (-0.1~-0.04 m^3/m^3), implying an improved dry bias. For SMAP L2 (Figure 8C), the peaks of its difference with 5, 5.08, and 10 cm in situ data were around 0~0.04 and 0.04~0.1 m^3/m^3, where the dry bias started to change to a wet bias. For SMAP L4 (Figure 8D), the difference around 0.1~0.3 m^3/m^3 was suppressed, and the wet bias was weakened, while the difference around -0.1~0 m^3/m^3 was enhanced, and the dry bias was strengthened.

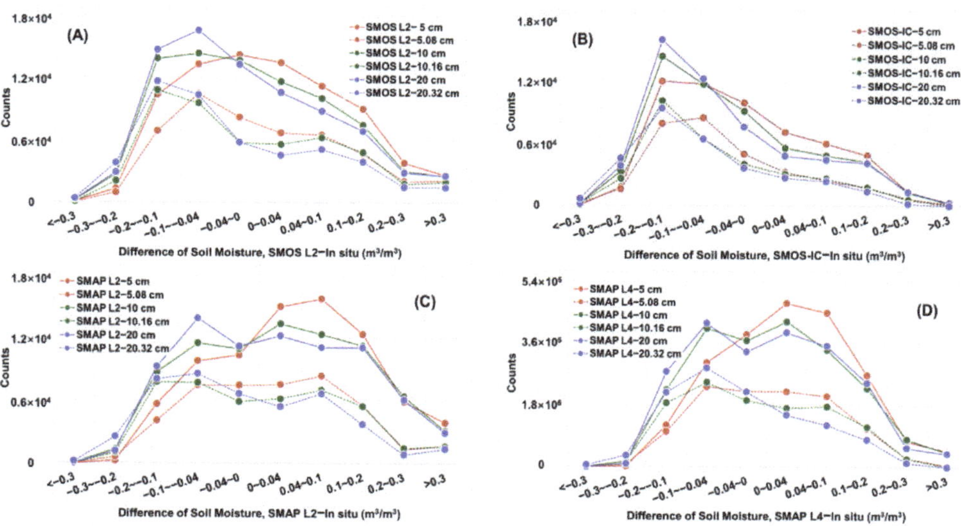

Figure 8. Soil moisture difference between satellite and in situ, separate comparison: (**A**) the difference between SMOS L2 and in situ; (**B**) the difference between SMOS-IC and in situ; (**C**) the difference between SMAP L2 and in situ; (**D**) the difference between SMAP L4 and in situ.

There seemed to be a turning point in the distribution of the difference between the four satellite soil moisture products and the three layers of in situ data. For the 5/10/20 cm group, the turning point was around $-0.04 \sim 0$ m^3/m^3 except for SMOS-IC, and for the 5.08/10.16/20.32 cm group, the turning point was around $-0.1 \sim -0.04$ m^3/m^3 except for SMAP L4. The order of sample size from large to small was 20/20.32, 10/10.16, and 5/5.08 cm in areas where the difference was below the inflection point, while above the inflection point, the order was reversed. In general, the difference between the satellite products and the 5/5.08 cm in situ data was not similar to the other two layers, with SMOS L2 and SMOS-IC soil moisture lower than the in situ data and SMAP L2 and L4 soil moisture higher than the in situ data.

The numerical difference between the satellite products and in situ data was further explored in groups. The first group was based on land cover, sand fraction, and clay fraction. For each condition, the mean positive and negative difference was calculated separately, as well as the difference in sample size on both sides, and the results are shown in Figure 9.

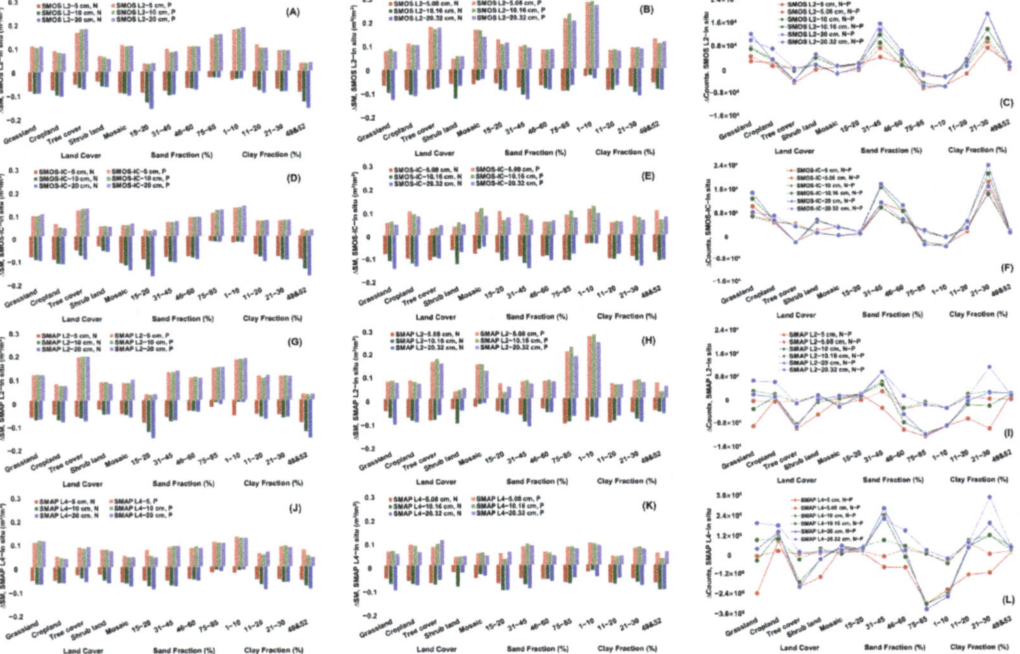

Figure 9. (**A–L**) Soil moisture difference between satellite and in situ, grouped by land cover, sand fraction, and clay fraction, separate comparison; P and N refer to positive and negative, and N − P refers to positive minus negative.

When comparing the satellite products with the 5/10/20 cm in situ data, the difference was significantly different in the cropland, tree cover, and mosaic conditions. The largest negative and positive differences were observed for SMOS-IC in the mosaic condition (Figure 9D) and SMAP L2 in the tree cover condition (Figure 9G). The negative difference decreased, and the positive difference increased with the increase in the sand fraction, while this trend was completely reversed with the increase in the clay fraction. The largest negative difference was contributed by SMOS-IC in conditions with the "15~20" sand fraction and "49&52" clay fraction, and the largest positive difference was contributed by SMOS L2 in conditions with the 75~85 sand fraction and 1~10 clay fraction (Figure 9A).

When compared to the 5.08/10.16/20.32 cm in situ data, none of the differences between the four satellite products and the in situ data were similar, especially in the tree cover and mosaic conditions. The negative difference in SMOS-IC (Figure 9E) and the positive difference in SMOS L2 (Figure 9B) appeared to be higher than those of the other products. The trend of increasing and decreasing negative and positive differences could still be found with variations in sand and clay fractions, but the pattern was not as clear and consistent. In conditions where the sand fraction was very high, and the clay fraction was very low, the positive difference in SMOS L2 and SMAP L2 (Figure 9H) increased to about 0.2~0.3 m^3/m^3, which can be considered anomalies. In conclusion, regardless of the group with which the comparison was carried out, the negative difference between the satellite products and 5/5.08 cm in situ data was the smallest, and it was the largest with 20/20.32 cm; however, a similar pattern of a positive difference could only be found for SMOS L2, SMOS-IC, and SMAP L2 with their comparison to the 5/10/20 cm group.

The difference in sample size shown in Figure 9 is also revealing. Compared with the 5/10/20 cm in situ data, the sample size distributions of the four satellite products looked very different in the grassland condition but appeared similar in the mosaic condition. SMOS-IC (Figure 9F) and SMAP L4 (Figure 9L) were similar in the cropland condition, with a significantly higher negative than positive sample size, whereas in the tree cover and shrubland conditions, the sample size bias showed similarities within the SMOS (SMOS L2 and SMOS-IC) and SMAP (SMAP L2 and SMAP L4) groups, as well as differences between the groups. In terms of soil properties, the negative bias gradually became positive as the sand fraction increased, whereas the opposite trend was observed as the clay fraction increased, with exceptions where the sand fraction was very low (15~20) and the clay fraction was very high (49&52). Compared with the 5.08/10.16/20.32 cm in situ data, SMOS-IC in cropland and tree cover conditions and SMAP L4 in grassland conditions appeared to be significantly different from the other satellite products. The shift in dominance was still clearly discernible as the sand and clay fraction increased, and its magnitude slowed down but became more uniform for the SMAP group (SMAP L2, L4). There was a general pattern in which the difference between negative and positive sample sizes increased with depth.

The second group was based on the in situ data, and the results are shown in Figure 10. The mean negative difference increased with the increase in soil moisture, and the order from largest to smallest was SMOS-IC, SMOS L2, SMAP L2, and SMAP L4. The satellite products had the smallest negative difference with the 5/5.08 cm in situ data and the largest with the 20/20.32 cm. The descending order of the positive difference was SMOS L2, SMAP L2, SMOS-IC, and SMAP L4. A trend of decreasing positive difference with the increase in soil moisture can be found for SMAP L2 and SMAP L4, especially when comparing SMAP L2 and 5.08/10.16/20.32 cm in situ data (Figure 10J). However, SMOS L2 and SMOS-IC did not show such a trend, and the peak of their positive difference occurred mainly around 0.3~0.4 m^3/m^3, where the soil moisture was at a higher level. In most cases, the positive difference between the satellite products and 5/5.08 cm in situ data was the smallest.

As the soil moisture increased, the difference in sample size showed a basic pattern in which the negative difference gradually exceeded the positive one, peaking at about 0.3~0.4 m^3/m^3. The sample sizes on both sides became comparable when the soil moisture was higher than 0.4 m^3/m^3, but their difference remained positive. However, the comparison with the 20 cm in situ data seemed to be quite different from the others, as the difference between negative and positive values reached a maximum at around 0.1~0.2 m^3/m^3, and then the gap between the two sides narrowed with the increase in soil moisture, but it did not cross the 0 line. The performance of SMOS-IC was also somewhat peculiar in that the difference in sample size remained above the 0 line (excluding 20 cm), which meant that the magnitude of the negative difference was always greater than that of the positive one. In contrast, SMOS-IC had the least variation in the difference in sample size, while SMAP L4 had the most; if the degree of variation in the difference was to be ranked from small to large, the order was 5/5.08, 10/10.16, and 20/20.32 cm.

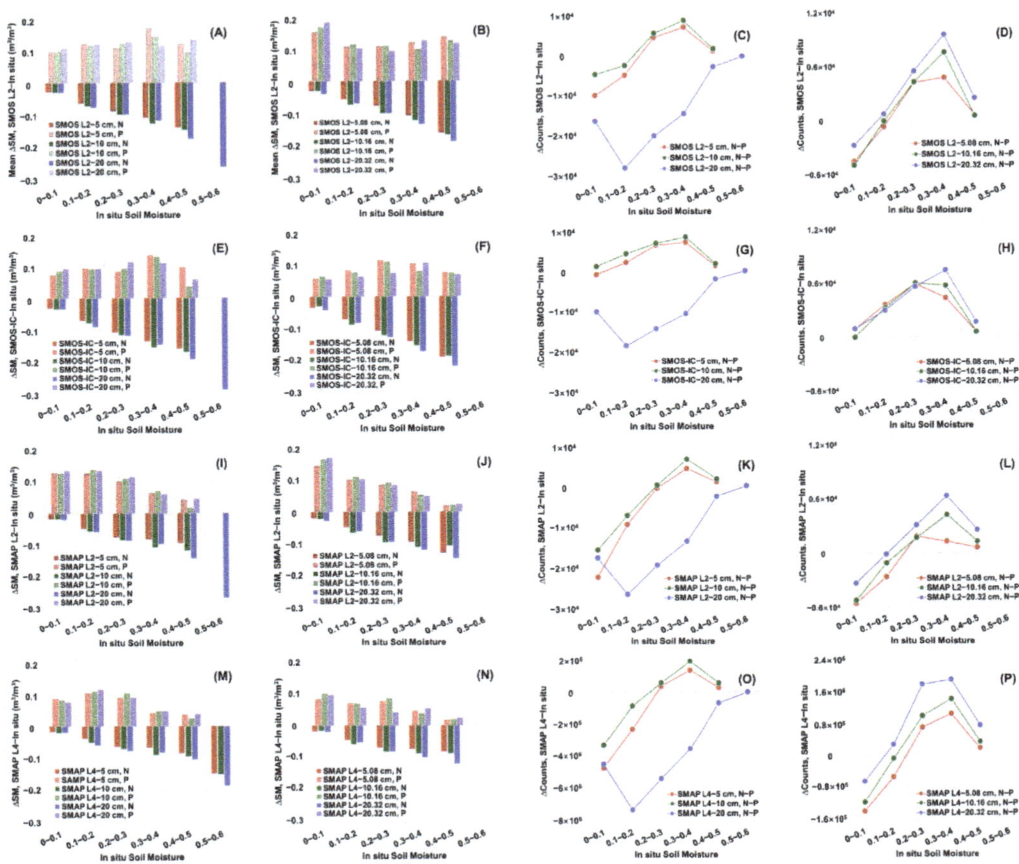

Figure 10. (**A–P**) Soil moisture difference between satellite and in situ, grouped according to in situ soil moisture, separate comparison; P and N refer to positive and negative, and N − P refers to positive minus negative.

3.2.2. Simultaneous Comparison

The SMOS L2, SMOS-IC, SMAP L2, and SMAP L4 soil moisture products were simultaneously compared with the in situ data at 5, 5.08, 10, 10.16, 20, and 20.32 cm, and their correlation coefficients and numerical differences are shown in Table 3 and Figures 11–13. It should be noted that the representativeness of the results may be limited, as the sample size was only 7848 under strict temporal matching (Table 1).

Table 3. Correlation coefficients of satellite soil moisture products and multilayer in situ measurements, simultaneous comparison.

R	5 cm	10 cm	20 cm	5.08 cm	10.16 cm	20.32 cm
SMOS L2	0.535	0.557	0.463	0.479	0.381	0.453
SMOS IC	0.685	0.614	0.608	0.549	0.510	0.519
SMAP L2	0.692	0.647	0.617	0.592	0.519	0.555
SMAP L4	0.700	0.693	0.635	0.629	0.541	0.623

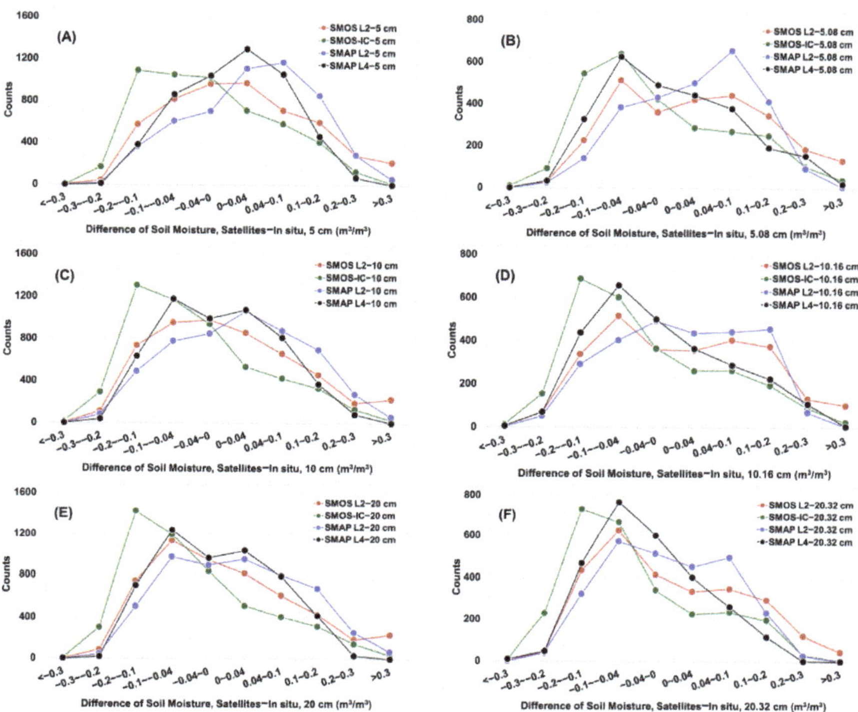

Figure 11. (**A–F**) Soil moisture differences between satellite and in situ data, simultaneous comparison.

The trends of the separate comparison (Table 2) are also presented in Table 3. Within the groups, the correlation coefficient decreased with the increase in depth. Between the groups, the SMOS-IC, SMAP L2, and SMAP L4 products had a higher correlation coefficient with the 5/10 cm in situ data than with the 5.08/10.16 cm. The ranking of the satellite products from small to large remained SMOS L2, SMOS-IC, SMAP L2, and SMAP L4, but they all had a higher correlation coefficient with the 5/5.08 cm in situ data.

The numerical difference between the four satellite products and the in situ data of each layer is shown in Figure 11, which shows the characteristics of each satellite product more clearly.

Compared with the 5 cm in situ data (Figure 11A), for SMOS L2, the difference concentrated within -0.1~0.1 m^3/m^3, and the negative was slightly higher than the positive. For SMOS-IC, the difference concentrated within -0.2~0.04 m^3/m^3, there was a peak around -0.2~-0.1 m^3/m^3, and the negative was much higher than the positive. The difference for SMAP L2 seemed to be the opposite of SMOS-IC: It concentrated within -0.04~0.2 m^3/m^3, and the peak was around 0.04~0.1 m^3/m^3, with the positive difference significantly higher than the negative. SMAP L4 seemed to have a normal distribution, as the difference was concentrated within -0.1~0.1 m^3/m^3, with a peak around 0~0.4 m^3/m^3, and the positive was slightly higher than the negative, probably due to some calibration of the simulation when the soil moisture was high.

Compared with the 5.08 cm in situ data (Figure 11B), the dry bias of SMOS L2 would probably disappear since the size of the positive difference exceeded the negative, while the dry bias of SMOS-IC seemed to become stronger, with the difference narrowly concentrated within -0.2~0 m^3/m^3, and the size of the negative difference much higher than the positive. For SMAP L2, the difference remained positive without weakening. SMAP L4 was also found to have a remarkable dry bias, with the negative difference taking over and peaking at around -0.1~-0.04 m^3/m^3.

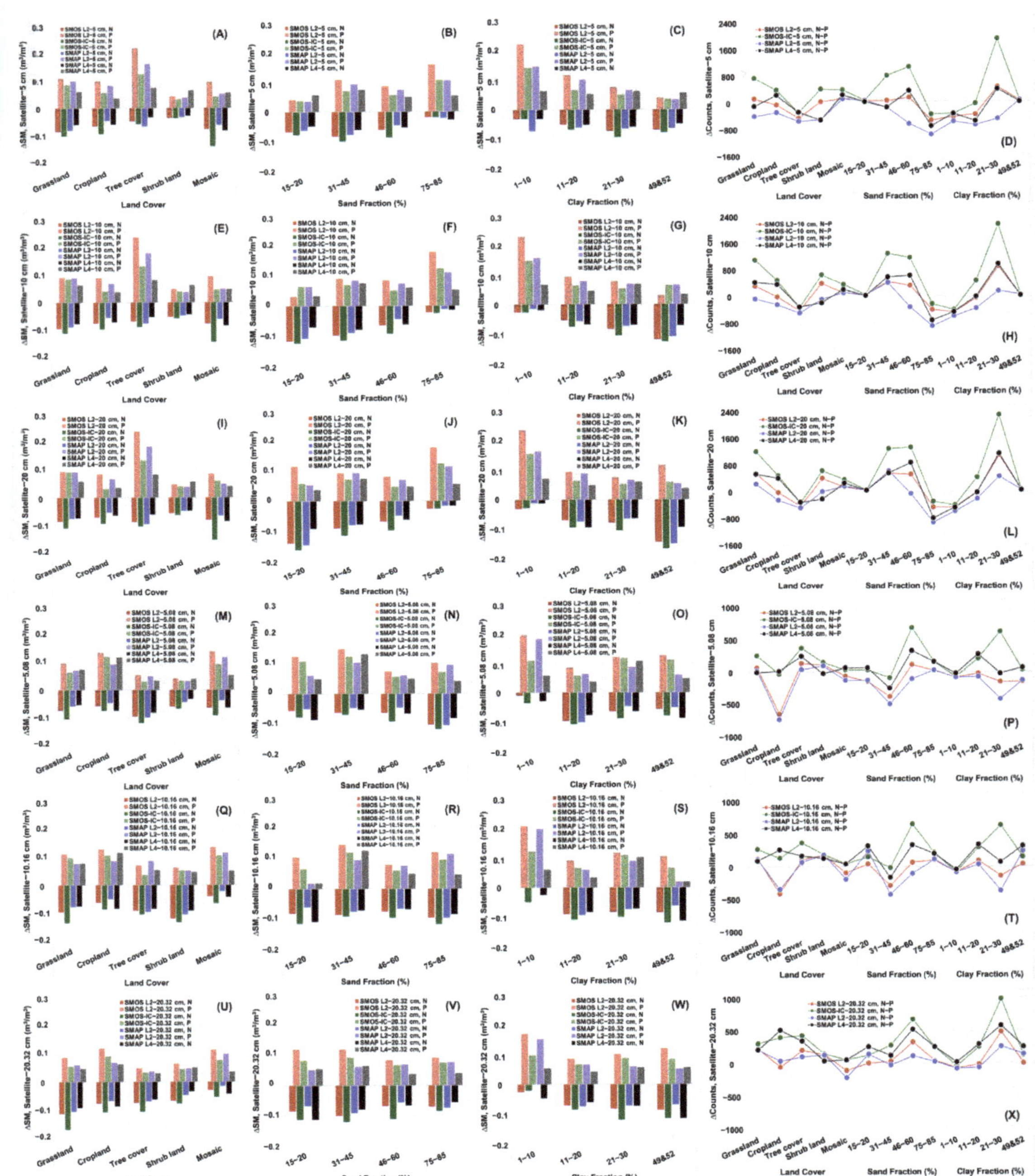

Figure 12. (**A–X**) Soil moisture differences between satellite and in situ data, grouped according to land cover and sand and clay fractions, simultaneous comparison; P and N refer to positive and negative, and N − P refers to positive minus negative.

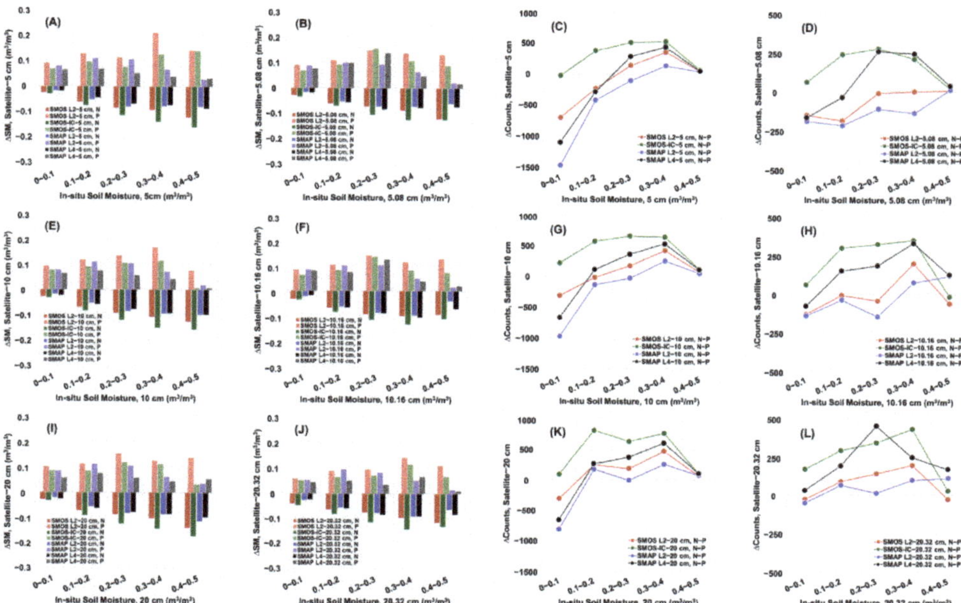

Figure 13. (**A–L**) Soil moisture differences between satellite and in situ data, grouped according to in situ soil moisture, simultaneous comparison; P and N refer to positive and negative, and N − P refers to positive minus negative.

Taking the comparison with the 5/5.08 cm in situ data as a reference, the differences between the four satellite products all moved progressively into the negative direction with the increase in depth, and the dry bias became stronger, and the distributions of their differences became more similar (Figure 11E,F). In addition, regardless of the depth to which the comparison was performed, the descending order of negative differences below the range of -0.04~$0\ m^3/m^3$ was SMOS-IC, SMAP L4, SMOS L2, and SMAP L2, and when the differences were above this range, SMOS L2 had the largest scale of positive difference and SMOS-IC the smallest.

The differences between the satellite products and in situ data in a simultaneous comparison were also analyzed in terms of land cover, sand fraction, and clay fraction. The differences between the four satellite products varied in terms of land cover. In the comparison with the 5/10/20 cm in situ data (Figure 12A–L), the difference was largest in the tree cover and smallest in the shrubland, and there was little change in the positive difference with the increase in depth, but the negative difference gradually increased. The comparison with the 5.08/10.16/20.32 cm in situ data (Figure 12M–X) seemed to lack regularity, as there was a large negative difference in the tree cover, shrubland (Figure 12Q), and grassland (Figure 12U) but a large positive difference in the cropland and mosaic. The tendency for the negative difference to increase and the positive difference to decrease with the increase in depth could only be observed under grassland and cropland, with no common change for the others, and the comparison with the 10.16 cm in situ data seemed to show a large difference on both sides.

In the grouping of the sand and clay fractions, the trend in which the negative difference decreased and increased, respectively, as the two parameters increased remained highly significant compared with the 5/10/20 cm in situ data, and the opposite trend of the positive difference could also be distinguished. With the increase in depth, the negative difference continued to increase and reached a large magnitude with a low sand content and a high clay content (Figure 12J,K), while the positive difference was very high with a high sand content and a low clay content but did not show a clear pattern of variation with depth. In comparison with the 5.08/10.16/20.32 in situ data, the trend of variation differed in each range of sand and clay fractions. There was a large negative difference in 31~45 (Figure 12V) and 75~85 (Figure 12N,R) sand fractions and a large positive difference in 31~45 sand fractions (Figure 12N,R) and 1~10 clay fractions, whereas the difference did not show a distinctive pattern of variation with sand and clay fractions but was found to increase in the negative difference and decrease in the positive difference with the increase in depth.

Of the four satellite products, SMOS L2 and SMOS-IC had the largest positive and negative differences, respectively, while SMAP L4 had the smallest positive and negative differences. The difference in sample size indicates that the deviation between the two sides can be arranged in descending order as SMOS-IC, SMAP L4, SMOS L2, and SMAP L2, with SMOS-IC mostly above the 0 line and SMAP L2 remaining below. Some cases are worth noting: Compared with the 5/10/20 cm in situ data (Figure 12D,H,L), SMAP L4, SMOS L2, and SMAP showed an increase and a reverse trend in the grassland and 31~45 sand fraction, and compared with the 5.08/10.16/20.32 cm in situ data (Figure 12P,T,X), there was a large decrease and a reverse trend in the cropland and 21~30 clay fraction. With the increase in depth, the distribution became closer to the 0 line and the fluctuation became weaker, which corresponds well to the trend in Figure 11 in which the magnitude of the negative difference increased and the predominance of the positive difference decreased.

The difference in depth between 5/10/20 cm and 5.08/10.16/20.32 cm was mainly due to the different unit settings of the observation depth, i.e., one was in centimeters and the other in inches. This 1.6% difference is difficult to detect in practice and may therefore be of little significance at a distance. The fundamental difference lies in the soil conditions and the type of land cover on which they rest, which will lead to not only an absolute difference between the networks but also a relative difference between stations within the network; in a sense, the difference between the satellite products and the two sets of in situ data may not be comparable. As mentioned before, land cover and soil properties are interdependent, and together, they drive the distribution characteristics of soil moisture in the vertical direction. The variety and variation in land cover in terms of temporal and spatial variables will probably be stronger and faster than those of the sand and clay fractions, and thus it has a greater influence on soil moisture. To some extent, this also indicates that the satellite retrieval of soil moisture should be more focused on land cover, especially the response and interaction with meteorological conditions of transient conditions.

The differences in the simultaneous comparison were also grouped according to the in situ data, and the results are shown in Figure 13. With the increase in soil moisture, the negative difference continued to increase, whereas the positive difference first increased and then decreased, peaking at around 0.3~0.4 m^3/m^3. SMOS-IC and SMOS L2 had the highest negative and positive differences, respectively, while SMAP L4 still remained the smallest on both sides. With the increase in depth, the negative difference showed an increasing trend, whereas most of the positive differences decreased. On the other hand, the distribution gradually approached or even crossed the 0 line with the increase in depth, indicating that the quantitative advantage of the negative difference constantly increased. SMOS-IC was above the 0 line and had a more negative difference, while SMAP L2 remained below this line and had a more positive difference, which is consistent with the results in Figure 11A,B and again confirms the numerical characteristics of the four satellite products.

3.2.3. The Depth Mismatch

To evaluate the depth mismatch, the mean difference (*MD*, Equation (1)) and mean absolute difference (*MAD*, Equation (2)) between the satellite soil moisture products and the multilayer in situ soil moisture data were calculated, and the results are presented in Tables 4 and 5.

Table 4. Differences between satellite data and in situ data: interlayer differences and separate comparison.

(m^3/m^3)	*MD*				*MAD*			
	SMOS L2	**SMOS-IC**	**SMAP L2**	**SMAP L4**	**SMOS L2**	**SMOS-IC**	**SMAP L2**	**SMAP L4**
Satellite−5 cm	0.018	−0.027	0.055	0.034	0.092	0.086	0.098	0.075
Satellite−10 cm	−0.001	−0.045	0.037	0.016	0.097	0.097	0.100	0.079
Satellite−20 cm	−0.008	−0.052	0.031	0.011	0.099	0.103	0.100	0.081
(Satellite−10 cm) − (Satellite−5 cm)	−0.019	−0.018	−0.018	−0.018	0.005	0.011	0.002	0.004
(Satellite−20 cm) − (Satellite−10 cm)	−0.007	−0.007	−0.006	−0.005	0.002	0.006	0	0.002
(Satellite−20 cm) − (Satellite−5 cm)	−0.026	−0.025	−0.024	−0.023	0.007	0.017	0.002	0.006
Satellite−5.08 cm	0.012	−0.049	0.025	0.005	0.097	0.093	0.086	0.067
Satellite−10.16 cm	−0.006	−0.064	0.009	−0.011	0.110	0.105	0.098	0.076
Satellite−20.32 cm	−0.031	−0.088	−0.016	−0.034	0.115	0.117	0.098	0.078
(Satellite−10.16 cm) − (Satellite−5.08 cm)	−0.018	−0.015	−0.016	−0.016	0.013	0.012	0.012	0.009
(Satellite−20.32 cm) − (Satellite−10.16 cm)	−0.025	−0.024	−0.025	−0.023	0.005	0.012	0	0.002
(Satellite−20.32 cm) − (Satellite−5.08 cm)	−0.043	−0.039	−0.041	−0.039	0.018	0.024	0.012	0.011

Table 5. Differences between satellite data and in situ data: interlayer difference and simultaneous comparison.

(m^3/m^3)	*MD*				*MAD*			
	SMOS L2	**SMOS-IC**	**SMAP L2**	**SMAP L4**	**SMOS L2**	**SMOS-IC**	**SMAP L2**	**SMAP L4**
Satellite−5 cm	0.030	−0.027	0.039	0.009	0.092	0.083	0.080	0.058
Satellite−10 cm	0.012	−0.045	0.021	−0.009	0.094	0.094	0.081	0.065
Satellite−20 cm	0.009	−0.048	0.018	−0.012	0.094	0.097	0.080	0.064
(Satellite−10 cm) − (Satellite−5 cm)	−0.018	−0.018	−0.018	−0.018	0.002	0.011	0	0.007
(Satellite−20 cm) − (Satellite−10 cm)	−0.003	−0.003	−0.003	−0.003	0	0.003	−0.001	−0.001
(Satellite−20 cm) − (Satellite−5 cm)	−0.021	−0.021	−0.021	−0.021	0.002	0.014	0	0.006
Satellite−5.08 cm	0.038	−0.020	0.031	0.001	0.098	0.095	0.075	0.078
Satellite−10.16 cm	0.020	−0.039	0.012	−0.017	0.100	0.100	0.080	0.083
Satellite−20.32 cm	−0.001	−0.060	−0.008	−0.038	0.090	0.102	0.069	0.068
(Satellite−10.16 cm) − (Satellite−5.08 cm)	−0.018	−0.019	−0.019	−0.018	0.002	0.005	0.005	0.005
(Satellite−20.32 cm) − (Satellite−10.16 cm)	−0.021	−0.021	−0.020	−0.021	−0.010	0.002	−0.011	−0.015
(Satellite−20.32 cm) − (Satellite−5.08 cm)	−0.039	−0.040	−0.039	−0.039	−0.008	0.007	−0.006	−0.010

In the separate comparison, *MD* reflected the numerical characteristics of each satellite product well. It continued to grow in a negative direction with the increase in depth, regardless of whether it started out positive or negative. The dry bias of SMOS L2, the enhanced dry bias of SMOS-IC, the strong wet bias of SMAP L2, and the modified wet bias of SMAP L4 were clearly visible. The depth difference between 10 and 5 cm ($-0.19\sim-0.18$ m^3/m^3) was much larger than that between 20 and 10 cm ($-0.07\sim-0.05$ m^3/m^3), while the difference between 20.32 and 10.16 cm ($-0.25\sim-0.23$ m^3/m^3) was somewhat larger than that between 10.16 and 5.08 cm ($-0.18\sim-0.15$ m^3/m^3), also reflecting the stratification characteristics of soil moisture. The *MAD* is actually the mean absolute cumulative difference, which increased slightly with depth. Focusing only on the first two layers, SMAP L4 always had the smallest *MAD*, while the largest *MAD* values were observed for SMAP L2 in the 5/10 group and SMOS L2 in the 5.08/10.16 group, respectively; the difference between 5/5.08 cm and 10/10.16 cm was slightly larger than that between 10/10.16 and 20/20.32 cm.

These results were further confirmed in the simultaneous comparison. *MD* also showed negative growth with depth, but the four satellite products behaved somewhat differently than in the separate comparison. SMOS L2 turned the dry bias into a wet bias, while SMAP L4 showed the opposite trend in the 5/10/20 cm group. SMOS-IC weakened the dry bias in the 5.08/10.16/20.32 cm group, and SMAP L2 weakened its wet bias in the 5/10/20 cm group. However, the interlayer difference remained stable, suggesting that, although the samples were screened in strict temporal matching, their inherent pattern did not change. *MAD* appeared to be slightly smaller in the simultaneous comparison, a ranking of the four satellite products could also be established, but there was still a lack of regularity.

4. Discussion

4.1. The Vertical Distribution Pattern of Surface Soil Moisture

The stratification characteristics of soil moisture (5/10/20, 5.08/10.16/20.32 cm) were studied from three aspects: single-layer distribution, interlayer correlation, and interlayer difference. The fact that soil moisture in the upper layers was less than that in the lower layers seemed to be a stable distribution pattern, as the negative difference (upper–lower) dominated, and to some extent, this can be regarded as a natural response to gravity. The small increase in the mean positive difference (0.020/0.024/0.028 vs. 0.037/0.036/0.040 m^3/m^3) should be noted, as it probably indicated that the soil moisture was close to or at saturation, in other words, that the maximum water capacity of this layer had been reached. The reverse growth reflected by the positive difference could be caused by external random conditions such as precipitation and can be considered an unconventional distribution pattern. Land cover and soil properties appeared to be the main determinants of the vertical distribution of soil moisture, particularly for shallow layers, where the effect of land cover may be greater. These two static variables were coupled and together determine the water-holding capacity of the soil. In conclusion, the absolute values of the positive and negative differences in soil moisture between the layers were very close to or even greater than 0.04 m^3/m^3, indicating that there was significant stratification in the vertical direction and that the effect of depth mismatch on the validation and comparison of satellite soil moisture products should be carefully considered.

4.2. The Difference between the Satellite Products and the In Situ Data

Land cover and soil properties of the sand and clay fractions were considered static variables and were used as the key parameters in the soil moisture retrieval algorithm. Quantification of the difference between the satellite soil moisture products and multilayer in situ measurements under these conditions is expected to provide references for data validation and algorithm optimization.

According to the separate comparison, the numerical difference showed that the satellite soil moisture retrievals had lower values than the in situ measurements. The dominance of the negative difference was likely to be the norm, and the background causing the positive difference could also be precipitation, as it occurred randomly and was mostly a persistent process, leading to an inverse distribution of soil moisture in the vertical direction. Such cases complicate the setting of dynamic conditions and ancillary information such as precipitation, temperature, and wetness, which in turn complicates the retrieval of soil moisture. Therefore, the retrieval optimization should more focus on soil moisture at higher levels, especially when the surface layer is high. It can be seen that the differences between all four satellite products and the 5/5.08 cm in situ data were smaller than the differences between the four satellite products and 10/10.16 and 20/20.32 cm in situ data. A common pattern can be observed in which both the correlation coefficient and the numerical difference increased with the increase in depth.

In terms of simultaneous comparison, it is worth noting that, under each condition, namely, land cover, sand fraction, clay fraction, and soil moisture background, the difference between each satellite product varied with depth, but the order between them was roughly the same at all depths. In each of the products, unique strategies are used for setting these conditions in the soil moisture retrieval algorithm, which ultimately led to different results. The depth mismatch can be related to two aspects in the validation of the satellite products. The first was for the comparison between the satellite products and multilayer in situ data; their difference varied with depth, and the effect of the mismatch was observed. The second was for the comparison between the multisource satellite products; there was no significant change in the relative magnitude of their difference when they were all compared to the same in situ data at a given depth, and the mismatch effect may not be of concern.

In fact, the brightness temperature (TB, L1) was the common source of the soil moisture product at higher levels (L2 and L4). The reasons for the difference between the TB observations of SMOS and SMAP may be mainly due to their detection mechanism, hardware implementation, and reconstruction methods. However, the results of this study showed that the pattern of difference between the four satellite products and the multilayer in situ data did not change significantly with land cover, soil properties, and soil moisture background, which meant that the difference in penetration depth due to the observation conditions may not be large enough to cause the difference between the satellite soil moisture products.

5. Conclusions

Based on the ISMN multilayer in situ data (5, 10, 20, 5.08, 10.16, and 20.32 cm), the stratification characteristics of soil moisture were studied in this paper, and then SMOS (SMOS L2 and SMOS-IC) and SMAP (SMAP L2 and SMAP L4) soil moisture products were compared with the in situ data.

It was found that the soil moisture in the lower layers was usually higher than that in the upper layers, and there was a very significant hierarchical distribution in the vertical direction. The negative and positive differences of soil moisture between the layers were $-0.042/-0.67\sim-0.024/-0.44$ and $0.020/0.036\sim0.028/0.040$ m^3/m^3, respectively, which were close to or even greater than the nominal retrieval accuracy of 0.04 m^3/m^3 of SMOS and SMAP. The comparison showed that the correlation coefficient between the satellite products and the 5/5.08 cm in situ data was the highest, and their numerical difference was the smallest. The mismatch induced by using the 10/10.16 or 20/20.32 cm in situ data as a substitute was about $-0.019\sim-0.018/-0.18\sim-0.015$ m^3/m^3 and $-0.026\sim-0.023/-0.043\sim-0.039$ m^3/m^3 in the mean difference, respectively.

The mismatch of multisource data was mainly in the form of temporal, spatial, and depth mismatch. In previous studies, the influence of the temporal mismatch of SMOS and SMAP was found to be much smaller than the nominal retrieval accuracy of the satellites and can be safely ignored. The depth mismatch was analyzed in this study. It appeared to be larger than the temporal mismatch, according to the numerical differences.

Some shortcomings need to be mentioned. First, under the strict temporal matching, the sample size was too small to support a comparison of the sensitivity to the depth mismatch between satellite products. Second, the comparison between satellite products and multilayer in situ data was only formal, and their numerical differences could be due to multiple effects caused by external conditions such as precipitation, temperature, and wind, leaving much room for further research.

Author Contributions: Conceptualization, methodology, and formal analysis, N.Y.; software, validation, and investigation, F.X.; resources and data curation, H.Z.; writing—original draft preparation, N.Y.; writing—review and editing, F.X.; visualization, H.Z.; project administration and funding acquisition, N.Y. All authors have read and agreed to the published version of the manuscript.

Funding: This work was funded by the State Key Project of the National Natural Science Foundation of China—Key Projects of Joint Fund for Regional Innovation and Development (Grant Number U21A20108) and by the Double First-Class Project Cultivation Special Project—Key Technology for Intelligent Equipment and Intelligent Processing of Spatiotemporal Information (Grant Number 722403/067/004).

Data Availability Statement: No new data were created or analyzed in this study. Data sharing is not applicable to this article.

Acknowledgments: The authors would like to thank the International Soil Moisture Network (ISMN) for making available the field observations on soil moisture. We also wish to thank the European Astronomy Centre (ESAC) SMOS Data Processing Ground Segment (DPGS) for providing SMOS Level 2 data; INRAE BORDEAUX remote sensing products for providing SMOS-IC product; and the National Aeronautics and Space Administration (NASA) Distributed Active Archive Center (DAAC) at National Snow and Ice Data Center (NSIDC) for providing SMAP Level 2 and Level 4 data. We thank the Editorial Review Board Members and reviewers of *Remote Sensing*, for the time they took to review the manuscript and for their valuable feedback.

Conflicts of Interest: The authors declare no conflict of interest.

References

1. Shen, X.J.; Walker, J.P.; Ye, N.; Wu, X.L.; Boopathi, N.; Yeo, I.Y.; Zhang, L.L.; Zhu, L.J. Soil Moisture Retrieval Depth of P- and L-Band Radiometry: Predictions and Observations. *IEEE Trans. Geosci. Remote Sens.* **2021**, *59*, 6814–6822. [CrossRef]
2. Konkathi, P.; Karthikeyan, L. Error and uncertainty characterization of soil moisture and VOD retrievals obtained from L-band SMAP radiometer. *Remote Sens. Environ.* **2022**, *280*, 113146. [CrossRef]
3. Zhang, P.; Zheng, D.H.; van der Velde, R.; Wen, J.; Ma, Y.M.; Zeng, Y.J.; Wang, X.; Wang, Z.L.; Chen, J.L.; Su, Z.B. A dataset of 10-year regional-scale soil moisture and soil temperature measurements at multiple depths on the Tibetan Plateau. *Earth Syst. Sci. Data* **2022**, *14*, 5513–5542. [CrossRef]
4. Hu, F.M.; Wei, Z.S.; Yang, X.N.; Xie, W.J.; Li, Y.X.; Cui, C.L.; Yang, B.B.; Tao, C.X.; Zhang, W.; Meng, L.K. Assessment of SMAP and SMOS soil moisture products using triple collocation method over Inner Mongolia. *J. Hydrol.-Reg. Stud.* **2022**, *40*, 101027. [CrossRef]
5. Wigneron, J.-P.; Jackson, T.J.; O'Neill, P.; De Lannoy, G.; de Rosnay, P.; Walker, J.P.; Ferrazzoli, P.; Mironov, V.; Bircher, S.; Grant, J.P.; et al. Modelling the passive microwave signature from land surfaces: A review of recent results and application to the L-band SMOS & SMAP soil moisture retrieval algorithms. *Remote Sens. Environ.* **2017**, *192*, 238–262.
6. Chan, S.K.; Bindlish, R.; O'Neill, P.; Jackson, T.; Kerr, Y. Development and assessment of the SMAP enhanced passive soil moisture product. *Remote Sens. Environ.* **2018**, *204*, 2539–2542. [CrossRef]
7. Zheng, D.H.; Li, X.; Wang, X.; Wang, Z.L.; Wen, J.; van der Velde, R.; Schwank, M.; Su, Z.B. Sampling depth of L-band radiometer measurements of soil moisture and freeze-thaw dynamics on the Tibetan Plateau. *Remote Sens. Environ.* **2019**, *226*, 16–25. [CrossRef]
8. Kerr, Y.H.; Al-Yaari, A.; Rodriguez-Fernandez, N.; Parrens, M.; Molero, B.; Leroux, D.; Bircher, S.; Mahmoodi, A.; Mialon, A.; Richaume, P.; et al. Overview of SMOS performance in terms of global soil moisture monitoring after six years in operation. *Remote Sens. Environ.* **2016**, *180*, 40–63. [CrossRef]
9. Zeng, J.; Chen, K.-S.; Bi, H.; Quan, Q. A Preliminary Evaluation of the SMAP Radiometer Soil Moisture Product Over United States and Europe Using Ground-Based Measurements. *IEEE Trans. Geosci. Remote Sens.* **2016**, *54*, 4929–4940. [CrossRef]
10. Al-Yaari, A.; Wigneron, J.-P.; Kerr, Y.; Rodriguez-Fernandez, N.; O'Neill, P.E.; Jackson, T.J.; De Lannoy, G.J.M.; Al Bitar, A.; Mialon, A.; Richaume, P.; et al. Evaluating soil moisture retrievals from ESA's SMOS and NASA's SMAP brightness temperature datasets. *Remote Sens. Environ.* **2017**, *193*, 257–273. [CrossRef] [PubMed]
11. Burgin, M.S.; Colliander, A.; Njoku, E.G.; Chan, S.K.; Cabot, F.; Kerr, Y.H.; Bindlish, R.; Jackson, T.J.; Entekhabi, D.; Yueh, S.H. A Comparative Study of the SMAP Passive Soil Moisture Product With Existing Satellite-Based Soil Moisture Products. *IEEE Trans. Geosci. Remote Sens.* **2017**, *55*, 2959–2971. [CrossRef]
12. Das, N.N.; Entekhabi, D.; Dunbar, R.S.; Colliander, A.; Chen, F.; Crow, W.; Jackson, T.J.; Berg, A.; Bosch, D.D.; Caldwell, T.; et al. The SMAP mission combined active-passive soil moisture product at 9 km and 3 km spatial resolutions. *Remote Sens. Environ.* **2018**, *211*, 204–217. [CrossRef]
13. Walker, V.A.; Hornbuckle, B.K.; Cosh, M.H. A Five-Year Evaluation of SMOS Level 2 Soil Moisture in the Corn Belt of the United States. *IEEE J. Sel. Top. Appl. Earth Obs. Remote Sens.* **2018**, *11*, 4664–4675. [CrossRef]
14. Colliander, A.; Cosh, M.H.; Misra, S.; Jackson, T.J.; Crow, W.T.; Powers, J.; Mcnairn, H.; Bullock, P.; Berg, A.; Magagi, R. Comparison of high-resolution airborne soil moisture retrievals to SMAP soil moisture during the SMAP validation experiment 2016 (SMAPVEX16). *Remote Sens. Environ.* **2019**, *227*, 137–150. [CrossRef]

15. Li, X.; Al-Yaari, A.; Schwank, M.; Fan, L.; FFrappart Swenson, J.; Wigneron, J.-P. Compared performances of SMOS-IC soil moisture and vegetation optical depth retrievals based on Tau-Omega and Two-Stream microwave emission models. *Remote Sens. Environ.* **2020**, *236*, 111502. [CrossRef]
16. Ma, H.; Zeng, J.; Chen, N.; Zhang, X.; Wang, W. Satellite surface soil moisture from SMAP, SMOS, AMSR2 and ESA CCI: A comprehensive assessment using global ground-based observations. *Remote Sens. Environ.* **2019**, *231C*, 111215. [CrossRef]
17. Mousa, B.G.; Hong, S. Spatial Evaluation and Assimilation of SMAP, SMOS, and ASCAT Satellite Soil Moisture Products over Africa Using Statistical Techniques. *Earth Space Sci.* **2020**, *7*, e2019EA000841. [CrossRef]
18. Pan, M.; Cai, X.; Chaney, N.W.; Entekhabi, D.; Wood, E.F. An initial assessment of SMAP soil moisture retrievals using high-resolution model simulations and in situ observations. *Geophys. Res. Lett.* **2016**, *43*, 9662–9668. [CrossRef]
19. Wang, Z.; Che, T.; Liou, Y.A. Global Sensitivity Analysis of the L-MEB Model for Retrieving Soil Moisture. *IEEE Trans. Geosci. Remote Sens.* **2016**, *54*, 2949–2962. [CrossRef]
20. Dong, J.; Crow, W.T.; Bindlish, R. The Error Structure of the SMAP Single and Dual Channel Soil Moisture Retrievals. *Geophys. Res. Lett.* **2017**, *45*, 758–765. [CrossRef]
21. Li, D.; Jin, R.; Zhou, J.; Kang, J. Analysis and Reduction of the Uncertainties in Soil Moisture Estimation With the L-MEB Model Using EFAST and Ensemble Retrieval. *IEEE Geosci. Remote Sens.* **2017**, *12*, 1337–1341.
22. Chen, Q.; Zeng, J.; Cui, C.; Li, Z.; Chen, K.S.; Bai, X.; Xu, J. Soil Moisture Retrieval From SMAP: A Validation and Error Analysis Study Using Ground-Based Observations Over the Little Washita Watershed. *IEEE Trans. Geosci. Remote Sens.* **2018**, *56*, 1394–1408. [CrossRef]
23. Chaubell, M.J.; Asanuma, J.; Berg, A.A.; Bosch, D.D.; O'Neill, P.E. Improved SMAP Dual-Channel Algorithm for the Retrieval of Soil Moisture. *IEEE Trans. Geosci. Remote Sens.* **2020**, *58*, 3894–3905. [CrossRef]
24. Mialon, A.; Richaume, P.; Leroux, D.; Bircher, S.; Bitar, A.A.; Pellarin, T.; Wigneron, J.; Kerr, Y.H. Comparison of Dobson and Mironov Dielectric Models in the SMOS Soil Moisture Retrieval Algorithm. *IEEE Trans. Geosci. Remote Sens.* **2015**, *53*, 3084–3094. [CrossRef]
25. Ebrahimi-Khusfi, M.; Alavipanah, S.K.; Hamzeh, S.; Amiraslani, F.; Samany, N.N.; Wigneron, J.P. Comparison of soil moisture retrieval algorithms based on the synergy between SMAP and SMOS-IC. *Int. J. Appl. Earth Obs. Geoinf.* **2018**, *67*, 148–160. [CrossRef]
26. Zheng, D.; Rogier, V.; Wen, J.; Wang, X.; Ferrazzoli, P.; Schwank, M.; Colliander, A.; Bindlish, R.; Su, Z. Assessment of the SMAP Soil Emission Model and Soil Moisture Retrieval Algorithms for a Tibetan Desert Ecosystem. *IEEE Trans. Geosci. Remote Sens.* **2018**, *56*, 3786–3799. [CrossRef]
27. Kang, C.S.; Kanniah, K.D.; Kerr, Y.H. Calibration of SMOS Soil Moisture Retrieval Algorithm: A Case of Tropical Site in Malaysia. *IEEE Trans. Geosci. Remote Sens.* **2019**, *57*, 3827–3839. [CrossRef]
28. Khazal, A.; Richaume, P.; Cabot, F.; Anterrieu, E.; Mialon, A.; Kerr, Y.H. Improving the Spatial Bias Correction Algorithm in SMOS Image Reconstruction Processor: Validation of Soil Moisture Retrievals with In Situ Data. *IEEE Trans. Geosci. Remote Sens.* **2019**, *57*, 277–290. [CrossRef]
29. Zheng, D.H.; Li, X.; Zhao, T.J.; Wen, J.; van der Velde, R.; Schwank, M.; Wang, X.; Wang, Z.L.; Su, Z.B. Impact of Soil Permittivity and Temperature Profile on L-Band Microwave Emission of Frozen Soil. *IEEE Trans. Geosci. Remote Sens.* **2021**, *59*, 4080–4093. [CrossRef]
30. Yee, M.S.; Walker, J.P.; Monerris, A.; Rüdiger, C.; Jackson, T.J. On the identification of representative in situ soil moisture monitoring stations for the validation of SMAP soil moisture products in Australia. *J. Hydrol.* **2016**, *537*, 367–381. [CrossRef]
31. González-Zamora, Á.; Sánchez, N.; Pablos, M.; Martínez-Fernández, J. CCI soil moisture assessment with SMOS soil moisture and in situ data under different environmental conditions and spatial scales in Spain. *Remote Sens. Environ.* **2019**, *225*, 469–482. [CrossRef]
32. Whitcomb, J.; Clewley, D.; Colliander, A.; Cosh, M.H.; Moghaddam, M. Evaluation of SMAP Core Validation Site Representativeness Errors Using Dense Networks of In Situ Sensors and Random Forests. *IEEE J. Sel. Top. Appl. Earth Obs. Remote Sens.* **2020**, *13*, 6457–6472. [CrossRef]
33. Zhang, P.; Zheng, D.H.; van der Velde, R.; Wen, J.; Zeng, Y.J.; Wang, X.; Wang, Z.L.; Chen, J.L.; Su, Z.B. Status of the Tibetan Plateau observatory (Tibet-Obs) and a 10-year (2009–2019) surface soil moisture dataset. *Earth Syst. Sci. Data* **2021**, *13*, 3075–3102. [CrossRef]
34. Yang, N.; Tang, Y.; Chen, Y.; Xiang, F. Study on Stability of Surface Soil Moisture and Other Meteorological Variables within Time Intervals of SMOS and SMAP. *IEEE Geosci. Remote Sens. Lett.* **2021**, *18*, 1911–1915. [CrossRef]
35. Dorigo, W.; Himmelbauer, I.; Aberer, D.; Schremmer, L.; Sabia, R. The International Soil Moisture Network: Serving Earth system science for over a decade. *Hydrol. Earth Syst. Sci.* **2021**, *25*, 5749–5804. [CrossRef]
36. Yi, C.X.; Li, X.J.; Zeng, J.Y.; Fan, L.; Xie, Z.Q.; Gao, L.; Xing, Z.P.; Ma, H.L.; Boudah, A.; Zhou, H.W.; et al. Assessment of five SMAP soil moisture products using ISMN ground-based measurements over varied environmental conditions. *J. Hydrol.* **2023**, *619*, 129325. [CrossRef]
37. Colliander, A.; Kerr, Y.; Wigneron, J.P.; Al-Yaari, A.; Rodriguez-Fernandez, N.; Li, X.; Chaubell, J.; Richaume, P.; Mialon, A.; Asanuma, J.; et al. Performance of SMOS Soil Moisture Products Over Core Validation Sites. *IEEE Geosci. Remote Sens. Lett.* **2023**, *20*, 2502805. [CrossRef]

38. Pascal, C.; Ferrant, S.; Rodriguez-Fernandez, N.; Kerr, Y.; Selles, A.; Merlin, O. Indicator of Flood-Irrigated Crops From SMOS and SMAP Soil Moisture Products in Southern India. *IEEE Geosci. Remote Sens. Lett.* **2023**, *20*, 4500205. [CrossRef]
39. Fernandez-Moran, R.; Al-Yaari, A.; Mialon, A.; Mahmoodi, A.; Al Bitar, A.; De Lannoy, G.; Rodriguez-Fernandez, N.; Lopez-Baeza, E.; Kerr, Y.; Wigneron, J.P. SMOS-IC: An Alternative SMOS Soil Moisture and Vegetation Optical Depth Product. *Remote Sens.* **2017**, *9*, 457. [CrossRef]
40. Wigneron, J.P.; Li, X.; Frappart, F.; Fan, L.; Moisy, C. SMOS-IC data record of soil moisture and L-VOD: Historical development, applications and perspectives. *Remote Sens. Environ.* **2021**, *254*, 112238. [CrossRef]
41. van der Velde, R.; Colliander, A.; Pezij, M.; Benninga, H.J.F.; Bindlish, R.; Chan, S.K.; Jackson, T.J.; Hendriks, D.M.D.; Augustijn, D.C.M.; Su, Z.B. Validation of SMAP L2 passive-only soil moisture products using upscaled in situ measurements collected in Twente, the Netherlands. *Hydrol. Earth Syst. Sci.* **2021**, *25*, 473–495. [CrossRef]
42. Du, J.Y.; Kimball, J.S.; Chan, S.K.; Chaubell, M.J.; Bindlish, R.; Dunbar, R.S.; Colliander, A. Assessment of Surface Fractional Water Impacts on SMAP Soil Moisture Retrieval. *IEEE J. Sel. Top. Appl. Earth Obs. Remote Sens.* **2023**, *16*, 4871–4881. [CrossRef]
43. Purdy, A.J.; Fisher, J.B.; Goulden, M.L.; Colliander, A.; Halverson, G.; Tu, K.; Farniglietti, J.S. SMAP soil moisture improves global evapotranspiration. *Remote Sens. Environ.* **2018**, *219*, 1–14. [CrossRef]
44. Tavakol, A.; Rahmani, V.; Quiring, S.M.; Kumar, S.V. Evaluation analysis of NASA SMAP L3 and L4 and SPoRT-LIS soil moisture data in the United States. *Remote Sens. Environ.* **2019**, *229*, 234–246. [CrossRef]
45. Li, S.P.; Sawada, Y. Soil moisture-vegetation interaction from near-global in-situ soil moisture measurements. *Environ. Res. Lett.* **2022**, *17*, 114028. [CrossRef]
46. Kivi, M.; Vergopolan, N.; Dokoohaki, H. A comprehensive assessment of in situ and remote sensing soil moisture data assimilation in the APSIM model for improving agricultural forecastingacross the US Midwest. *Hydrol. Earth Syst. Sci.* **2023**, *27*, 1173–1199. [CrossRef]
47. Fan, X.W.; Zhao, X.S.; Pan, X.; Liu, Y.W.; Liu, Y.B. Investigating multiple causes of time-varying SMAP soil moisture biases based on core validation sites data. *J. Hydrol.* **2022**, *612*, 128151. [CrossRef]
48. Gupta, D.K.; Srivastava, P.K.; Pandey, D.K.; Chaudhary, S.K.; Prasad, R.; O'Neill, P.E. Passive Only Microwave Soil Moisture Retrieval in Indian Cropping Conditions: Model Parameterization and Validation. *IEEE Trans. Geosci. Remote Sens.* **2023**, *61*, 4400412. [CrossRef]
49. Hong, Z.; Moreno, H.A.; Li, Z.; Li, S.; Greene, J.S.; Hong, Y.; Alvarez, L.V. Triple Collocation of Ground-, Satellite- and Land Surface Model-Based Surface Soil Moisture Products in Oklahoma-Part I: Individual Product Assessment. *Remote Sens.* **2022**, *14*, 5641. [CrossRef]
50. Zhu, L.Y.; Li, W.J.; Wang, H.Q.; Deng, X.D.; Tong, C.; He, S.; Wang, K. Merging Microwave, Optical, and Reanalysis Data for 1 Km Daily Soil Moisture by Triple Collocation. *Remote Sens.* **2023**, *15*, 159. [CrossRef]

Article

A Performance Analysis of Soil Dielectric Models over Organic Soils in Alaska for Passive Microwave Remote Sensing of Soil Moisture

Runze Zhang [1,*], Steven Chan [2], Rajat Bindlish [3] and Venkataraman Lakshmi [1]

1 Department of Engineering Systems and Environment, University of Virginia, Charlottesville, VA 22904, USA; vlakshmi@virginia.edu
2 NASA Jet Propulsion Laboratory, California Institute of Technology, Pasadena, CA 91109, USA; steventsz.k.chan@jpl.nasa.gov
3 NASA Goddard Space Flight Center, Greenbelt, MD 20771, USA; rajat.bindlish@nasa.gov
* Correspondence: rz4pd@virginia.edu

Abstract: Passive microwave remote sensing of soil moisture (SM) requires a physically based dielectric model that quantitatively converts the volumetric SM into the soil bulk dielectric constant. Mironov 2009 is the dielectric model used in the operational SM retrieval algorithms of the NASA Soil Moisture Active Passive (SMAP) and the ESA Soil Moisture and Ocean Salinity (SMOS) missions. However, Mironov 2009 suffers a challenge in deriving SM over organic soils, as it does not account for the impact of soil organic matter (SOM) on the soil bulk dielectric constant. To this end, we presented a comparative performance analysis of nine advanced soil dielectric models over organic soil in Alaska, four of which incorporate SOM. In the framework of the SMAP single-channel algorithm at vertical polarization (SCA-V), SM retrievals from different dielectric models were derived using an iterative optimization scheme. The skills of the different dielectric models over organic soils were reflected by the performance of their respective SM retrievals, which was measured by four conventional statistical metrics, calculated by comparing satellite-based SM time series with in-situ benchmarks. Overall, SM retrievals of organic-soil-based dielectric models tended to overestimate, while those from mineral-soil-based models displayed dry biases. All the models showed comparable values of unbiased root-mean-square error (ubRMSE) and Pearson Correlation (R), but Mironov 2019 exhibited a slight but consistent edge over the others. An integrated consideration of the model inputs, the physical basis, and the validated accuracy indicated that the separate use of Mironov 2009 and Mironov 2019 in the SMAP SCA-V for mineral soils (SOM $<15\%$) and organic soils (SOM $\geq 15\%$) would be the preferred option.

Keywords: soil moisture; dielectric models; SMAP; soil organic matter

Citation: Zhang, R.; Chan, S.; Bindlish, R.; Lakshmi, V. A Performance Analysis of Soil Dielectric Models over Organic Soils in Alaska for Passive Microwave Remote Sensing of Soil Moisture. *Remote Sens.* **2023**, *15*, 1658. https://doi.org/10.3390/rs15061658

Academic Editors: Jiangyuan Zeng, Jian Peng, Wei Zhao, Chunfeng Ma and Hongliang Ma

Received: 28 January 2023
Revised: 16 March 2023
Accepted: 17 March 2023
Published: 19 March 2023

1. Introduction

Passive microwave remote sensing is considered the most suitable tool for mapping spatial soil wetness, owing to the negligible atmospheric influence and less interference from canopy and surface roughness [1,2]. The remarkable performance of soil moisture (SM) retrievals from spaceborne L-band radiometers (i.e., soil moisture and ocean salinity (SMOS) [3] and soil moisture active passive (SMAP) [4]) has been substantiated by a number of validation studies [5–9]. The mechanism that physically bridges the surface emission at microwave bands and surface SM is based on the contrasting difference between the dielectric constants of liquid water (~80) and dry soil (~4) [10]. The dielectric model that quantitatively links the SM with the bulk dielectric constant of the soil–water–air system is therefore critical in the retrieval algorithms of SMOS and SMAP.

Recently, numerous dielectric models were developed and applied for both spaceborne microwave radiometers and in-situ electromagnetic sensors [11]. An ideal dielectric model is envisioned, to accurately account for the dielectric response of wet soils as a function of all the relevant factors, including soil compaction, soil composition, the fraction of bound and free water, salinity, soil temperature, soil particle size distribution, and observation frequency, etc. [12]. However, the practical dielectric models are often established on a limited set of soil properties and are unable to approximate proper dielectric constants for all the surface conditions. Previous studies found that applying mineral-soil-based dielectric models over organic soils could lead to a substantial underestimation of SM [11]. [13] revealed a significant drop in SMAP retrieval quality in regions with soil organic carbon (SOC) exceeding 8.72%. Given that Mironov 2009 [14], currently used in the SMOS and SMAP operation algorithms, was developed exclusively on samples of mineral soils, an update on the dielectric model that incorporates the effect of soil organic matter (SOM) is pressingly required for areas with organic-rich soils.

The influence of SOM on the bulk dielectric constant of the soil–water system is often summarized in two aspects. First, organic substrates have larger specific surface areas than minerals, indicating that organic soil has a higher fraction of bound water relative to mineral soil, when they contain the same amount of water [11,15,16]. As such, at the same moisture, the dielectric constant of organic soil tends to be lower than that of mineral soil, as the dielectric constant of bound water is much smaller than that of free water. Second, organic soil is often marked by a larger porosity than mineral soil, due to its complex structure [11,15–17]. Based on these principles, several organic-soil-based dielectric models have been developed in recent years.

Although model developers pointed out the potential applicability of their models in the retrieval of SM, assessment of the efficacy of these newly developed organic-soil-based dielectric models in the derivation of passive microwave remote sensing of SM has not been widely carried out. In light of these considerations, nine advanced dielectric mixing models were selected and tested in the context of the SMAP single-channel algorithm at vertical polarization (SCA-V) [18]. This study has two major objectives: (1) present the differences between the available mineral- and organic-soil-based models, in describing the complex dielectric behaviors of wet soils under various SOM conditions; and (2) evaluate their performance in organic-rich soils. The latter was achieved by comparing the SCA-V SM retrievals from different models against in-situ measurements scattered over Alaska, where the soils are identified with a noticeably higher SOM (~25%) relative to the global average level (Figure A1). The dielectric models considered here have been classified as mineral-soil-based dielectric models, including Wang 1980 [19], the semi-empirical Dobson 1985 modified by Peplinski 1995 [12,20] (hereafter Dobson 1985), the prevalent Mironov 2009 [14], Mironov 2012 [21], and Park 2017 [22], and organic-soil-based dielectric models, including the natural log fitting model in [11] (hereafter Bircher 2016), Mironov 2019 [23], Park 2019 [16], and Park 2021 [24].

As introduced earlier, five mineral-soil-based dielectric models were selected for a comprehensive survey of diverse models in the framework of the SMAP SCA-V algorithm over organic-rich soils. Two of them, Mironov 2013 and Park 2017, have not been widely examined under the SMOS and SMAP schemes [22,25]. In contrast, the other three classic models have been extensively assessed in wide domains covered by mineral soils [26–28]. However, their performances over regions with high SOM proportions have not been well-studied and compared with those of dedicated organic-soil-based models. In addition to water volume, mineral-soil-based models primarily focus on the influence of soil texture, commonly characterized by sand, clay, and silt. Yet, organic-soil-based models place a greater emphasis on the SOM effect. Mironov 2019, for example, describes all parameters as functions of SOM rather than the clay percentage used in Mironov 2009 [23]. Therefore, incorporating more mineral- and organic-soil-based models may also help to construct an impression of their systematic differences when describing the dielectric behaviors of organic soils.

The paper is organized as follows. In Section 2, all the data sets and preprocessing steps are presented. Next are the workflow of in-situ measurements screening and the partial SMAP SCA-V retrieval process used to derive the SM from the identical observations and different models (Section 3). The results of the synthetic experiments, validation consequences over Alaska, and a detailed discussion are subsequently displayed in Section 4. Finally, the conclusions are followed by a brief summary presented in Section 5.

2. Data

2.1. SMAP L2 Radiometer Half-Orbit 36 km EASE-Grid Soil Moisture, Version 8

Launched on 31 January 2015, the SMAP mission was designed to map high-resolution SM and freeze/thaw state by combining the attributes of L-band radar and radiometer. However, the SMAP SM products presently rely on radiometer observations alone, due to an unexpected malfunction of the SMAP radar in July 2015. With an average revisit frequency of two to three days, the SMAP sensors cross the Equator at the local solar times of 6 a.m. and 6 p.m.

SMAP L2 Radiometer Half-Orbit 36 km EASE-Grid Soil Moisture, Version 8 (SMAP V8) [29] was adopted in this study. Here, we only used the descending (6 a.m.) SM retrievals derived using the SCA-V algorithm. A series of masking procedures were utilized to avoid the application of SM retrievals of low accuracy and high uncertainty. Specifically, only the retrievals flagged as the "recommended quality" were retained and employed in the later analysis. Given Alaska, the focused region of this study, is located in the high-latitude portion with a long-term frozen duration, we only considered those qualified SM retrievals within the time intervals from June to August, between 2015 and 2021.

One noticeable improvement in SMAP V8 (relative to the older version) is the update and extension of gridded soil parameters, ranging from SOC, silt and sand fractions to bulk density. These newly added soil attributes originate from the SoilGrid 250 m [30] and replace the earlier patched version composed of the National Soil Data Canada (NSDC), the State Soil Geographic Database (STATSGO), the Australia Soil Resources Information System (ASRIS), and the Harmonized World Soil Database (HWSD) [31]. Since these soil attributes are often necessary inputs for dielectric models of soil, they were also extracted from the SMAP V8.

2.2. In-Situ Soil Moisture Measurements

Ground-based SM measurements over Alaska were employed as benchmarks to assess the skills of the diverse dielectric mixing models. Historical files of soil water content observed by in-situ sensors were first downloaded from the Natural Resources Conservation Service (NRCS), the National Water and Climate Center (NWCC) homepage (https://www.nrcs.usda.gov/wps/portal/wcc/home (accessed on 7 April 2022)). At present, there are more than 40 operating stations from the Snow Telemetry (SNOTEL) [32] and Soil Climate and Analysis Network (SCAN) [33]. These stations are able to monitor the sub-daily variations of SM and many other climatic variables in near real time.

However, some typical errors [34] of in-situ SM readings, such as breaks and plateaus, were found before their application. As a response, the other authoritative data source of in-situ SM, the International Soil Moisture Network (ISMN) [35,36], was also considered, aiming at incorporating its flag information. Given the limited stations in Alaska, it is expected that SM data from the above two sources (NWCC and ISMN) are mostly from the same set of stations. Additionally, for the same station, the observed SM time series from the NWCC and ISMN should be identical, as the ISMN only gathers data and harmonizes them in units and time steps, without extra data processing. Given the frequently abnormal SM readings (even after adopting the quality flag) and the necessity of checking the consistency of SM measurements from two different sources, several rigorous pre-checking procedures were applied (as described in Section 3.1) to filter out those suspicious observations where possible in advance.

3. Methodology

3.1. Preliminary Examination of In-Situ Measurements

The quality of in-situ SM data is of great importance, as these ground measurements are generally seen as the benchmark for evaluating remotely sensed and/or modeled SM data sets [5–7]. However, monitoring SM dynamics over high-latitude regions is still challenging, due to the long-term frozen periods and harsh environments. Such difficulties have been reflected by the flat limbs and breaks frequently occurring in the SM time series from the Alaskan stations. Given those, a careful examination of in-situ SM measurements is necessary.

The general workflow of the preliminary examination steps is delineated in Figure 1. Specifically, the in-situ SM data measured at the local time of 6 a.m. and 6 p.m. (temporally align with the SMAP overpass time) were first extracted from the NWCC and ISMN stations. SM measurements with the corresponding land surface temperature below 4 °C were excluded, as [6] demonstrates that some sensors begin to behave abnormally under this temperature. Meanwhile, the utilization of such a threshold would also be helpful to filter out those SM measurements likely obtained during a period of active thawing and re-freezing, where SM fluctuations are excessively unstable (e.g., Figure 3c in [34]). Additionally, stations with a distance shorter than 36 km to large water bodies or oceans were also masked, as the SMAP SM over those regions is likely influenced by water contamination. The flag information from the ISMN was also incorporated to filter the in-situ data of low quality.

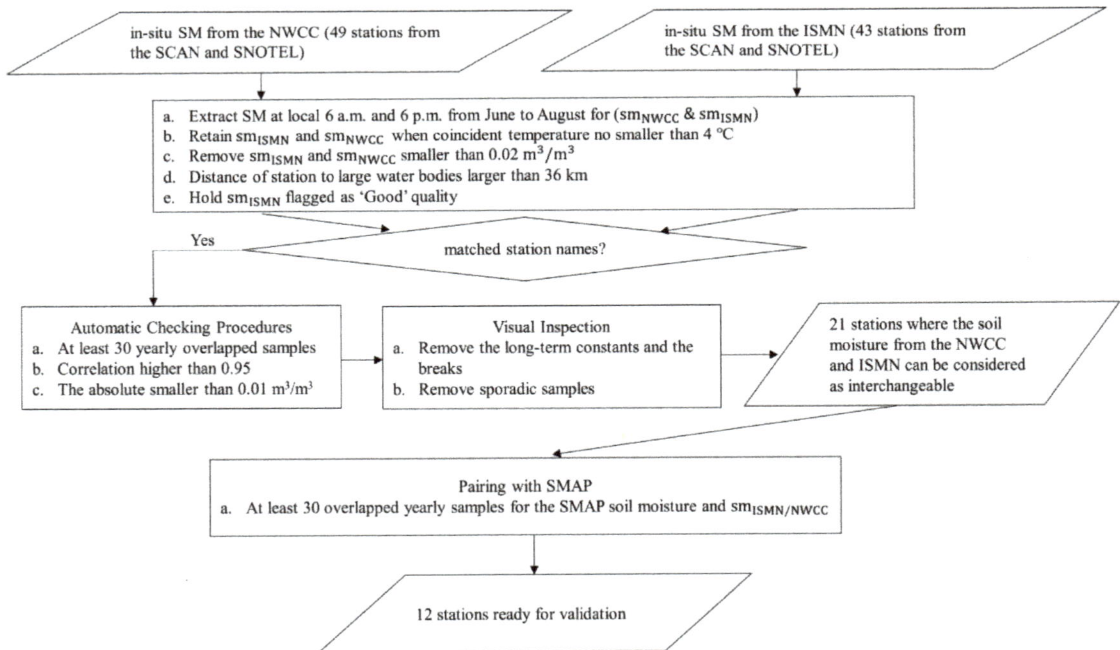

Figure 1. Flow chart of the preliminary examination of Alaskan in-situ soil moisture obtained from the NWCC and ISMN.

The matched SM data of the overlapped stations from the NWCC and ISMN are anticipated, and this greater consistency further enhances the reliability of these benchmarks. Therefore, an automatic consistency checking procedure, constrained by three requirements, was applied. Since breaks and plateaus still appeared on the SM time series after consistency checking, a manual visual inspection was then performed to screen these suspicious measurements. After those, there were 21 qualified stations left, and we assumed that their SM data from the NWCC and ISMN are interchangeable. Furthermore, pairing with the SMAP observations removed nine stations, and the remaining 12 stations (Figure S1) were used in the later validation steps.

3.2. Derivation of Soil Moisture from Various Dielectric Models

In the SCA-V algorithm, the SMAP SM value is determined when there is a minimized difference between the simulated and the observed reflectivity (r_{smap}) (reflectivity = 1 − emissivity) of smooth soil. At each temporal step, the value of r_{smap} over a pixel is fixed, as the SMAP SCA algorithm determines the radiative contribution from the canopy layer and the impact of surface roughness before subtracting them from SMAP observed surface brightness temperature (T_B). Hence, the influence of adopting different dielectric constant models on SM retrievals can be examined using the iterative feedback-loop procedure, to minimize the difference between the simulated reflectivity (r_{est}) and r_{smap}, and without the need to construct the whole process from SM to T_B, in consideration of simplicity.

However, r_{smap} is an intermediate product and unavailable in the original SMAP data set. Given this, the values of r_{smap} were first estimated leveraging SMAP SM and Mironov 2009. With these benchmarks, the SM retrievals of other dielectric models were then acquired based on the optimization flow described in Figure 2. Notably, the SM retrieval at a given time point is reproducible when the identical r_{smap} and model are used.

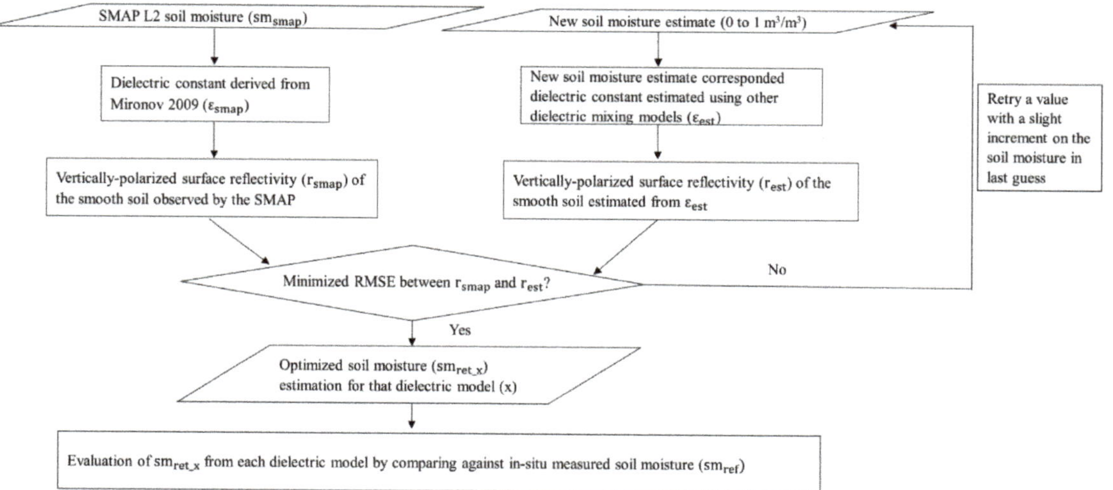

Figure 2. Flow chart that describes the retrieval of soil moisture using different dielectric models, based on identical SMAP observations.

3.3. Performance Metrics

The capability of the remote sensing SM data set has been described by four conventional metrics, which are bias, root-mean-square error (RMSE), unbiased root-mean-square error (ubRMSE), and the Pearson Correlation (R) [37]. These metrics could effectively reflect the discrepancies in terms of magnitude, as well as the links of the temporal evolutions between the SM estimations and the ground truth. The formulas used to compute these metrics are shown in Equations (1)–(4), where E [. . .] represents the arithmetic mean; and σ_{opt} and σ_{ref} denote the standard deviations of SM retrievals of the respective dielectric models and in-situ measured SM.

$$\text{bias} = E[sm_{ret}] - E[sm_{ref}] \quad (1)$$

$$\mathrm{RMSE} = \sqrt{\mathrm{E}\left[(\mathrm{sm_{ret}} - \mathrm{sm_{ref}})^2\right]} \quad (2)$$

$$\mathrm{ubRMSE} = \sqrt{\mathrm{RMSE}^2 - \mathrm{bias}^2} \quad (3)$$

$$\mathrm{R} = \frac{\mathrm{E}[(\mathrm{sm_{ret}} - \mathrm{E}[\mathrm{sm_{ret}}])(\mathrm{sm_{ref}} - \mathrm{E}[\mathrm{sm_{ref}}])]}{\sigma_{\mathrm{ret}}\sigma_{\mathrm{ref}}} \quad (4)$$

4. Results and Discussion

4.1. Simulated Brightness Temperature of Smooth Soil through Synthetic Experiments

Synthetic experiments have the capability to afford complete dielectric responses to a whole SM range, by artificially controlling all the inputs required for the dielectric models (Table 1). With the SOM increasing from 0% to 75% at a step of 15%, the differences between the dielectric constants estimated by mineral- and organic-soil-based dielectric models were explored. These various dielectric responses were further transferred to their corresponding thermal radiations of smooth soils, represented by the vertically polarized T_B.

Table 1. Input variables required for the nine dielectric models.

Model Inputs	Mineral Soil Based Models					Organic Soil Based Models			
	Wang 1980	**Dobson 1985**	**Mironov 2009**	**Mironov 2013**	**Park 2017**	**Bircher 2016**	**Mironov 2019**	**Park 2019**	**Park 2021**
Soil Moisture	Volumetric Soil Moisture (m^3/m^3)	Volumetric Soil Moisture (m^3/m^3)	Volumetric Soil Moisture (m^3/m^3)	Volumetric Soil Moisture (m^3/m^3)	Volumetric Soil Moisture (m^3/m^3)	Volumetric Soil Moisture (m^3/m^3)	Gravimetric Soil Moisture (g/g)	Volumetric Soil Moisture (m^3/m^3)	Volumetric Soil Moisture (m^3/m^3)
Soil Organic Matter	/	/	/	/	/	/	Gravimetric Soil Organic Matter (%)	Gravimetric Soil Organic Matter (%)	Gravimetric Soil Organic Matter (%)
Clay	Gravimetric Clay Fraction (0–1)	Gravimetric Clay Fraction (0–1)	Gravimetric Clay Fraction (%)	Gravimetric Clay Fraction (%)	Volumetric Clay Fraction (0–1)	/	/	Volumetric Clay Fraction (0–1)	Volumetric Clay Fraction (0–1)
Sand	Gravimetric Sand Fraction (0–1)	Gravimetric Sand Fraction (0–1)	/	/	Volumetric Sand Fraction (0–1)	/	/	Volumetric Sand Fraction (0–1)	Volumetric Sand Fraction (0–1)
Silt	/	/	/	/	Volumetric Silt Fraction (0–1)	/	/	Volumetric Silt Fraction (0–1)	Volumetric Silt Fraction (0–1)
Bulk Density	Bulk Density (g/cm^3)	Bulk Density (g/cm^3)	/	/	/	/	Bulk Density (g/cm^3)	/	/
Frequency	/	Frequency (Hz)	Frequency (Hz)	/	Frequency (Hz)	/	/	Frequency (Hz)	Frequency (Hz)
Salinity	/	/	/	/	Salinity (‰)	/	/	Salinity (‰)	Salinity (‰)
Soil Temperature	/	Soil Temperature (°C)	/	Soil Temperature (°C)	Soil Temperature (°C)	/	Soil Temperature (°C)	Soil Temperature (°C)	Soil Temperature (°C)
Total Number of Inputs	4	6	3	3	7	1	4	8	8

Figure 3 presents the T_B curves derived using different dielectric models, across the range of SM from 0 to 0.8 m^3/m^3. Generally, the T_B values estimated using organic-soil-based models are greater than those derived using the mineral-soil-based models, particularly when SOM exceeds 15% and the SM is higher than 0.1 m^3/m^3. In other words, the SM retrievals from organic-soil-based models tend to be wetter than the SM retrievals from mineral-soil-based models (e.g., Mironov 2009) given the same surface reflectivity (or T_B) of bare, smooth soil. The discrepancies between the simulated T_B magnitudes from mineral- and organic-soil-based models further grow with the increase of SOM (Figure 3). However, it should be noted that the estimated dielectric constants and their subsequent T_B values from mineral-soil-based models do not vary with SOM. The higher SM estimations of organic-soil-based models relative to mineral-soil-based models could be attributed to the fact that these organic-soil-based models assume a higher volumetric proportion of bound water [11,15,16]. When the SOM is at 15% (and below), the simulated T_B curves from all the considered models are clustered together, bounded by Dobson 1985 and Bircher 2016 (Figure 3b). Therefore, the SOM of 15% might be treated as an appropriate demarcation point for the separate use of mineral- and organic-soil-based dielectric models over mineral soils and organic soils.

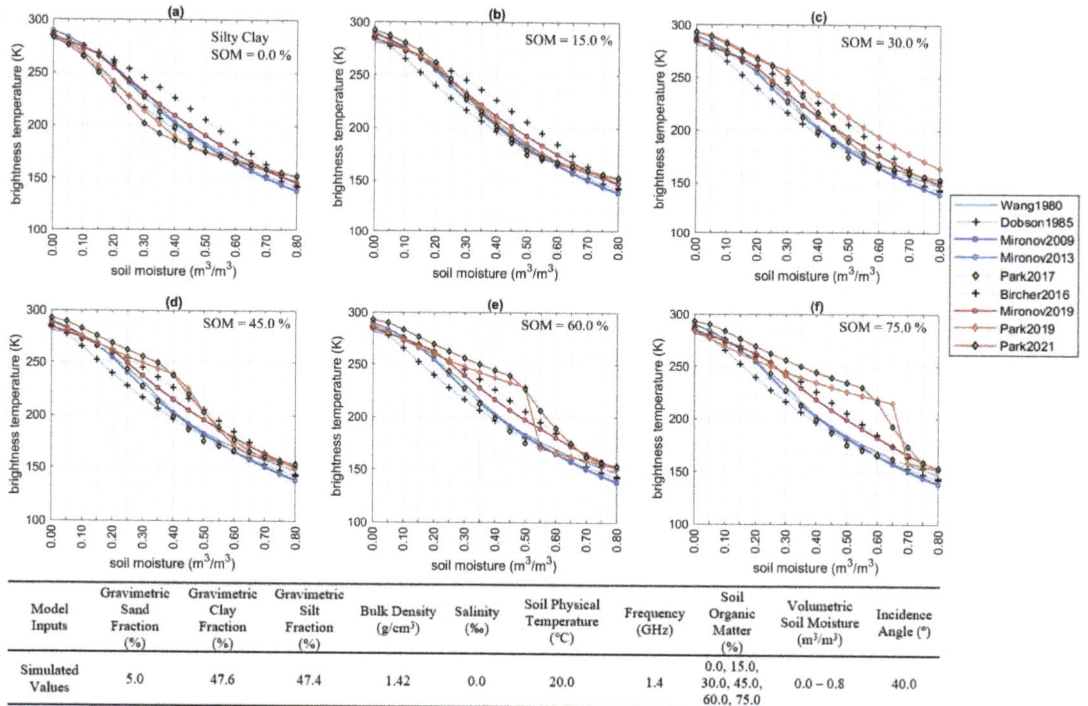

Model Inputs	Gravimetric Sand Fraction (%)	Gravimetric Clay Fraction (%)	Gravimetric Silt Fraction (%)	Bulk Density (g/cm³)	Salinity (‰)	Soil Physical Temperature (°C)	Frequency (GHz)	Soil Organic Matter (%)	Volumetric Soil Moisture (m³/m³)	Incidence Angle (°)
Simulated Values	5.0	47.6	47.4	1.42	0.0	20.0	1.4	0.0, 15.0, 30.0, 45.0, 60.0, 75.0	0.0 – 0.8	40.0

Figure 3. Simulated brightness temperature of a silty clay with various soil organic matter, and the accompanying table displays all the input values, where most soil parameters were directly taken from the sample of silty clay used in [38]. (**a**–**f**) represent the simulated brightness temperature curves variations across various soil organic matter with an increase step of 15%.

Moreover, similar features of the T_B curves of those considered dielectric models have been observed when a sandy sample is tested (Figure S2). Such a stable-magnitude discrepancy between the red curves (organic models) and the blue curves (mineral models) under contrasting textures (sandy and clay soils) can be attributed to the insensitivity of the organic-soil-based dielectric models to soil texture. For example, Mironov 2019 only accounts for the effects of soil moisture, SOM, and soil temperature on the dielectric permittivity of organic soils (Table 1). Although Park 2019 and Park 2021 incorporate both textural and SOM information, the differences in their estimated T_B values from sandy and clay samples seem insignificant under the same SOM level (Figures 3 and S2).

Compared to Mironov 2019, the influence of organic content on the simulated T_B magnitude seem more pronounced for Park 2019 and Park 2021. When the SOM increases from 0% to 75% and the SM values are smaller than 0.5 m^3/m^3, the T_B curve of Park 2021 jumps from the bottom to the top line, with a varying amplitude on the order of tens of Kelvins (Figure 3). In contrast, as a response to the growing SOM, the estimations from Mironov 2019 slowly move upward, approaching the T_B curve of Bircher 2016. According to Figure 3e,f, there is a rapidly dropping segment on the T_B curve of Park 2019. Such abnormal dielectric behavior can be attributed to the improper formulas used to calculate the wilting point and porosity, with a detailed explanation in Section 4.4.

4.2. Evaluation of Dielectric Models over In-Situ Sites in Alaska

Here, SM measurements from 12 sites served as benchmarks to evaluate the skills of the multiple dielectric models in the setting of SMAP observations and the SCA-V algorithm. Before inter-comparison, it was found that the assessment metrics of the satellite-based SM retrievals over the same pixel could vary a lot in different years. Using the time series in Monument Creek as an instance (Figure 4), the R values ranged from 0.18 (2017) to 0.69 (2015). Hence, the obtained metrics (Tables 2–4) averaged over multiple years of each station might be underrated, as they may have been compromised by abnormal behavior in one year. Additionally, the amplitudes and frequencies of in-situ SM variations are often more pronounced relative to the SM retrievals, as the latter reflects the changes over a coarse spatial extent (Figure 4). SM variations at local scales often cannot be captured by the 36 km-scale SM retrievals, due to the omission of spatial variability within the footprint-scale area. As noted by [39], spatial mismatching between satellite SM retrievals and point-scale in-situ measurements could adversely impact the perceived accuracy of SMAP observations.

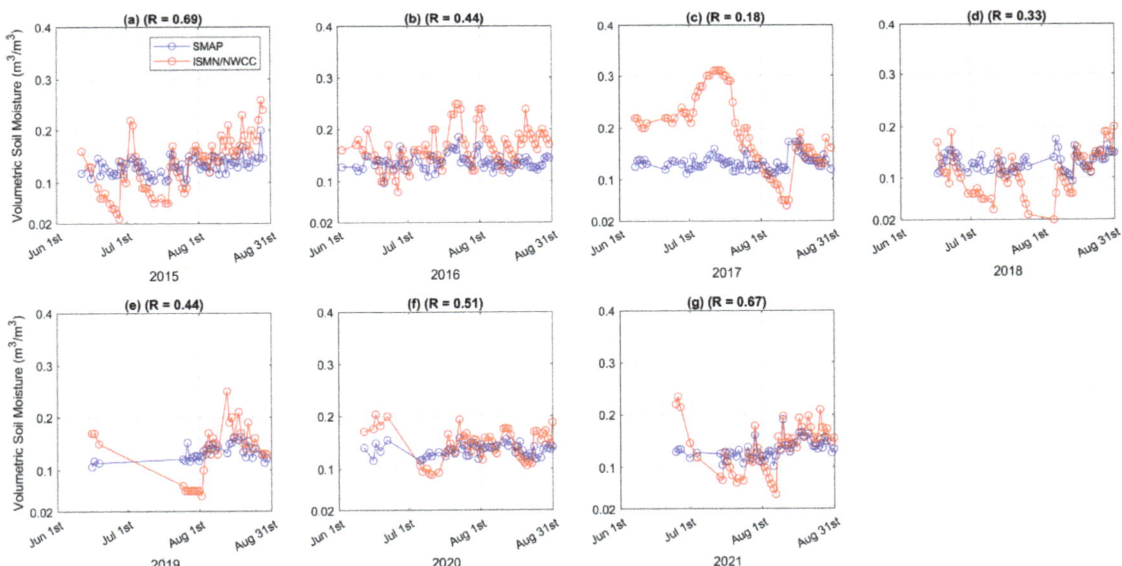

Figure 4. Time series of soil moisture derived from satellite observations and in-situ measurements at Monument Creek (65.18° N, 145.87° W). (**a–g**) describe the soil moisture variations of SMAP retrievals and ground measurements from 2015 to 2021.

Table 2. Bias of soil moisture retrievals using various dielectric models over in-situ sites in Alaska, where biases from mineral- and organic-soil-based models tend to underestimate and overestimate relative to in-situ measurements.

Station/Bias (m^3/m^3)	N	Mineral Soil Based Models					Organic Soil Based Models			
		Wang 1980	Dobson 1985	Mironov 2009	Mironov 2013	Park 2017	Bircher 2016	Mironov 2019	Park 2019	Park 2021
Gulkana River	72	0.058	**0.025**	0.046	0.044	0.039	0.195	0.142	0.104	0.085
Spring Creek	37	−0.108	−0.153	−0.137	−0.137	−0.139	**−0.022**	−0.051	−0.105	−0.109
Atigun Pass	81	0.047	**−0.002**	0.015	0.016	0.009	0.092	0.092	0.044	0.061
Coldfoot	156	−0.085	−0.133	−0.121	−0.121	−0.124	**−0.030**	−0.036	−0.083	−0.067
Eagle Summit	320	−0.028	−0.068	−0.062	−0.061	−0.068	**0.014**	0.017	−0.033	−0.015
Gobblers Knob	262	0.031	−0.010	**−0.003**	−0.003	−0.007	0.096	0.083	0.039	0.055
Monahan Flat	121	−0.047	−0.093	−0.076	−0.077	−0.081	0.035	**0.009**	−0.029	−0.029
Monument Creek	405	0.018	−0.022	**−0.014**	−0.014	−0.016	0.091	0.073	0.029	0.041
Mt. Ryan	194	0.114	**0.078**	0.082	0.082	0.080	0.196	0.172	0.132	0.142
Munson Ridge	383	0.018	−0.019	**−0.015**	−0.015	−0.016	0.096	0.075	0.034	0.045
Tokositna Valley	253	0.014	−0.008	**−0.006**	−0.008	−0.008	0.147	0.093	0.062	0.046
Upper Nome Creek	283	−0.138	−0.180	−0.171	−0.171	−0.176	**−0.086**	−0.091	−0.138	−0.120
Mean	214	−0.009	−0.049	−0.038	−0.039	−0.042	0.069	0.048	**0.005**	0.011

Where the column of the number in bold font represents the dielectric model with the smallest absolute bias in that station or mean, and 'N' in the second column represents the total number of paired SMAP retrievals and in-situ SM measurements used to calculate the bias for each station.

Table 3. ubRMSE of soil moisture retrievals using various dielectric models over in-situ sites in Alaska.

Station/ubRMSE (m^3/m^3)	N	Mineral Soil Based Models					Organic Soil Based Models			
		Wang 1980	Dobson 1985	Mironov 2009	Mironov 2013	Park 2017	Bircher 2016	Mironov 2019	Park 2019	Park 2021
Gulkana River	72	**0.0132**	0.0164	0.0156	0.0154	0.0152	0.0209	0.0180	0.0169	0.0138
Spring Creek	37	0.0460	0.0457	0.0452	0.0454	0.0455	**0.0408**	0.0428	0.0446	0.0462
Atigun Pass	81	0.0311	0.0311	0.0311	0.0311	0.0311	0.0317	0.0311	0.0310	**0.0310**
Coldfoot	156	0.0736	0.0736	0.0736	0.0736	**0.0736**	0.0743	0.0737	0.0739	0.0737
Eagle Summit	320	0.0487	0.0490	0.0487	0.0487	0.0487	0.0480	**0.0477**	0.0482	0.0481
Gobblers Knob	262	0.0665	0.0663	0.0660	0.0662	0.0662	**0.0622**	0.0643	0.0628	0.0637
Monahan Flat	121	0.0722	0.0721	0.0720	0.0721	0.0721	**0.0714**	0.0718	0.0715	0.0722

Table 3. *Cont.*

Station/ubRMSE (m^3/m^3)	N	Mineral Soil Based Models					Organic Soil Based Models			
		Wang 1980	**Dobson 1985**	**Mironov 2009**	**Mironov 2013**	**Park 2017**	**Bircher 2016**	**Mironov 2019**	**Park 2019**	**Park 2021**
Monument Creek	405	0.0510	0.0509	0.0508	0.0508	0.0508	0.0505	0.0503	0.0504	**0.0503**
Mt. Ryan	194	**0.0163**	0.0177	0.0173	0.0172	0.0173	0.0262	0.0186	0.0237	0.0187
Munson Ridge	383	0.0499	0.0492	0.0490	0.0492	0.0492	**0.0465**	0.0475	0.0467	0.0478
Tokositna Valley	253	0.1295	0.1296	0.1295	0.1295	0.1296	0.1298	**0.1294**	0.1296	0.1296
Upper Nome Creek	283	**0.0122**	0.0126	0.0124	0.0123	0.0126	0.0196	0.0129	0.0163	0.0160
Mean	214	0.0509	0.0512	0.0509	0.0510	0.0510	0.0518	**0.0507**	0.0513	0.0509

Where the column of the number in bold font represents the dielectric model with the best ubRMSE in that station or mean, and 'N' in the second column represents the total number of paired SMAP retrievals and in-situ SM measurements used to calculate the ubRMSE for each station.

Table 4. R of soil moisture retrievals using various dielectric models over in-situ sites in Alaska.

Station/R	N	Mineral Soil Based Models					Organic Soil Based Models			
		Wang 1980	**Dobson 1985**	**Mironov 2009**	**Mironov 2013**	**Park 2017**	**Bircher 2016**	**Mironov 2019**	**Park 2019**	**Park 2021**
Gulkana River	72	0.605	0.596	0.607	0.604	0.599	0.608	**0.621**	0.603	0.601
Spring Creek	37	0.757	0.737	0.758	0.752	0.745	0.757	**0.805**	0.752	0.746
Atigun Pass	81	0.342	**0.348**	0.344	0.344	0.344	0.341	0.333	0.347	0.347
Coldfoot	156	0.205	0.205	0.204	0.204	0.205	0.206	0.199	0.202	**0.208**
Eagle Summit	320	0.375	0.353	0.372	0.376	0.368	0.376	**0.429**	0.368	0.372
Gobblers Knob	262	0.571	0.557	0.571	0.570	0.564	0.571	**0.603**	0.575	0.577
Monahan Flat	121	0.276	0.273	0.275	0.274	0.274	0.277	0.275	**0.284**	0.276
Monument Creek	405	0.407	0.401	0.406	0.405	0.404	0.409	0.413	0.406	**0.418**
Mt. Ryan	194	0.604	0.595	0.604	0.601	0.599	0.605	**0.624**	0.604	0.601
Munson Ridge	383	0.608	0.597	0.606	0.604	0.602	0.610	**0.624**	0.611	0.611
Tokositna Valley	253	0.177	0.171	0.174	0.172	0.170	0.172	**0.176**	0.172	0.171
Upper Nome Creek	283	0.416	0.398	0.418	0.420	0.410	0.416	**0.477**	0.421	0.416
Mean	214	0.445	0.436	0.445	0.444	0.440	0.446	**0.465**	0.445	0.445

Where the column of the number in bold font represents the dielectric model with the best R in that station or mean, and 'N' in the second column represents the total number of paired SMAP retrievals and in-situ SM measurements used to calculate the R for each station.

Assessment metrics of the SM retrievals derived using identical r_{smap} values and different dielectric models were computed by their temporally paired in-situ measurements. According to Table 2, the SM estimates from mineral-soil-based models tend to underestimate, while the organic-soil-based models generally exhibit wet biases compared to the ground recordings. In terms of both ubRMSE and R (Tables 3 and 4), all the models show comparable accuracy levels, similar to the previous results in [27], whereas Mironov 2019 displays a slight but consistent edge over the other models. Compared to the other dielectric models, the modest improvement in R of Mironov 2019 was likely due to its simultaneous consideration of bulk density and SOM effects [23].

The other aspect that we attempted to evaluate for the predictive power of various dielectric models was checking the correlations between the SM retrievals of different models and SMAP observed vertically polarized T_B. If the higher absolute R values between the time series of SM and SMAP vertically polarized T_B are assumed as a criterion that reflects the better skill of a dielectric mixing model, Mironov 2019 presents an overwhelming superiority over the other models in the 765 Alaskan pixels (Figure 5). Table S2 displays that in-situ measured SM usually has a lower correlation with SMAP vertically polarized T_B relative to the correlations between satellite-based SM retrievals and SMAP T_B. However, it should be noted that such correlation-based results were inconclusive and functioned as a reference only, since the impacts of vegetation disturbance and surface roughness were entirely ignored.

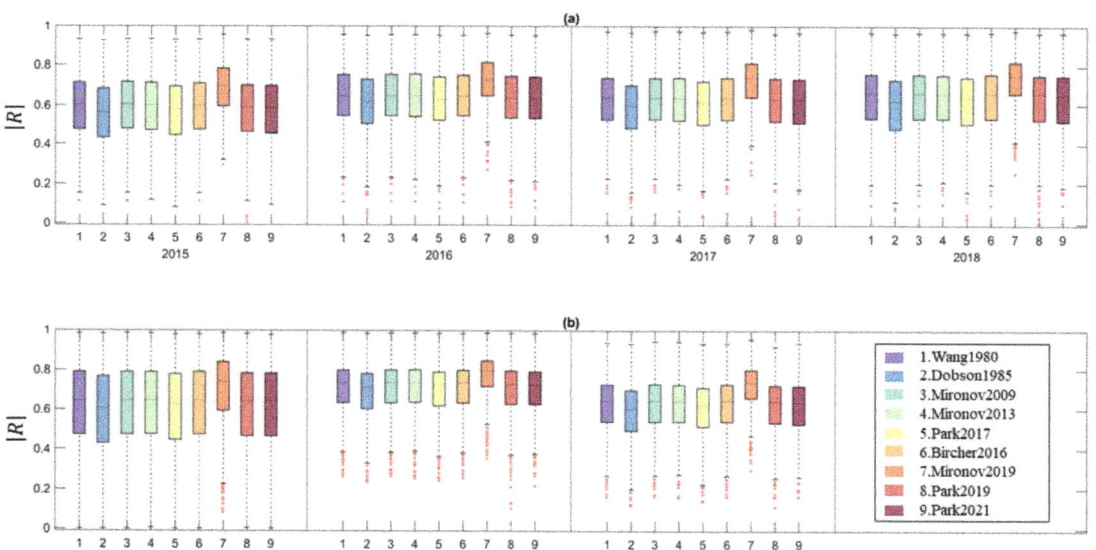

Figure 5. Boxplots of the absolute correlations between the soil moisture retrievals from various dielectric mixing models and the SMAP vertically polarized brightness temperature over the 765 pixels in Alaska. (**a**) and (**b**) represent the boxplots of absolute R values from 2015 to 2018 and 2019 to 2021, respectively.

4.3. *A Global Intercomparison between Mironov 2009 and Mironov 2019*

Mironov 2009 and Mironov 2019 were selected as the representatives for mineral- and organic-soil-based dielectric models and were then compared with each other at the global scale using one-week SMAP observations from 2 July 2018 to 8 July 2018. The one-week SM retrievals of Mironov 2009 and Mironov 2019 were analyzed over more regions with abundant SOM and were also used to acquire performance clues for applying Mironov 2019 to mineral soils.

According to Figure 6a,b, satellite-based SM data are usually unavailable in many areas characterized by organic-rich soils, likely owing to dense boreal forests, steep surface roughness, as well as permanently frozen soils on the land surface [11,40]. The magnitude differences between Mironov 2009 and Mironov 2019 yielded SM retrievals are commonly above 0.05 m^3/m^3 generally when the SOM is over 10% (Figure 6b,e). In the case of extreme dryness (SM < 0.1 m^3/m^3) over mineral soils (SOM < 5%), the SM retrievals from Mironov 2019 are likely lower than those from Mironov 2009. As illustrated in Figure 6d, there is a limb where the SM retrievals of Mironov 2019 are nearly constant, while those from Mironov 2009 vary, possibly because of the soil texture.

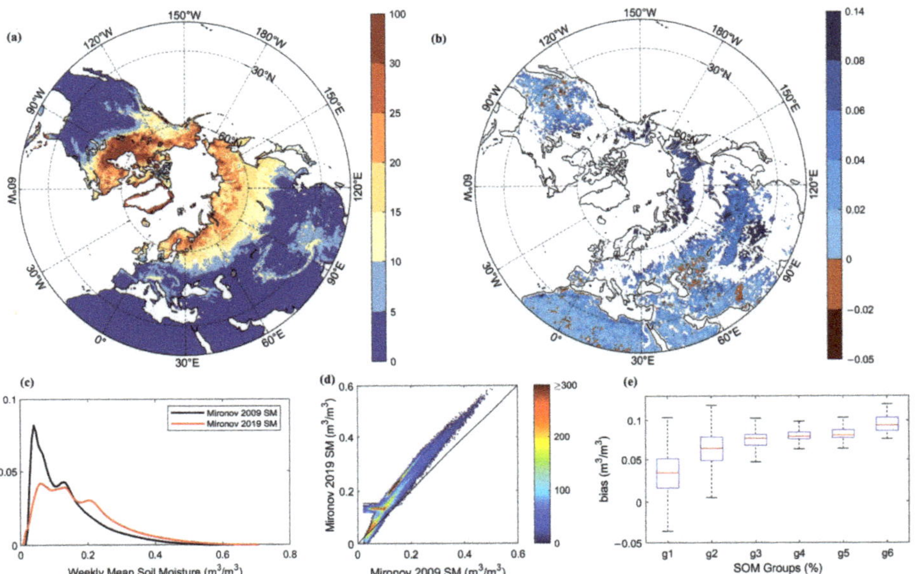

Figure 6. A global intercomparison of soil moisture retrievals from Mironov 2009 and Mironov 2019: (**a**) the spatial distribution of soil organic matter (SOM) in percentage from a north polar view, (**b**) the spatial distribution of mean differences between soil moisture estimations using Mironov 2009 and Mironov 2019 (bias = $SM_{Mironov2019} - SM_{Mironov2009}$), (**c**) the probability distribution function of weekly mean soil moistures derived using the above two models, (**d**) scatterplot of soil moisture using both models across the globe, where the color bar shows the number of pixels, and (**e**) boxplot that describes the bias variations along with the increase of SOM that was organized into 6 groups (g1–g6). The organic range of each group is 0–5% (g1), 5–10% (g2), 10–15% (g3), 15–20% (g4), 20–30% (g5), and >30% (g6).

4.4. Discussion

4.4.1. The Applicable Range of Dielectric Models

Although the above validation results over in-situ sites in Alaska demonstrated the slightly better performance of Mironov 2019 over the other models, it may be not the best model across all landscapes and climatic conditions. The accuracy of a dielectric model heavily depends on its respective applicable range. A dielectric model is likely to acquire a better performance score when being applied over the samples used to develop it. In other scenarios, potential degradation of the model skill can be expected. For instance, when Dobson 1985 is adopted in soils that fall beyond the prototypal soils on which Dobson 1985 was established, some unrealistic dielectric constants were yielded [14]. According to SMAP configurations and parameters, the frequency is confined to 1.4 GHz, while most pixels in Alaska show SOM values spanning from 15% to 30%. However, it should be noted that Mironov 2019 was designed for a surface soil layer with SOM ranging from 35% to 80% [23]. Meanwhile, the natural log calibration function from [11] was proposed for highly organic soils and the Decagon 5TE (in-situ sensor), which is operated at 70 MHz. Such imperfect alignments between the applicable ranges of dielectric models and the actual settings are surprisingly common, possibly leading to underestimations of the quality of these dielectric models.

4.4.2. Organic-Soil-Based Dielectric Models

Similar to other empirical dielectric models [41–46] accounting for the influence of SOM, SOM itself is not treated as a necessary input in Bircher 2016 to derive the dielectric constants of organic soils. Mironov 2019, however, incorporates the dielectric impacts of SOM and soil bulk density, while omitting the clay fraction. In contrast, Park 2019 and Park 2021 consider both mineralogy and SOM. Though comprehensive, the confidence in representing the dielectric interactions among various soil properties and the quality of those global-scale soil databases greatly limit the practical uses of Park models. For example, SOM, as the most critical index for classifying mineral and organic soils, was estimated by multiplying the SOC content by a fixed factor of 1.724 [23,47]. However, the conversion factor between SOC and SOM is unlikely a global constant, while [47] pointed out that this conversion factor would vary from 1.4 to 2.5 across different geographical regions.

Additionally, mineral-soil-based dielectric models are usually based on the assumption that the soil is composed of sand, silt, and clay, and thus the summation of their fractions is 100% [12,19,22]. However, this assumption is likely inappropriate over organic-rich soils, where SOM has a great gravimetric contribution. Here, the texture fractions extracted from the SoilGrids250m [30] were normalized. As a result, the summation of minerals and SOM currently exceeds 100%, while a further re-normalization is difficult to proceed with, as the SOM contents (sometimes over 100%) were empirically estimated. Despite these issues, at this time, these data sets might be the most practical sources to support running those dielectric models over a wide spatial coverage. Therefore, a soil property data set that can accurately describe the gravimetric relationship among sand, silt, clay, and SOM is pressingly needed.

4.4.3. Limitations of In-Situ Benchmarks

Besides the limits of the model applicable range and the quality of input data sets of soil properties, the other critical factor that directly affects the assessment results is the quality of the benchmarks, i.e., in-situ SM measurements. As mentioned, breaks, missing values, and jumps were commonly found during the examination of the in-situ SM time series. Furthermore, many of the calibration functions used to deduce in-situ SM values were designed for mineral soils only, due to the unavailability of organic-soil-based calibration functions over those regions. As a result, in-situ SM values might have an underestimation issue.

Due to the limited availability of in-situ measurements over Alaska, only one ground station was selected as the regional benchmark for each validation pixel. However, the estimated SMAP retrieval performance over these areas was likely degraded given the unmatched spatial representatives and measuring depths between the passive microwave SM derivations and ground measurements [39]. Additionally, inconsistent SM variations from the radiometer snapshots and the ground sensors may have arisen during the transition period between two years (e.g., from the end of August 2015 to the beginning of June 2016), adversely affecting the validation metrics. In spite of these factors, this study presents an evaluation that maximizes the use of existing data sets and can serve as a valuable reference for further investigations as more data become available.

4.4.4. Characteristics of Park Models

Compared to the other conventional semi-empirical dielectric models [12,16,19,21–23], Park models describe the fractions of bound water and free water differently [16,22,24]. First, Park models use the wilting point as the beginning point where free water starts to occur, whereas other models set that value using an independent term, named maximum bound water fraction. When the volumetric SM is between the maximum bound water fraction and porosity, most dielectric models fix the bound water content and the dielectric contribution of bound water. However, in the same SM range, Park models assume that the content of bound water and free water alters with the volumetric SM. Specifically, SM is treated as a weighted summation of the bound water and free water, where the sum of the weights of bound water (w_b) and free water (w_f) is constrained as one. It is assumed that w_b is one when SM is equal to the wilting point. On the contrary, w_b declines to zero when SM reaches porosity.

According to Figure 3e,f, there are a few rapid drops in the curves of Park 2019 and Park 2021 when the SOM exceeds 60%. Such scenarios could be explained by the wilting-point and porosity calculation equations used in Park 2019 and Park 2021. As shown in Figure S3, the porosity equation of Park 2019 could lead to a porosity greater than $1m^3/m^3$ when SOM ranges from 30% to 35%. Meanwhile, in Park 2019, the derived wilting point could surpass the porosity when the SOM is over 60%. Although the above issues were substantially mitigated for Park 2021 with valid magnitudes of its derived porosity and wilting point, an evident bending near the wilting point could still be observed in its simulated T_B curves at highly organic soils. Therefore, caution should be paid when applying Park 2019 and Park 2021 over organic-rich soils.

4.4.5. Selection of a Globally Optimal Combination of Dielectric Models

In general, Mironov 2019 can be concluded as the prime dielectric model for use in the SMAP SCA-V algorithm over organic-rich soils. Similar to [27], such a determination was not only yielded from the validation results, but also incorporated the input parameters and configurations of various models. Specifically, Mironov 2019 requires fewer input parameters compared to Park 2019 and Park 2021, making it less susceptible to the uncertainties introduced by different soil property data sources, while accounting for the SOM effects. Additionally, Mironov 2019 was developed based on a physically refractive mixing dielectric model, where the parameters were calibrated and validated across several soil samples, with a SOM ranging from 35% to 80% [23]. In contrast, Bircher 2016 was derived from straightforward regression analyses between two measured variables, while Park 2019 and Park 2021 lack effective calibration [11,16,24]. Furthermore, Mironov 2019 consistently demonstrated a slight edge over the other models, in terms of the averaged ubRMSE and R. This accuracy advantage of Mironov 2019 would likely extend to other regions with organic-rich soils (Figure A1), given similar climatic conditions and vegetation types with Alaska [48,49].

While the operational SMAP retrieval algorithms apply a single dielectric model globally [50], finding a universal dielectric model that outperforms the other models across all possible conditions seems overambitious. As described above, mineral-soil-based dielectric models do not include the SOM effect on soil dielectric constants, whereas organic-soil-based models often ignore the influence of soil texture. Although Park 2019 and Park 2021 consider both soil texture and SOM, they are prone to higher errors, due to a few improper formulations and excessive uncertainties introduced by various input data sources. Hence, based on the previous studies [15,27] and the results obtained here, the separate use of Mironov 2009 and Mironov 2019 in the SMAP SCA-V algorithm over mineral and organic soils is proposed. The selection of utilizing Mironov 2009 is somewhat arbitrary, as Mironov 2009 has not been comprehensively assessed against Mironov 2013 and Park 2017 over mineral soils. The applicability of Mironov 2009 has been extensively validated, and the use of Mironov 2009 will not further degrade the retrieval quality.

The simultaneous use of Mironov 2009 and Mironov 2019 requires a sophisticated SOM threshold that can demarcate mineral and organic soils. However, there is presently no rigorous set of rules for this threshold. [23] state that soil can be categorized into organic soil if the SOM is more than 20%, whereas [51] and [52] declare that organic soil should contain a SOM of at least 30% [11]. According to the results of the synthetic experiments, a SOM of 15% might be an optimal threshold for distinguishing soil types, as the T_B curves of different models are closely clustered and the divergence between mineral- and organic-soil-based models seems to start after a SOM exceeding 15% (Figure 3). Such a threshold conforms to [53] who classifies soils into organic soil or highly organic soil when the SOM is more than 15%.

The utilization of an optimal organic-soil-based dielectric model (i.e., Mironov 2019 here) is anticipated to improve the overall precision of SMAP SM retrievals over organic soils. Since SM is a crucial factor in determining carbon fluxes in boreal regions [18], having precise knowledge of SM variations can effectively monitor the health of local ecosystems and predict the trends in carbon storage. In the current context of global warming, the snow extent has rapidly dropped in the Northern Hemisphere [54]. Consequently, more snow-covered regions become bare soils, and the period of thawing seasons tends to last longer. Hence, decreasing SM retrieval uncertainties over these high-SOM areas would greatly aid in tracking the potential significant hydrologic shifts triggered by climate change and permafrost thawing [55,56].

Meanwhile, the deficiencies in the quality of soil property products and in-situ data sets in the Northern environment have been identified. For instance, the universal conversion formula between SOC and SOM is still rudimentary, occasionally leading to an estimation over 100%. As such, the limitations discovered in this study offer a strong motivation and direction for developing soil property data sets with better applicability. Additionally, the necessity for accurate SM in high-latitude areas highlights the need for more ground stations and dense SM observation networks over the circumpolar zone.

4.4.6. Future Work

Here, the determination of the SOM threshold at 15%, based solely on synthetic experiments, likely caused spatial inconsistencies at the boundary of the mineral and organic soils. Hence, location/time-dependent SOM thresholds may be necessary to produce smooth SM maps in high-latitude regions. An alternative approach would be the mixed use of mineral- and organic-soil-based models over each pixel, provided that an accurate relative proportion of SOM and clay is available in advance.

Although this study evaluated various dielectric models under the SMAP SCA-V algorithm, their use in other radiative transfer model-based algorithms and with observations from different polarizations, angles, and frequencies remains to be investigated. Of particular interest is the dual-channel algorithm (DCA), the current SMAP baseline algorithm, which exhibited moderate edges over agricultural sites [18]. The objective of the DCA algorithm is to achieve the optimal vegetation optical depth (VOD) and SM simultaneously, by minimizing the aggregated differences between the simulated and observed brightness temperatures at both horizontal and vertical polarizations. Thus, the alternation of the dielectric model could indirectly affect the derived vegetation water content. In addition to passive microwave remote sensing, the dielectric mixing model is also critical for other fields, such as SMAP L4 and the European Centre for Medium-Range Weather Forecasts (ECMWF) Community Microwave Emission Model (CHEM) [57,58]. Radar sensors also require a dielectric model to simulate the backscatter coefficients [59]. However, there is currently no clear consensus on the best dielectric model for these platforms, making further investigations necessary and valuable.

5. Conclusions

In this study, the skills of nine dielectric models over organic soil in Alaska were evaluated and compared in the context of the SMAP SCA-V algorithm. Four out of nine models carefully account for the SOM effect on the complex dielectric constant of the soil–water mixtures, while the remaining models were designed for use in mineral soils. The dielectric responses (expressed in a form of T_B) of those models to the increasing SOM were comprehensively investigated through artificially controlling input values. At a given SM over 0.1 m^3/m^3 and a SOM higher than 15%, the simulated T_B values from organic-soil-based dielectric models were higher than those estimated from the mineral-soil-based dielectric models. In other words, relative to mineral-soil-based dielectric models, organic-soil-based models are inclined to obtain higher SM estimates from identical observed radiations. The different magnitudes from the above two types of dielectric model were relatively stable across soil textures (e.g., silty, clay, and sandy loam), as organic-soil-based models are less sensitive to the proportions of sand, silt, and clay content. Furthermore, a SOM threshold of 15% was suggested for the separate use of mineral- and organic-soil-based dielectric models in the retrieval algorithm, as the divergence of T_B curves of mineral- and organic-soil models was observed when the SOM exceeded 15%.

The predictive power of each dielectric model was represented using several statistic metrics computed by comparing the SM retrievals with in-situ measurements. Compared to satellite products reflecting SM variations over a large spatial extent, in-situ point-based SM measurements exhibited more temporal variability. Additionally, even over the same location, the annual correlations between satellite-based SM retrievals and in-situ data fluctuated a lot. Consistent with the results from the synthetic experiments, organic- and mineral-soil-based models tended to induce wet and dry biases. In an integrated evaluation, Mironov 2019 presented a slightly, but consistently, better performance over the other dielectric models, which showed a mean ubRMSE of 0.0507 m^3/m^3 and a mean R of 0.465.

Furthermore, an inter-comparison between the SM retrievals within a one-week time interval from mineral- and organic-soil-based dielectric models was conducted at a global scale. Such a comparison would be useful to capture clues about the performance of organic-soil-based models over mineral soils. Mironov 2009 and Mironov 2019 were elected as the representatives of mineral- and organic-soil-based models, respectively. As a result, SM estimates from Mironov 2019 were at least 0.05 m^3/m^3 higher than those from Mironov 2009. When the SM was below 0.1 m^3/m^3, the SM retrievals from Mironov 2019 were occasionally smaller than the SM retrievals from Mironov 2009 in mineral soils.

It should be noted that the performance of each dielectric model heavily depends on its designed application range, the quality of the input data sets, as well as the accuracy of in-situ benchmarks. Different assessment results might be obtained with the updating of the dielectric models, in-situ measurements, and soil parameters. Given the contrasting sensitivity of mineral- and organic-soil-based models to soil texture and SOM, it is of great importance to ensure a consistent source of soil ancillary data. As such, a routine evaluation study that incorporates all the potential dielectric models and the most recent soil auxiliary data sets is recommended. In an integrated consideration of model inputs, the model physical foundation, and the practical accuracy, the separate use of Mironov 2009 and Mironov 2019 in the SMAP SCA-V algorithm for mineral soils (SOM $<$ 15%) and organic soils (SOM $\geq$ 15%) would be the optimal option at this time. Considering the SOM magnitudes at the 36 km scale, developing a sophisticated dielectric model accounting for a variable SOM from 10% to 30% is required for passive microwave remote sensing of SM.

Supplementary Materials: The following supporting information can be downloaded at: https://www.mdpi.com/article/10.3390/rs15061658/s1, Figure S1: The geographical distributions of all the 12 stations finally used for validation. Figure S1: Simulated brightness temperature of a sandy loam with various soil organic matter, and the accompanied table displays all the input values where most of soil parameters are directly taken from the sample of sandy loam used in [38]. (a)–(f) represent the simulated brightness temperature curves variations across various soil organic matter with an increase step of 15%. Figure S2: Variations of wilting point and porosity estimated from Park 2019 and Park 2021 with increasing soil organic matter with assumed volumetric textural compositions. Table S1: Detailed information of all in-situ stations investigated in this study. Table S2: Annual R values between soil moisture retrievals from various dielectric models and in-situ measurements and the SMAP vertically polarized brightness temperature.

Author Contributions: Conceptualization, R.Z, S.C., R.B. and V.L.; methodology, R.Z., S.C. and R.B.; data analysis, R.Z. and S.C.; writing—original draft preparation, R.Z.; writing—review and editing, S.C., R.B. and V.L. All authors have read and agreed to the published version of the manuscript.

Funding: This investigation was funded as a university subcontract under the NASA Making Earth System Data Records for USE in Research Environments (MEaSUREs) Program.

Data Availability Statement: Publicly available data sets were analyzed in this study. SMAP L2 data were downloaded from National Snow and Ice Data Center (https://nsidc.org/data/data-access-tool/SPL2SMP/versions/8, access date: 14 April 2022). In-situ soil moisture measurements are freely available on the Natural Resources Conservation Service (NRCS), the National Water and Climate Center (NWCC) homepage (https://www.nrcs.usda.gov/wps/portal/wcc/home, access date: 7 April 2022), and the International Soil Moisture Network (ISMN) (https://ismn.earth/en/networks, access data: 10 April 2022), respectively.

Acknowledgments: We thank Chang-Hwan Park (Ajou University) for providing his model scripts with detailed explanations. We are also grateful to all contributors to the data sets used in this study.

Conflicts of Interest: The authors declare no conflict of interest.

Appendix A

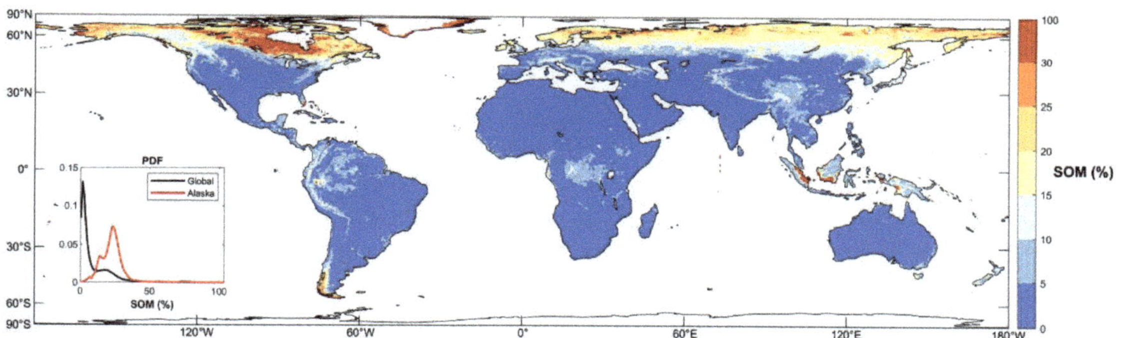

Figure A1. Global distribution of soil organic matter (SOM), where the inset describes the probability distribution function (PDF) of SOM at the global scale and in Alaska.

References

1. Njoku, E.G.; Entekhabi, D. Passive microwave remote sensing of soil moisture. *J. Hydrol.* **1996**, *184*, 101–129. [CrossRef]
2. De Jeu, R.A.; Wagner, W.; Holmes, T.; Dolman, A.; Van De Giesen, N.; Friesen, J. Global soil moisture patterns observed by space borne microwave radiometers and scatterometers. *Surv. Geophys.* **2008**, *29*, 399–420. [CrossRef]
3. Kerr, Y.H.; Waldteufel, P.; Wigneron, J.-P.; Martinuzzi, J.; Font, J.; Berger, M. Soil moisture retrieval from space: The Soil Moisture and Ocean Salinity (SMOS) mission. *IEEE Trans. Geosci. Remote Sens.* **2001**, *39*, 1729–1735. [CrossRef]
4. Entekhabi, D.; Njoku, E.G.; O'Neill, P.E.; Kellogg, K.H.; Crow, W.T.; Edelstein, W.N.; Entin, J.K.; Goodman, S.D.; Jackson, T.J.; Johnson, J. The soil moisture active passive (SMAP) mission. *Proc. IEEE* **2010**, *98*, 704–716. [CrossRef]
5. Chan, S.K.; Bindlish, R.; O'Neill, P.E.; Njoku, E.; Jackson, T.; Colliander, A.; Chen, F.; Burgin, M.; Dunbar, S.; Piepmeier, J. Assessment of the SMAP passive soil moisture product. *IEEE Trans. Geosci. Remote Sens.* **2016**, *54*, 4994–5007. [CrossRef]
6. Colliander, A.; Jackson, T.J.; Bindlish, R.; Chan, S.; Das, N.; Kim, S.; Cosh, M.; Dunbar, R.; Dang, L.; Pashaian, L. Validation of SMAP surface soil moisture products with core validation sites. *Remote Sens. Environ.* **2017**, *191*, 215–231. [CrossRef]
7. Chan, S.K.; Bindlish, R.; O'Neill, P.; Jackson, T.; Njoku, E.; Dunbar, S.; Chaubell, J.; Piepmeier, J.; Yueh, S.; Entekhabi, D. Development and assessment of the SMAP enhanced passive soil moisture product. *Remote Sens. Environ.* **2018**, *204*, 931–941. [CrossRef]
8. Kim, H.; Wigneron, J.-P.; Kumar, S.; Dong, J.; Wagner, W.; Cosh, M.H.; Bosch, D.D.; Collins, C.H.; Starks, P.J.; Seyfried, M. Global scale error assessments of soil moisture estimates from microwave-based active and passive satellites and land surface models over forest and mixed irrigated/dryland agriculture regions. *Remote Sens. Environ.* **2020**, *251*, 112052. [CrossRef]
9. Zhang, R.; Kim, S.; Sharma, A.; Lakshmi, V. Identifying relative strengths of SMAP, SMOS-IC, and ASCAT to capture temporal variability. *Remote Sens. Environ.* **2021**, *252*, 112126. [CrossRef]
10. Ulaby, F.T.; Moore, R.K.; Fung, A.K. *Radar Remote Sensing and Surface Scattering and Emission Theory*; Artech House: Norwood, MA, USA, 1986; Volume II.
11. Bircher, S.; Andreasen, M.; Vuollet, J.; Vehviläinen, J.; Rautiainen, K.; Jonard, F.; Weihermüller, L.; Zakharova, E.; Wigneron, J.-P.; Kerr, Y.H. Soil moisture sensor calibration for organic soil surface layers. *Geosci. Instrum. Methods Data Syst.* **2016**, *5*, 109–125. [CrossRef]

12. Dobson, M.C.; Ulaby, F.T.; Hallikainen, M.T.; El-Rayes, M.A. Microwave dielectric behavior of wet soil-Part II: Dielectric mixing models. *IEEE Trans. Geosci. Remote Sens.* **1985**, *23*, 35–46. [CrossRef]
13. Zhang, R.; Kim, S.; Sharma, A. A comprehensive validation of the SMAP Enhanced Level-3 Soil Moisture product using ground measurements over varied climates and landscapes. *Remote Sens. Environ.* **2019**, *223*, 82–94. [CrossRef]
14. Mironov, V.L.; Kosolapova, L.G.; Fomin, S.V. Physically and mineralogically based spectroscopic dielectric model for moist soils. *IEEE Trans. Geosci. Remote Sens.* **2009**, *47*, 2059–2070. [CrossRef]
15. Wigneron, J.-P.; Jackson, T.; O'neill, P.; De Lannoy, G.; de Rosnay, P.; Walker, J.; Ferrazzoli, P.; Mironov, V.; Bircher, S.; Grant, J. Modelling the passive microwave signature from land surfaces: A review of recent results and application to the L-band SMOS & SMAP soil moisture retrieval algorithms. *Remote Sens. Environ.* **2017**, *192*, 238–262.
16. Park, C.H.; Montzka, C.; Jagdhuber, T.; Jonard, F.; De Lannoy, G.; Hong, J.; Jackson, T.J.; Wulfmeyer, V. A dielectric mixing model accounting for soil organic matter. *Vadose Zone J.* **2019**, *18*, 190036. [CrossRef]
17. O'Neill, P.; Jackson, T. Observed effects of soil organic matter content on the microwave emissivity of soils. *Remote Sens. Environ.* **1990**, *31*, 175–182. [CrossRef]
18. O'Neill, P.; Bindlish, R.; Chan, S.; Chaubell, J.; Colliander, A.; Njoku, E.; Jackson, T. Algorithm Theoretical Basis Document Level 2 & 3 Soil Moisture (Passive) Data Products, Revision G, 12 October 2021, SMAP Project, JPL D-66480, Jet Propulsion Laboratory, Pasadena, CA. Available online: https://nsidc.org/sites/nsidc.org/files/technical-references/L2_SM_P_ATBD_rev_G_final_Oct2021.pdf (accessed on 12 May 2022).
19. Wang, J.R.; Schmugge, T.J. An empirical model for the complex dielectric permittivity of soils as a function of water content. *IEEE Trans. Geosci. Remote Sens.* **1980**, *18*, 288–295. [CrossRef]
20. Peplinski, N.R.; Ulaby, F.T.; Dobson, M.C. Dielectric properties of soils in the 0.3–1.3-GHz range. *IEEE Trans. Geosci. Remote Sens.* **1995**, *33*, 803–807. [CrossRef]
21. Mironov, V.; Kerr, Y.; Wigneron, J.-P.; Kosolapova, L.; Demontoux, F. Temperature-and texture-dependent dielectric model for moist soils at 1.4 GHz. *IEEE Geosci. Remote Sens. Lett.* **2012**, *10*, 419–423. [CrossRef]
22. Park, C.-H.; Behrendt, A.; LeDrew, E.; Wulfmeyer, V. New approach for calculating the effective dielectric constant of the moist soil for microwaves. *Remote Sens.* **2017**, *9*, 732. [CrossRef]
23. Mironov, V.L.; Kosolapova, L.G.; Fomin, S.V.; Savin, I.V. Experimental analysis and empirical model of the complex permittivity of five organic soils at 1.4 GHz in the temperature range from −30 °C to 25 °C. *IEEE Trans. Geosci. Remote Sens.* **2019**, *57*, 3778–3787. [CrossRef]
24. Park, C.-H.; Berg, A.; Cosh, M.H.; Colliander, A.; Behrendt, A.; Manns, H.; Hong, J.; Lee, J.; Zhang, R.; Wulfmeyer, V. An inverse dielectric mixing model at 50 MHz that considers soil organic carbon. *Hydrol. Earth Syst. Sci.* **2021**, *25*, 6407–6420. [CrossRef]
25. Yi, Y.; Chen, R.H.; Kimball, J.S.; Moghaddam, M.; Xu, X.; Euskirchen, E.S.; Das, N.; Miller, C.E. Potential Satellite Monitoring of Surface Organic Soil Properties in Arctic Tundra from SMAP. *Water Resour. Res.* **2022**, *58*, e2021WR030957. [CrossRef]
26. Suman, S.; Srivastava, P.K.; Pandey, D.K.; Prasad, R.; Mall, R.; O'Neill, P. Comparison of soil dielectric mixing models for soil moisture retrieval using SMAP brightness temperature over croplands in India. *J. Hydrol.* **2021**, *602*, 126673. [CrossRef]
27. Mialon, A.; Richaume, P.; Leroux, D.; Bircher, S.; Al Bitar, A.; Pellarin, T.; Wigneron, J.-P.; Kerr, Y.H. Comparison of Dobson and Mironov dielectric models in the SMOS soil moisture retrieval algorithm. *IEEE Trans. Geosci. Remote Sens.* **2015**, *53*, 3084–3094. [CrossRef]
28. Srivastava, P.K.; O'Neill, P.; Cosh, M.; Kurum, M.; Lang, R.; Joseph, A. Evaluation of dielectric mixing models for passive microwave soil moisture retrieval using data from ComRAD ground-based SMAP simulator. *IEEE J. Sel. Top. Appl. Earth Obs. Remote Sens.* **2014**, *8*, 4345–4354. [CrossRef]
29. O'Neill, P.; Chan, S.; Njoku, E.; Jackson, T.; Bindlish, R.; Chaubell, J. *L3 Radiometer Global Daily 36 km EASE-Grid Soil Moisture, Version 8.*; NASA National Snow and Ice Data Center Distributed Active Archive Center: Boulder, CO, USA, 2021. [CrossRef]
30. Hengl, T.; Mendes de Jesus, J.; Heuvelink, G.B.; Ruiperez Gonzalez, M.; Kilibarda, M.; Blagotić, A.; Shangguan, W.; Wright, M.N.; Geng, X.; Bauer-Marschallinger, B. SoilGrids250m: Global gridded soil information based on machine learning. *PLoS ONE* **2017**, *12*, e0169748. [CrossRef] [PubMed]
31. Das, N.N.; O'Neill, P. Soil Moisture Active Passive (SMAP) Ancillary Data Report, Soil Attributes, 15 August 2020, JPL D-53058, Version B, Jet Propulsion Laboratory, Pasadena, CA, USA. Available online: https://smap.jpl.nasa.gov/documents (accessed on 16 March 2023).
32. Schaefer, G.L.; Paetzold, R.F. SNOTEL (SNOwpack TELemetry) and SCAN (soil climate analysis network). *Autom. Weather. Station. Appl. Agric. Water Resour. Manag. Curr. Use Future Perspect.* **2001**, *1074*, 187–194.
33. Schaefer, G.L.; Cosh, M.H.; Jackson, T.J. The USDA natural resources conservation service soil climate analysis network (SCAN). *J. Atmos. Ocean. Technol.* **2007**, *24*, 2073–2077. [CrossRef]
34. Dorigo, W.; Xaver, A.; Vreugdenhil, M.; Gruber, A.; Hegyiova, A.; Sanchis-Dufau, A.; Zamojski, D.; Cordes, C.; Wagner, W.; Drusch, M. Global automated quality control of in-situ soil moisture data from the International Soil Moisture Network. *Vadose Zone J.* **2013**, *12*, 1–21. [CrossRef]
35. Dorigo, W.; Wagner, W.; Hohensinn, R.; Hahn, S.; Paulik, C.; Xaver, A.; Gruber, A.; Drusch, M.; Mecklenburg, S.; van Oevelen, P. The International Soil Moisture Network: A data hosting facility for global in-situ soil moisture measurements. *Hydrol. Earth Syst. Sci.* **2011**, *15*, 1675–1698. [CrossRef]

36. Dorigo, W.; Himmelbauer, I.; Aberer, D.; Schremmer, L.; Petrakovic, I.; Zappa, L.; Preimesberger, W.; Xaver, A.; Annor, F.; Ardö, J. The International Soil Moisture Network: Serving Earth system science for over a decade. *Hydrol. Earth Syst. Sci.* **2021**, *25*, 5749–5804. [CrossRef]
37. Entekhabi, D.; Reichle, R.H.; Koster, R.D.; Crow, W.T. Performance metrics for soil moisture retrievals and application requirements. *J. Hydrometeorol.* **2010**, *11*, 832–840. [CrossRef]
38. Hallikainen, M.T.; Ulaby, F.T.; Dobson, M.C.; El-Rayes, M.A.; Wu, L.-K. Microwave dielectric behavior of wet soil-part 1: Empirical models and experimental observations. *IEEE Trans. Geosci. Remote Sens.* **1985**, *23*, 25–34. [CrossRef]
39. Crow, W.T.; Berg, A.A.; Cosh, M.H.; Loew, A.; Mohanty, B.P.; Panciera, R.; de Rosnay, P.; Ryu, D.; Walker, J.P. Upscaling sparse ground-based soil moisture observations for the validation of coarse-resolution satellite soil moisture products. *Rev. Geophys.* **2012**, 50. [CrossRef]
40. Yi, Y.; Kimball, J.S.; Chen, R.H.; Moghaddam, M.; Miller, C.E. Sensitivity of active-layer freezing process to snow cover in Arctic Alaska. *Cryosphere* **2019**, *13*, 197–218. [CrossRef]
41. Topp, G.C.; Davis, J.; Annan, A.P. Electromagnetic determination of soil water content: Measurements in coaxial transmission lines. *Water Resour. Res.* **1980**, *16*, 574–582. [CrossRef]
42. Roth, C.; Malicki, M.; Plagge, R. Empirical evaluation of the relationship between soil dielectric constant and volumetric water content as the basis for calibrating soil moisture measurements by TDR. *J. Soil Sci.* **1992**, *43*, 1–13. [CrossRef]
43. Paquet, J.; Caron, J.; Banton, O. In-situ determination of the water desorption characteristics of peat substrates. *Can. J. Soil Sci.* **1993**, *73*, 329–339. [CrossRef]
44. Skierucha, W. Accuracy of soil moisture measurement by TDR technique. *Int. Agrophys.* **2000**, *14*, 417–426.
45. Kellner, E.; Lundin, L.-C. Calibration of time domain reflectometry for water content in peat soil. *Hydrol. Res.* **2001**, *32*, 315–332. [CrossRef]
46. Malicki, M.; Plagge, R.; Roth, C. Improving the calibration of dielectric TDR soil moisture determination taking into account the solid soil. *Eur. J. Soil Sci.* **1996**, *47*, 357–366. [CrossRef]
47. Pribyl, D.W. A critical review of the conventional SOC to SOM conversion factor. *Geoderma* **2010**, *156*, 75–83. [CrossRef]
48. Sulla-Menashe, D.; Gray, J.M.; Abercrombie, S.P.; Friedl, M.A. Hierarchical mapping of annual global land cover 2001 to present: The MODIS Collection 6 Land Cover product. *Remote Sens. Environ.* **2019**, *222*, 183–194. [CrossRef]
49. Beck, H.E.; Zimmermann, N.E.; McVicar, T.R.; Vergopolan, N.; Berg, A.; Wood, E.F. Present and future Köppen-Geiger climate classification maps at 1-km resolution. *Sci. Data* **2018**, *5*, 180214. [CrossRef] [PubMed]
50. O'Neill, P.; Chan, S.; Bindlish, R.; Chaubell, J.; Colliander, A.; Chen, F.; Dunbar, S.; Jackson, T.; Peng, J.; Mousavi, M.; et al. *Calibration and Validation for the L2/3_SM_P Version 8 and L2/3_SM_P_E Version 5 Data Products*; SMAP Project, JPL D-56297; Jet Propulsion Laboratory: Pasadena, CA, USA, 2021.
51. Broll, G.; Brauckmann, H.J.; Overesch, M.; Junge, B.; Erber, C.; Milbert, G.; Baize, D.; Nachtergaele, F. Topsoil characterization—Recommendations for revision and expansion of the FAO-Draft (1998) with emphasis on humus forms and biological features. *J. Plant Nutr. Soil Sci.* **2006**, *169*, 453–461. [CrossRef]
52. Zanella, A.; Jabiol, B.; Ponge, J.-F.; Sartori, G.; De Waal, R.; Van Delft, B.; Graefe, U.; Cools, N.; Katzensteiner, K.; Hager, H. European Humus Forms Reference Base. 2011. Available online: https://hal.science/hal-00541496/file/Humus_Forms_ERB_31_01_2011.pdf (accessed on 16 March 2023).
53. Huang, P.; Patel, M.; Bobet, A. FHWA/IN/JTRP-2008/2 Classification of Organic Soils. 2008. Available online: https://www.geostructures.com/library/technical-bulletins/pdf/Classification-of-Organic-Soils-FHWA-IN-JTRP-2008-2.pdf (accessed on 16 March 2023).
54. Mudryk, L.; Santolaria-Otín, M.; Krinner, G.; Ménégoz, M.; Derksen, C.; Brutel-Vuilmet, C.; Brady, M.; Essery, R. Historical Northern Hemisphere snow cover trends and projected changes in the CMIP6 multi-model ensemble. *Cryosphere* **2020**, *14*, 2495–2514. [CrossRef]
55. Vonk, J.E.; Tank, S.; Walvoord, M.A. Integrating hydrology and biogeochemistry across frozen landscapes. *Nat. Commun.* **2019**, *10*, 5377. [CrossRef] [PubMed]
56. Liljedahl, A.K.; Boike, J.; Daanen, R.P.; Fedorov, A.N.; Frost, G.V.; Grosse, G.; Hinzman, L.D.; Iijma, Y.; Jorgenson, J.C.; Matveyeva, N. Pan-Arctic ice-wedge degradation in warming permafrost and its influence on tundra hydrology. *Nat. Geosci.* **2016**, *9*, 312–318. [CrossRef]
57. Reichle, R.; De Lannoy, G.; Koster, R.D.; Crow, W.T.; Kimball, J.S.; Liu, Q.; Bechtold, M. *SMAP L4 Global 3-Hourly 9 km EASE-Grid Surface and Root Zone Soil Moisture Geophysical Data, Version 7*; NASA National Snow and Ice Data Center Distributed Active Archive Center: Boulder, CO, USA, 2022. [CrossRef]
58. Sabater, J.M.; De Rosnay, P.; Balsamo, G. Sensitivity of L-band NWP forward modelling to soil roughness. *Int. J. Remote Sens.* **2011**, *32*, 5607–5620. [CrossRef]
59. Karthikeyan, L.; Pan, M.; Wanders, N.; Kumar, D.N.; Wood, E.F. Four decades of microwave satellite soil moisture observations: Part 1. A review of retrieval algorithms. *Adv. Water Resour.* **2017**, *109*, 106–120. [CrossRef]

Article

Surface Soil Moisture Retrieval on Qinghai-Tibetan Plateau Using Sentinel-1 Synthetic Aperture Radar Data and Machine Learning Algorithms

Leilei Dong [1], Weizhen Wang [1,2,*], Rui Jin [1], Feinan Xu [1] and Yang Zhang [1]

[1] Key Laboratory of Remote Sensing of Gansu Province, Heihe Remote Sensing Experimental Research Station, Northwest Institute of Eco-Environment and Resources, Chinese Academy of Sciences, Lanzhou 730000, China

[2] Key Laboratory of Land Surface Process and Climate Change in Cold and Arid Regions, Chinese Academy of Sciences, Lanzhou 730000, China

* Correspondence: weizhen@lzb.ac.cn; Tel.: +86-931-4967243

Abstract: Soil moisture is a key factor in the water and heat exchange and energy transformation of the ecological systems and is of critical importance to the accurate obtainment of the soil moisture content for supervising water resources and protecting regional and global eco environments. In this study, we selected the soil moisture monitoring networks of Naqu, Maqu, and Tianjun on the Qinghai–Tibetan Plateau as the research areas, and we established a database of surface microwave scattering with the AIEM (advanced integral equation model) and the mathematical expressions for the backscattering coefficient, soil moisture, and surface roughness of the VV and VH polarizations. We proposed the soil moisture retrieval models of empirical and machine learnings algorithms (backpropagation neural network (BPNN), support vector machine (SVM), K-nearest neighbors (KNN), and random forest (RF)) for the ascending and descending orbits using Sentinel-1 and measurement data, and we also validated the accuracies of the retrieval model in the research areas. According to the results, there is a substantial logarithmic correlation among the backscattering coefficient, soil moisture, and combined roughness. Generally, we can use empirical models to estimate the soil moisture content, with an R^2 of 0.609, RMSE of 0.08, and MAE of 0.064 for the ascending orbit model and an R^2 of 0.554, RMSE of 0.086, and MAE of 0.071 for the descending orbit model. The soil moisture contents are underestimated when the volumetric water content is high. The soil moisture retrieval accuracy is improved with machine learning algorithms compared to the empirical model, and the performance of the RF algorithm is superior to those of the other machine learning algorithms. The RF algorithm also achieved satisfactory performances for the Maqu and Tianjun networks. The accuracies of the inversion models for the ascending orbit in the three soil moisture monitoring networks were better than those for the descending orbit.

Keywords: soil moisture; AIEM; machine learning algorithms; Sentinel-1; Qinghai–Tibetan Plateau

Citation: Dong, L.; Wang, W.; Jin, R.; Xu, F.; Zhang, Y. Surface Soil Moisture Retrieval on Qinghai-Tibetan Plateau Using Sentinel-1 Synthetic Aperture Radar Data and Machine Learning Algorithms. *Remote Sens.* **2023**, *15*, 153. https://doi.org/10.3390/rs15010153

Academic Editor: Dominique Arrouays

Received: 20 November 2022
Revised: 23 December 2022
Accepted: 24 December 2022
Published: 27 December 2022

1. Introduction

Soil moisture is a key factor in the water and heat transfer and energy transposition in land–atmosphere systems [1], and it is also vital to connecting the water of surface water, groundwater, and carbon cycles of terrestrial ecosystems [2]. As a crucial parameter in hydrology, meteorology, ecology, and agriculture, researchers use soil moisture in hydrologic modeling [3], numerical weather forecasting [4], and overland flow predictions [5]. Therefore, the accurate and dynamic monitoring of soil moisture is critical for environmental protection.

The main advantages of microwave remote sensing are its real-time detection, high penetrating power, and the fact that it is not easily influenced by cloudy weather. The soil volumetric water content has a substantial effect on the variation in the soil dielectric constant, and the soil dielectric properties are bound up with the brightness temperature and backscatter coefficient of the microwaves [6]. Consequently, microwave remote sensing technology is a potential method for soil moisture monitoring [7]. According to different energy sources, we can divide microwave remote sensing into two types: active and passive. The resolution of passive microwave radiometers is generally above 10 km, which is helpful for monitoring surface ecological environmental elements on a global scale and obtaining essential data for global change research [8]. However, passive microwave remote sensing cannot represent the changes in the local-scale soil moisture. Active microwave remote sensing makes up for these deficiencies with its high resolution. The Sentinel-1 can provide C-band SAR data with repeated observations, the revisit period is 6 days, the spatial resolution is 10 m, and it has considerable potential for soil moisture inversion [9,10].

The establishment of the microwave surface scattering model and an understanding of the influence of the soil volumetric moisture content on the SAR parameters are the prerequisites for soil moisture inversion using SAR data [11]. The interaction between the electromagnetic waves scattered by random surfaces and ground objects primarily depends on the system factors (frequency, polarization mode, and incidence angle) of the microwave sensor, and it is also closely related to the ground roughness and dielectric properties. Therefore, researchers have proposed empirical and theoretical models to reveal the relationship between the soil moisture content and SAR factors. The Oh model [12], Dubois model [13], and Shi model [14] are common empirical models. However, they are only suitable for special environments and lack universality due to their dependence on observation data. Researchers widely use theoretical models based on the electromagnetic wave radiation transfer equation to describe surface scattering due to it good physical basis. The early theoretical models include the SPM (small perturbation model) [15], GOM (geometrical optics model) [16], and POM (physical optics model) [17]; however, we can only apply these models within a certain ground roughness range. Fung developed the IEM (integral equation model) using the Maxwell equation of electromagnetic waves to broaden the model's application [18]. We can use the model to simulate surface scattering within a large ground roughness range. Chen [19] proposed the AIEM (advanced integral equation model), which has a higher accuracy and more compact form, by improving the IEM. We can use the model to simulate surface scattering due to the advantages of its higher theoretical foundation, clearer structure, and stronger universality. Baghdadi [20] proposed semiempirical calibration by using the IEM to better reconstruct the surface scattering characteristics of bare farmland. According to the results, the backscattering coefficient measured in the experiment coincided with that of the simulation of the semi-empirical model. The researchers validated the performance of the AIEM through different correlation length parameterizations [21]. According to the results, we can retrieve the soil moisture from SAR images based on the AIEM in semiarid districts. However, the IEM and AIEM achieve good satisfaction only in bare soil, and there are obvious errors in vegetation-covered areas [19].

Reducing the impacts of the roughness and vegetation on the surface backscattering is a critical issue in the process of soil moisture inversion using microwave images. Zribi [22] modified the geometrical features of the local soil framework based on the fractional Brownian model to estimate the backscattering coefficients of farmlands. The authors present the theoretical research on the generation and propagation of the roughness error, and according to the result, the profile extent, profile morphology, profile measurement number, and profile measurement precision in different directions are the major factors that affect roughness errors [23]. However, there are still uncertainties in the research on the roughness parameterization scheme.

Machine learning algorithms can describe the complicated relationships of variables and have been introduced to monitor soil moisture at different scales. A method using an Artificial Neural Network (ANN) has been put forward to model, test, and validate soil moisture for GMES Sentinel-1 [24]. The brightness temperature, soil moisture, surface soil temperature, and vegetation water content were employed to simulate global soil moisture by using a Neural Network technique by Kolassa [25]. The three machine learning algorithms of random forest (RF), support vector machine (SVM), and K-nearest neighbors (KNN) were used to research the soil moisture, thus downscaling the presence of seasonal differences [26]. The data fusion and random forest were used to generate surface soil moisture over the agricultural field [27]. A new approach combining machine learning and multi-sensor data was put forward to predict soil moisture in Australia [28], and the proposed model generated satisfactory performance compared to random forest regression, support vector machine, and CatBoost gradient boosting regression. Although machine learning algorithms can effectively explain non-linear problems, the lack of a physical foundation and the excessive dependence on training samples are their main disadvantages. Therefore, combining physical models and machine learning algorithms is a valid approach for modifying soil moisture inversion precision.

The Qinghai-Tibetan Plateau (QTP) is the highest and largest plateau in the world [29]. The QTP directly affects the local climate and environment via atmospheric circulation and hydrology procedures, and it also impacts climate change not only in China and Asia but also around the globe [30]. The soil moisture, as the critical surface element of the QTP, is of critical importance to predicting the atmospheric circulation and climate change of the plateau through the adjustment of the ground evaporation and infiltration, controlling the surface energy allocation, and influencing the soil freezing and thawing. The soil moisture also influences the monsoon climate and rainfall forms of the plateau. Therefore, the use of the active microwave technique to grasp the exact local soil moisture information of the QTP is essential for understanding the energy exchange of this district and its impacts on the environments of the surrounding areas.

Therefore, in this study, we selected three soil moisture observation networks in the QTP as the research areas: Naqu, Maqu, and Tianjun. We used the soil moisture measurement and Sentinel-1 data with the VV and VH polarizations of the ascending and descending orbits to model and retrieve the soil moisture. First, we analyzed the response of the soil moisture and surface roughness to the backscattering coefficient based on the AIEM, and we established the mathematical expressions for the backscattering coefficient, soil moisture, and surface roughness of the VV and VH polarizations. Subsequently, we proposed empirical and machine learning models for the soil moisture retrieval for the ascending and descending orbits by using the soil moisture measurement data and Sentinel-1 images from 2017–2019 of the Naqu station. Finally, we obtained the 2020 soil moisture results of the Naqu station based on the empirical model and machine learning models, and we also evaluated the accuracies of these models with measurement data. We also obtained the soil moisture results for Maqu in 2018 and Tianjun in 2020 to further verify the precision and applicability of the soil moisture retrieval models.

2. Materials and Methods

2.1. Soil Moisture Monitoring Networks

In order to obtain more accurate local soil moisture measurements in the QTP, we selected three soil moisture monitoring networks as the research areas: Naqu, Maqu, and Tainjun (Figure 1). The Naqu network is on the central QTP, the Maqu network is on the eastern QTP, and the Tianjun network is on the northeast QTP.

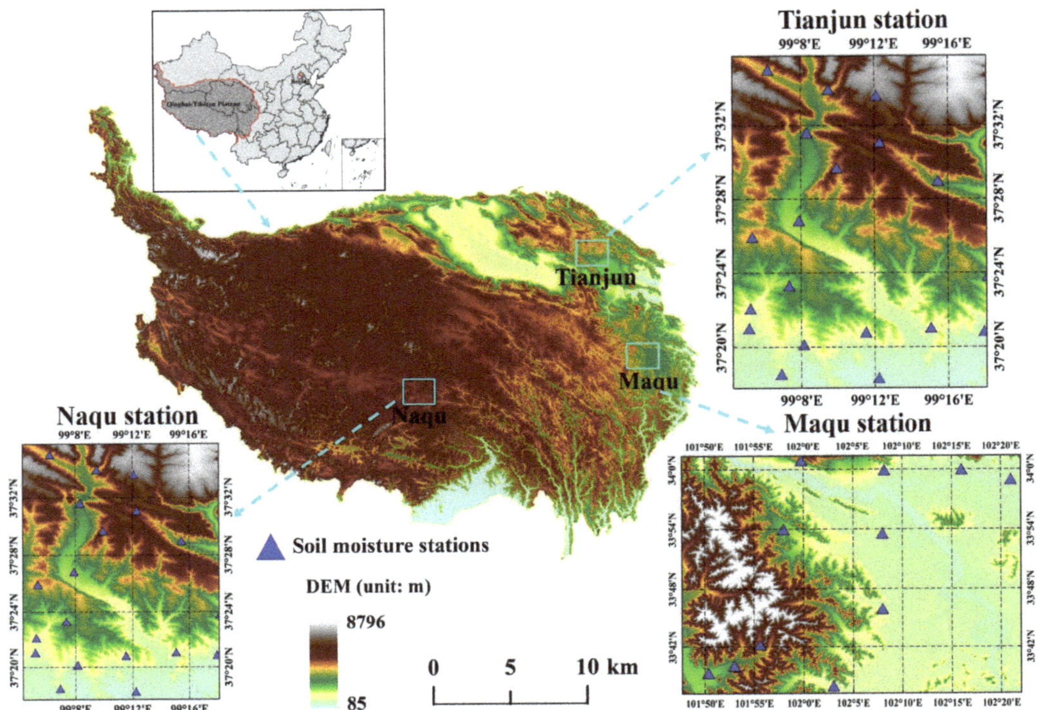

Figure 1. Locations of soil moisture monitoring networks on Qinghai–Tibetan Plateau: Naqu, Maqu, and Tianjun.

2.1.1. Naqu Soil Moisture Monitoring Network

The Naqu network was established in Naqu (29°55′–36°30′N, 83°55′–95°5′E), the Tibet Autonomous Region, China. The mean elevation is 4650 m, and the terrain is mountainous. The subrigid semiarid climate is the dominant climate type in the observation area. The average annual precipitation is about 500 mm, with 75% of the precipitation falling from May to October. The surface vegetation is mainly alpine grassland. The Naqu network consists of 56 soil moisture and temperature measurement stations, which were installed in three different networks to meet different spatial scale needs. At each station, soil moisture/temperature sensors were inserted horizontally at 5 cm, 10 cm, 20 cm, and 40 cm soil depths, respectively. The data collection interval is 30 min. The EC-TM and 5 TM capacitance probes manufactured by Decagon (United States) are used to establish the monitoring network. The sensors measure soil moisture according to the sensitivity of soil dielectric permittivity to liquid soil water. The 10 soil samples from different stations were collected to calibrate the sensor, the soil moisture is measured by the gravimetric method, and the soil dielectric permittivity is measured by the sensor simultaneously. A calibrated conversion between the measured soil moisture and the measured dielectric permittivity is then developed. The measured soil moisture turns out to be in the physical range after the calibration [31]. The measurements of soil moisture and temperature at different depths in the Naqu network from 2015 to 2021 are shown in Figure 2. The mean values of soil moisture and temperature during the observation period were 0.16 m^3/m^3 and 3.45 °C, respectively, and their trends were relatively similar. In order to match the Sentinel-1 data, we selected 23 soil moisture station measurements from the Naqu network for the soil moisture modeling and validation for 2017–2020.

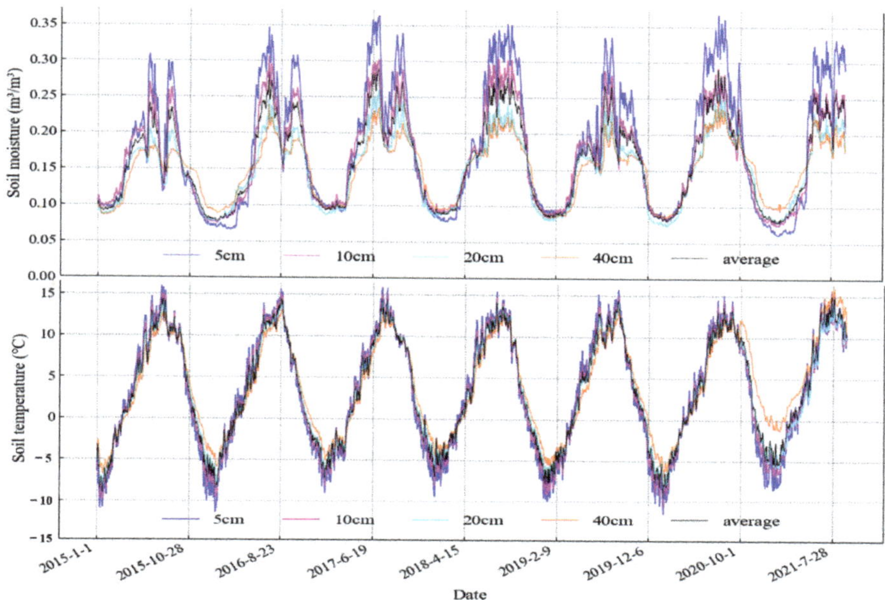

Figure 2. Naqu soil moisture and temperature measurement data for 2015–2021.

2.1.2. Maqu Soil Moisture Monitoring Network

The Maqu network was established in Maqu County (33°6′30″–34°30′15″N, 100°45′45″–102°29′E), the Ganan Tibetan Autonomous Prefecture, Gansu Province, China. The terrain of Maqu County is high in the northwest and low in the southeast, with elevations ranging from 3300 m to 4800 m. Maqu has a subrigid semihumid climate, the cold season is long and cold, and the warm season is short and mild. The average annual temperature and precipitation in the observation area are 2.9 °C and 611.9 mm, respectively. The surface vegetation is mainly low grassland. A total of 20 soil moisture and temperature measurement stations were installed in the Maqu network, and the soil moisture and temperature at depths of 5 cm, 10 cm, 40 cm, and 80 cm were observed at each station. The data collection interval is 60 min. Su [32] provides more detailed information on the Maqu soil moisture monitoring network. We selected 18 soil moisture station measurements from the Maqu network for the soil moisture validation for 2018.

2.1.3. Tianjun Soil Moisture Monitoring Network

The Tianjun network was established in Tianjun County (36°53′–48°39′12″N, 96°49′42″–99°41′48″E), the Haixi Mongolian and Tibetan Autonomous Prefecture, Qinghai Province, China. The mean elevation of Tianjun County is more than 4000 m in the territory. This region has a plateau continental climate with low temperatures and an uneven precipitation distribution. The alpine meadow is the main land cover type. The 58 soil moisture and temperature measurement stations were installed in the Tianjun network. The soil moisture and temperature at depths of 5 cm, 10 cm, and 30 cm were observed at each station. The data collection interval is 30 min. We selected 19 soil moisture station measurements from the Tianjun network for the soil moisture validation for 2020.

2.2. Remote Sensing Data

The Sentinel-1 is composed of two satellites (A and B), carrying a C-band synthetic aperture radar that provides continuous images. In this paper, the ground range detected (GRD) products from the interferometric wide swath (IW) mode in the VV and VH polarizations were employed to inverse the surface soil moisture. We performed the preprocessing steps (updating of orbit metadata, removal of border noise, removal of thermal noise removal, radiometric calibration, terrain correction, normalization of incident angle, and noise filtering) for Sentinel-1 on the Google Earth Engine (GEE) platform. We used the range-Doppler approach for the geometric terrain correction, and we introduced 7 × 7 Lee wave filtering to remove the noise.

2.3. Methods

2.3.1. Advanced Integral Equation Model (AIEM)

Although IEM can simulate real surface backscattering characteristics within a broad range of ground roughness, its main disadvantages are the dependence on the local incident angle and the inaccurate description of the actual surface roughness. Therefore, Chen proposed an AIEM by modifying the IEM. In this study, we established the surface microwave scattering database with the AIEM. The detailed expression of the AIEM is presented as follows:

$$\sigma_{\mathrm{pq}}^{0} = \frac{{k_1}^2}{2} \exp\left(-s^2\left({k_z}^2 + {k_{sz}}^2\right)\right) \sum_{n=1}^{\infty} \frac{s^{2n}}{n!} \left|I_{\mathrm{pp}}^{n}\right|^2 W^n\left(k_{sx} - k_x, k_{sy} - k_y\right) \quad (1)$$

$$I_{pq}^{n} = \left(k_{sz} + k_z\right)^n f_{pq} \exp\left(-s^2 k_z k_{sz}\right) + \frac{\left(k_{sz}\right)^n F_{pq}\left(-k_x, -k_y\right) + \left(k_z\right)^n F_{pq}\left(-k_{sz}, -k_{sy}\right)}{2} \quad (2)$$

where pq represents the polarization mode; k_1 is the free-space wave; s is the root-mean-square height; $W^n(k_{sx}\text{-}k_x, k_{sy}\text{-}k_y)$ is the n factorial Fourier transform of the surface correlation function ($k_z = kcos\theta_i$; $k_{sz} = kcos\theta_s$; $k_x = ksin\theta_i cos\varphi$; $k_{sx} = ksin\theta_s cos\varphi_s$; $k_y = ksin\theta_i sin\varphi$; $k_{sy} = ksin\theta_s sin\varphi_s$); φ is the incident azimuth; θ and φ_s are the scattering angle and scattering azimuth, respectively; F_{pq} and f_{pq} are the functions related to the Fresnel reflectance.

2.3.2. Machine Learning Algorithms

In this study, the four machine learning algorithms including backpropagation neural network (BPNN), support vector machine (SVM), K-nearest neighbor (KNN), and random forest (RF) are introduced to retrieve soil moisture.

The backpropagation neural network (BPNN) is one of the common neural networks, and it is a multilayer feedforward network that is trained by an error backpropagation algorithm [33]. A complete BPNN consists of three parts: the input layer, hidden layer, and output layer. The input layer receives the external massage and transports it to the hidden layer, where the message transformation process is achieved. The output layer outputs the result. The error backpropagation process is conducted when the actual output does not match the expected output. The BPNN is continuously adjusted until the variance in the initial system output and desired output is minimized.

The support vector machine (SVM) is a supervised learning approach that researchers commonly employ for classification analyses and regression modeling [34]. The principle is to construct the best fragmenting lineoid in the character interspace based on the framework risk minimization fundamentals, which globally optimizes the algorithm and places a particular limit on the anticipated risk in the entire example interspace. Generally, researchers use SVMs to solve the linear separability problem, for which the linearly inseparable sample of the lower-dimension input interspace is converted to the higher-dimension characteristic interspace based on a kernel function. The commonly used kernel functions are the polynomial, Gaussian, and radial basis kernel functions. In this study, we used the radial basis kernel function (RBKF) for the analysis because, according to previous results, it achieves more satisfactory effects than the other kernel functions [35].

The K-nearest neighbor (KNN) is a theoretically mature machine-learning algorithm [36]. The basic idea of this method is as follows: When the training data are certain, the K examples that are closest to the new input example are found in the training data. If the majority of these K examples are classified into a certain class, then the input example can also be classified into this class. In addition to classification, we can also use KNNs for regression. In the regression process, the K-nearest samples of the target sample are found, and the average value of these neighbor samples is assigned to the target sample.

Random forest (RF) is one of the typical ensemble algorithms. The samples are obtained from the raw data collection using the bootstrap resampling approach, and the decision tree is employed to calculate each bootstrap sample. Then, the prediction results of the multiple decision trees are combined, and finally, the predicted outcome is obtained by majority voting [37]. We can use the RF algorithm to solve multidimensional information and nonlinear issues without making feature selections, and it is also able to overcome noise and avoid the overfitting issue in practical applications.

2.3.3. Establishment of Surface Microwave Scattering Database with AIEM

The AIEM is deemed to be a theoretical model that can present the actual situation of the ground scattering well. Therefore, the numerical simulation using the AIEM is conducted to establish the database of the ground microwave scattering. The input parameters in the AIEM were as follows: a soil temperature of 20 °C; a frequency of 5.405 GHz; sand and clay contents of 40% and 10%, respectively; a soil moisture content range from 1% to 40%, with a step of 5%; an incident angle range from 20° to 50°, with a step of 5°; a root-mean-square height range from 0.1 cm to 2.9 cm, with a step of 0.4 cm; a correlation length range from 4 cm to 18 cm, with a step of 2 cm; and the surface autocorrelation function is the exponential autocorrelation function. The simulation of AIEM is shown in Figure 3. According to the results, there is a substantial logarithmic correlation among the backscattering coefficient, soil moisture, and combined roughness. In addition, if the surface roughness is given, then this logarithm relationship is only related to the incident angle. We present the detailed expression with different polarization patterns as follows:

$$\sigma^0_{pq} = A_{pq} \ln(M_v) + f(s,l) \tag{3}$$

where pq is the polarization pattern; A_{pq} is the coefficient that is not related to the surface roughness when the incident angle is known; $f(s,l)$ is the known surface roughness.

Ground roughness is one of the critical parameters in the process of microwave surface scattering, and it mainly includes two unknown parameters: the correlation length (l) and root-mean-square height (s). The backscattering coefficient is affected by both the land s, and it is difficult to distinguish between their influences on it. Therefore, researchers have proposed a new parameter that combines the l and s [38,39] to decrease the error of the soil moisture inversion. Zribi [40] found that the model outputs and backscattering had good consistency under different experimental conditions by combining parameters: the Zs ($Z_S = S^2/l$) and soil moisture. According to the result, there was a substantial logarithmic relationship between the backscattering coefficient and Zs in the VV and HH polarization patterns. The detailed expression is presented as follows:

$$\sigma^0_{pq} = B_{pq} \ln(Z_s) + f(m_v) \tag{4}$$

where pq represents the polarization pattern; B_{pq} is the coefficient that is not related to the soil moisture when the incident angle is known; $f(m_v)$ is the known soil moisture content.

The relationships among the soil moisture, combined roughness, and backscattering coefficient in the VV and VH polarizations are shown in Figure 4. According to the results, there was a substantial logarithmic correlation among the backscattering coefficient, soil moisture, and combined roughness. If the combined roughness is known, then the backscattering coefficient increases as the soil moisture increases. When the soil moisture is 30%, the change becomes stable, and the sensitivity of the backscattering coefficient to the soil moisture decreases. When the soil moisture is known, the trend of the backscattering coefficient increases as the combined roughness initially increases and then decreases. Consequently, the backscattering coefficient increases with the increase in the soil moisture content. Moreover, the sensitivity of the backscattering coefficient to the soil moisture gradually decreases with the increase in the combined surface roughness.

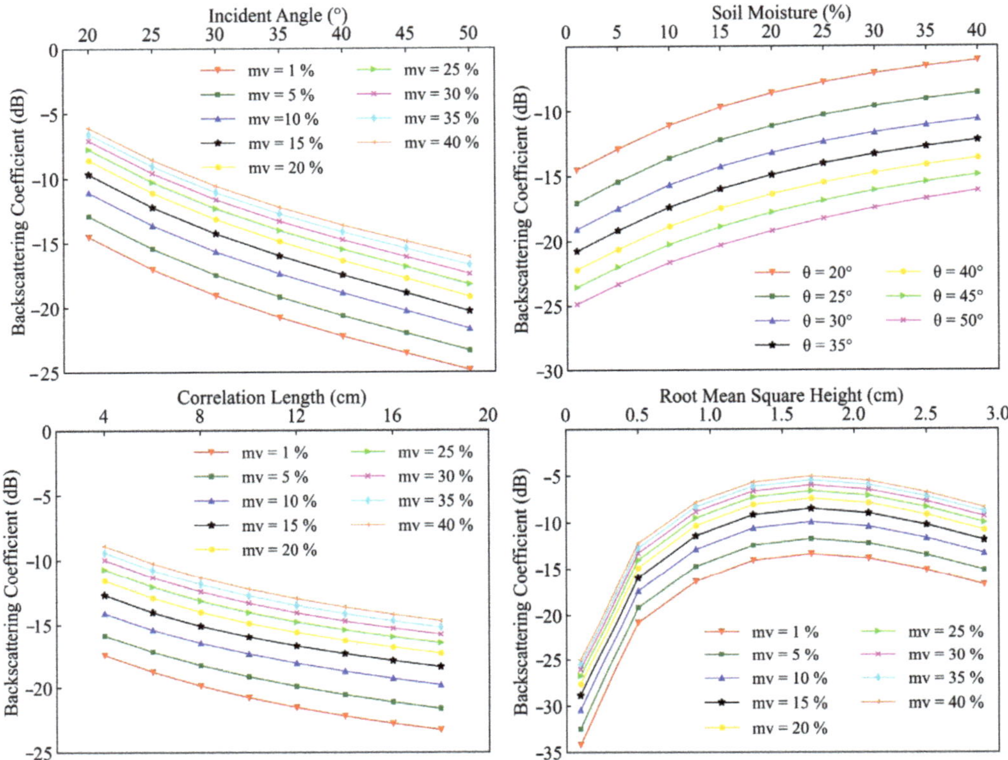

Figure 3. Response of backscattering coefficient to the surface parameters with AIEM (mv is soil moisture content).

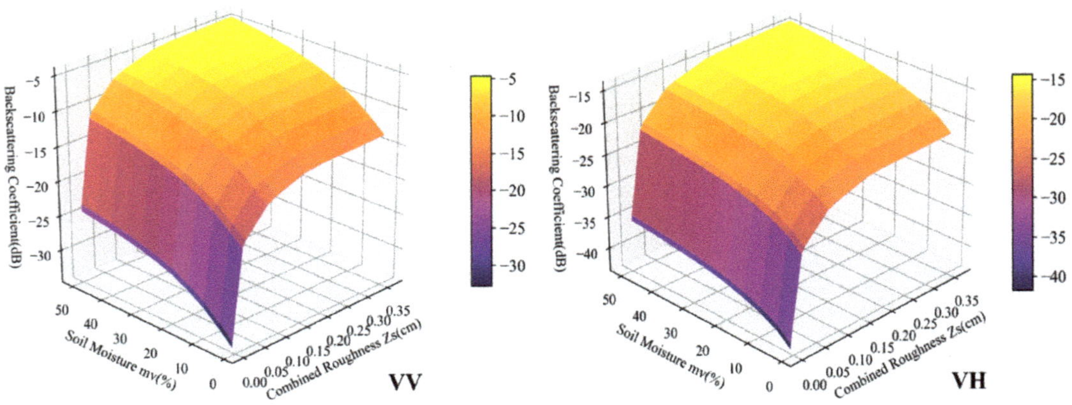

Figure 4. Relationships between soil moisture, combined roughness, and backscattering coefficient in VV and VH polarizations.

2.3.4. Construction of Empirical Model

Overall, the relationships between the soil moisture, combined roughness, and backscattering coefficient in the VV and VH polarization are as follows:

$$\sigma_{VV} = A_{VV}(\theta)\ln(m_v) + B_{VV}(\theta)\ln(Z_s) + C_{VV}(\theta) \quad (5)$$

$$\sigma_{VH} = A_{VH}(\theta)\ln(m_v) + B_{VH}(\theta)\ln(Z_s) + C_{VH}(\theta) \quad (6)$$

where σ represents the backscattering coefficient with different polarizations; $A(\theta)$, $B(\theta)$, and $C(\theta)$ are the coefficients that are only related to the incident angle (we obtained their values by simulating them in the AIEM database).

Although surface roughness is a critical parameter in soil moisture retrieval, it is difficult to measure the ground roughness in the actual application. Moreover, the measurement accuracy of the surface roughness also cannot be ensured. Therefore, if the surface roughness is replaced by other known parameters in the establishment of the empirical model, then this critical parameter has a substantial influence on and reduces soil moisture retrieval. In other words, the precision of soil moisture inversion will be improved by reducing the quantity of the unknown parameters or inaccuracy factors in the model. According to the simulation results of AIEM, the relationships between the backscattering coefficient, soil moisture, and surface roughness in the VV and VH polarizations are shown in Equations (5) and (6). When the backscattering coefficients of the VV and VH polarizations are known, the surface roughness *(Zs)* will be eliminated by combining Equations (5) and (6), and the final empirical model of the soil moisture retrieval can be obtained. The detailed expression is drawn as follows:

$$m_v = \mathrm{EXP}^{(A_{VVVH}\cdot\sigma_{VV} + B_{VVVH}\cdot\sigma_{VH} + C_{VVVH})} \quad (7)$$

where m_v is the soil moisture content; σ_{VV} and σ_{VH} are the backscattering coefficients of the VV and VH polarizations, respectively; A_{VVVH}, B_{VVVH}, and C_{VVVH} are the coefficients that are simulated by the modeling data.

3. Results

3.1. *Soil Moisture Retrieval Using the Empirical Model*

Although we collected soil moisture measurements from 2015 to 2021 at the Naqu station, we employed the Sentinel-1 synthetic aperture radar data from 2017 to 2019 and the soil moisture measurements from the corresponding time to establish the soil moisture retrieval models, which is because the Sentinel-1 images from 2015 to 2016 at the Naqu station were missing. Finally, we used 240 Sentinel-1 images of the VV and VH polarizations to construct the soil moisture retrieval models, and we obtained the soil moisture results from the Naqu station for 2020 from 2020 Sentinel-1 images by using retrieval models. In Section 3.2, we proposed the empirical models for the ascending and descending orbits based on the soil moisture measurement data of 5 cm and the backscattering coefficient of the VV and VH polarizations from Sentinel-1 images from 2017 to 2019 at the Naqu station. We introduced the least-squares method to calculate the A_{VVVH}, B_{VVVH}, and C_{VVVH} values. We present the detailed expressions in Equations (8) and (9), respectively:

$$m_v = \mathrm{EXP}^{(0.087\cdot\sigma_{VV} + 0.017\cdot\sigma_{VH} + 0.322)} \quad (8)$$

$$m_v = \mathrm{EXP}^{(0.071\cdot\sigma_{VV} + 0.016\cdot\sigma_{VH} - 0.071)} \quad (9)$$

The backscattering coefficients of the VV and VH polarizations from the Sentinel-1 images for 2020 from the Naqu station are put into the empirical models of the ascending and descending orbits to retrieve the soil moisture content, respectively. The inversion results of the soil moisture for the ascending and descending orbits at the Naqu station for 2020 are presented in Figures 5 and 6, respectively, and the comparisons of the soil moisture between the measured values and retrieved values for the ascending and descending orbits are shown in Figure 7. According to the results, we can use the empirical models to retrieve the surface soil moisture, with an R^2 of 0.609, RMSE of 0.08, and MAE of 0.064 for the ascending orbit model, and an R^2 of 0.554, RMSE of 0.086, and MAE of 0.071 for the descending orbit model. When the soil moisture is higher than 0.3 m^3/m^3, the empirical models underestimate the soil moisture so that it is markedly contrasted with the measured values. The simulation results of the ascending orbit are better than those of the descending orbit, which is also consistent with the results of Dabrowska-Zielinska [41], who retrieved the soil moisture from the Sentinel-1 imagery over wetlands and found that the retrieval result of the soil moisture achieved a satisfactory performance by using data from the ascending orbit of the Sentinel-1 images. The regions with high soil moisture are mainly distributed in mountainous areas, and the regions with low soil moisture are distributed among the flat terrain areas. The soil moisture contents in June, July, August, and September are substantially higher than in other months because the Naqu network climate is mainly influenced by the south Asian monsoon, and the precipitation falls between June and September.

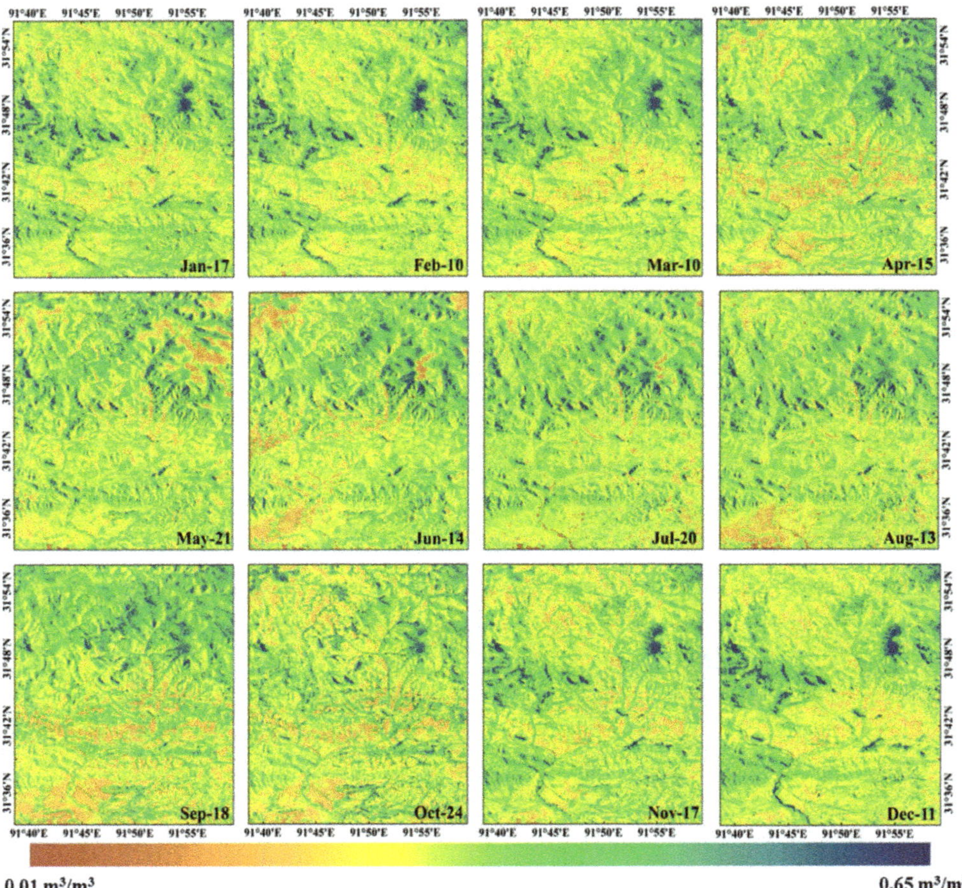

Figure 5. Soil moisture for ascending orbit of Naqu network for 2020.

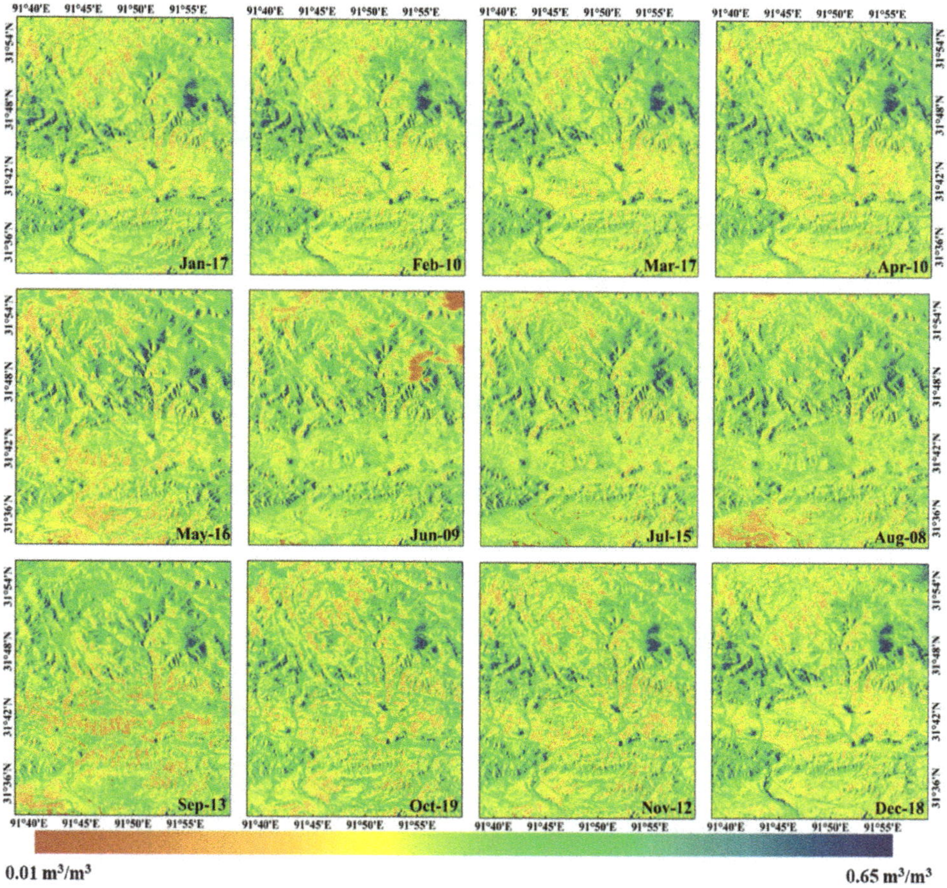

Figure 6. Soil moisture for descending orbit of Naqu network for 2020.

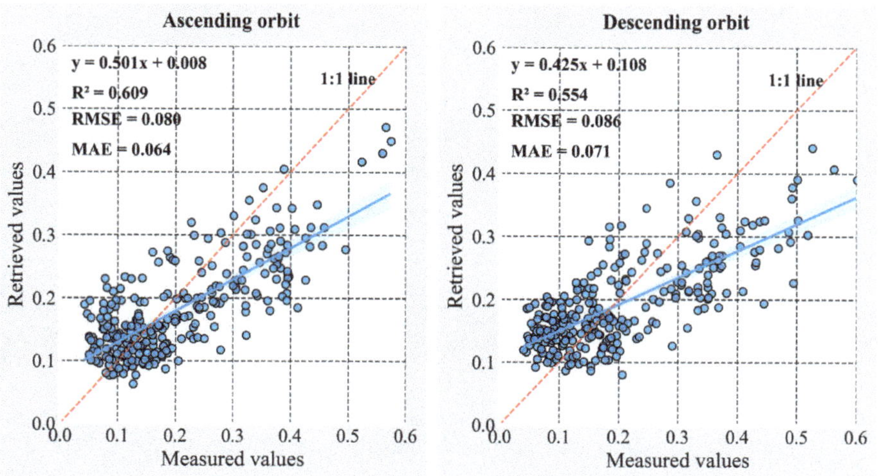

Figure 7. Comparison between measured and retrieved soil moisture values of Naqu network (The blue line is fitted line, and the red line is 1:1 line).

3.2. Soil Moisture Retrieval Using Machine Learning Algorithms

To further improve the accuracy of the soil moisture retrieval, the machine learning algorithms of SVM, BPNN, KNN, and RF are introduced to inverse the surface soil moisture of the Naqu network with the AIEM. In the process of machine learning modeling, the physically meaningful radar parameters in the AIEM are introduced to machine learning algorithms to establish the soil moisture retrieval model. The backscattering coefficient of the VV and VH polarizations and the incidence angle are the independent variables. The dependent variable is the soil moisture measurement data. Therefore, the measurements (soil moisture, backscattering coefficient of the VV and VH polarizations, and incidence angle) from the Naqu station for 2017–2019 are employed as the ensemble of training and testing samples, and the training and testing samples are set to 70% and 30% of the total number of samples, respectively. The model performances of the machine learning of the ascending and descending orbits are presented in Table 1. According to the results, the RF performance is better than those of the other machine learning algorithms, with an R^2 of 0.753, RMSE of 0.045, and MAE of 0.034 in the ascending orbit, and an R^2 of 0.671, RMSE of 0.049, and MAE of 0.038 in the descending orbit. In addition, the accuracies of the machine learning approaches in the ascending orbit are also better than those in the descending orbit. For the model application, the surface soil moisture contents for the ascending and descending orbits for 2020 from the Naqu network are retrieved by using different machine learning algorithms.

Table 1. Training accuracies of the different machine learning algorithms.

			R^2	RMSE	MAE	Bias
Ascending	SVM	Mean	0.634	0.057	0.046	0.015
		Std	0.025	0.005	0.013	0.008
	BPNN	Mean	0.614	0.058	0.047	0.018
		Std	0.029	0.006	0.015	0.009
	KNN	Mean	0.699	0.051	0.041	0.006
		Std	0.021	0.003	0.009	0.004
	RF	Mean	0.753	0.045	0.034	0.004
		Std	0.018	0.002	0.005	0.002
Descending	SVM	Mean	0.548	0.060	0.052	0.021
		Std	0.027	0.006	0.016	0.010
	BPNN	Mean	0.561	0.056	0.048	0.016
		Std	0.026	0.005	0.013	0.008
	KNN	Mean	0.616	0.053	0.042	0.007
		Std	0.023	0.005	0.044	0.007
	RF	Mean	0.671	0.049	0.038	0.006
		Std	0.020	0.004	0.007	0.003

The inversion results of the soil moisture retrieval with machine learning algorithms for the ascending and descending orbits of Naqu station for 2020 are presented in Figure 8, and we present the comparisons of the soil moisture retrieval of the different models for the ascending and descending orbits in Figure 9. The result indicates that the performances of the machine learning algorithms are substantially superior to the empirical model. For the ascending orbit, the coefficients (R^2) of the BPNN, KNN, SVM, RF, and EM (empirical) models are 0.615, 0.666, 0.626, 0.714, and 0.609, respectively. and the RMSE coefficients of these models are 0.076, 0.070, 0.078, 0.065, and 0.080, respectively. For the descending orbit, the coefficients (R^2) of the BPNN, KNN, SVM, RF, and EM (empirical) models are 0.590,

0.612, 0.588, 0.677, and 0.554, respectively, and the RMSE coefficients of these models were 0.080, 0.078, 0.083, 0.072, and 0.086, respectively. According to these results, the combination of the AIEM and machine learning algorithms can further enhance the precision of soil moisture retrieval. The inversion accuracies of the soil moisture with different machine learning algorithms in the ascending orbit are also better than those in the descending orbit. In addition, the accuracy of the RF algorithm is better than those of the BPNN, SVM, and KNN models.

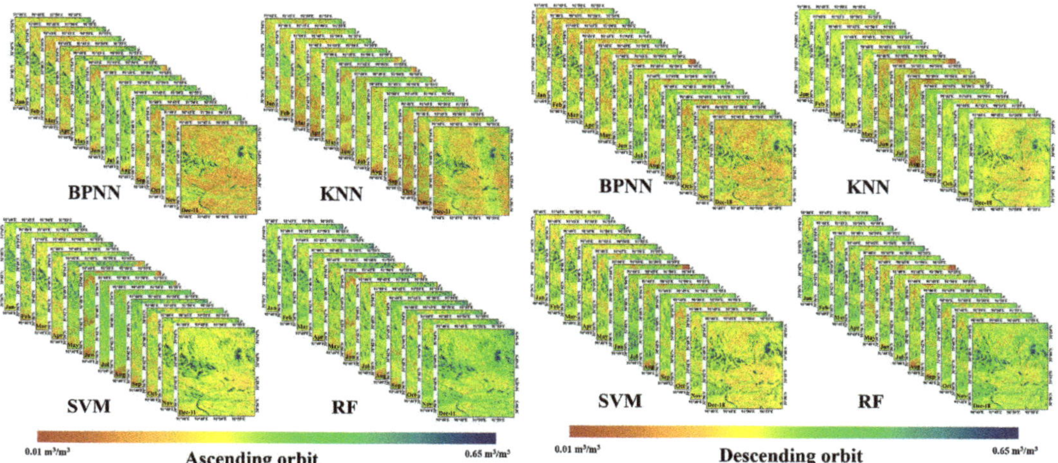

Figure 8. Soil moisture inversion results with machine learning algorithms for ascending and descending orbits of Naqu network for 2020.

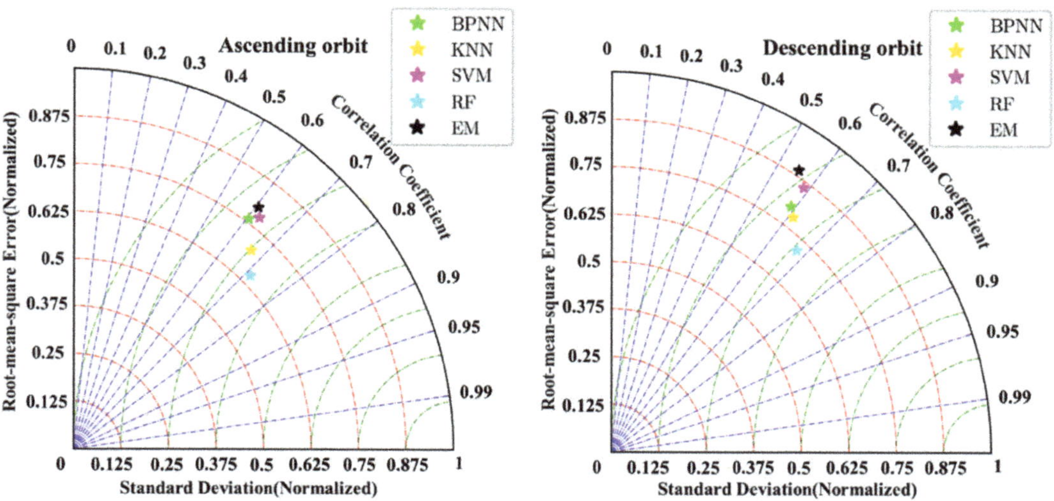

Figure 9. Comparison of soil moisture of different models for ascending and descending orbits.

Although the soil moisture inversion results with the RF in the Naqu network indicate a satisfactory performance, we also retrieve the soil moisture contents of the Maqu network in 2018 and of the Tianjun network in 2020 using the RF algorithm to further validate the precision of the soil moisture retrieval. Figures 10 and 11 present the soil moisture inversion result for the ascending and descending orbits of the Maqu network for 2018, respectively. The soil moisture inversion results for the ascending and descending orbits of the Tianjun network for 2020 are shown in Figures 12 and 13, respectively. The validations of soil moisture for the ascending and descending orbits in the Maqu and Tianjun networks are presented in Figures 14 and 15, respectively. The results indicate that the RF algorithm achieves a satisfactory performance for the ascending and descending orbits in both the Maqu and Tianjun networks. In the Maqu network, the R^2, RMSE, and MAE values for the ascending orbit are 0.696, 0.062, and 0.052, respectively, and the values of these coefficients for the descending orbit are 0.648, 0.075, and 0.064, respectively. In the Tianjun network, the R^2, RMSE, and MAE values for the ascending orbit are 0.709, 0.069, and 0.057, respectively, and the values of these coefficients for the descending orbit are 0.638, 0.074, and 0.063, respectively. Moreover, the inversion accuracies of the soil moisture for the ascending orbit are also higher than those for the descending orbit for both the Maqu and Tianjun networks.

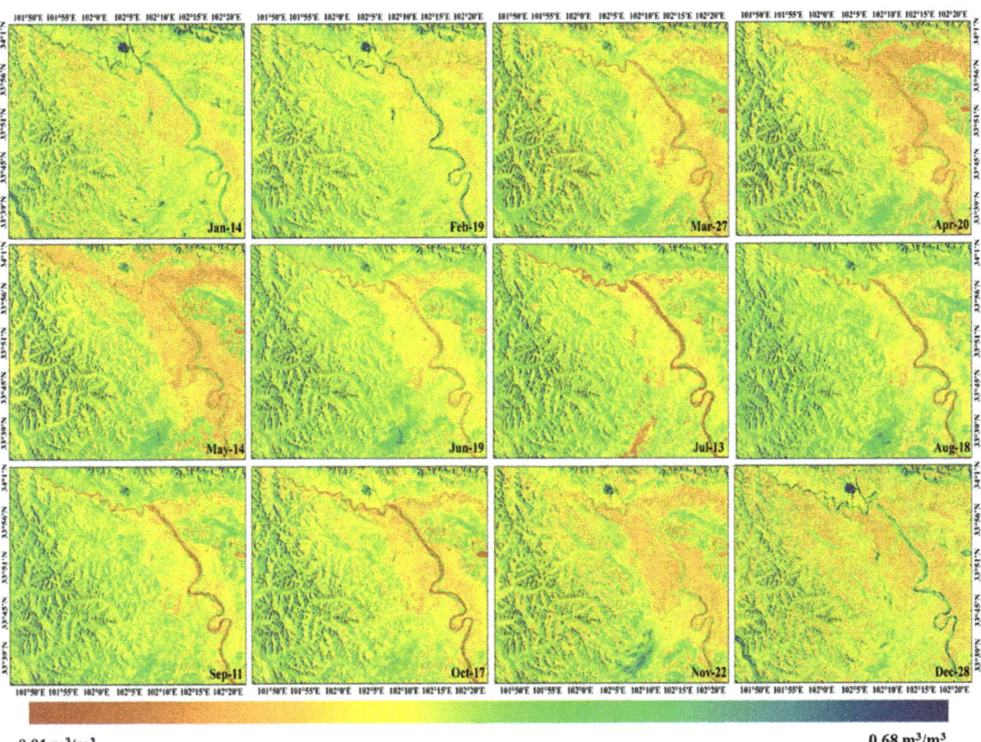

Figure 10. Soil moisture with RF for ascending orbit of Maqu network for 2018.

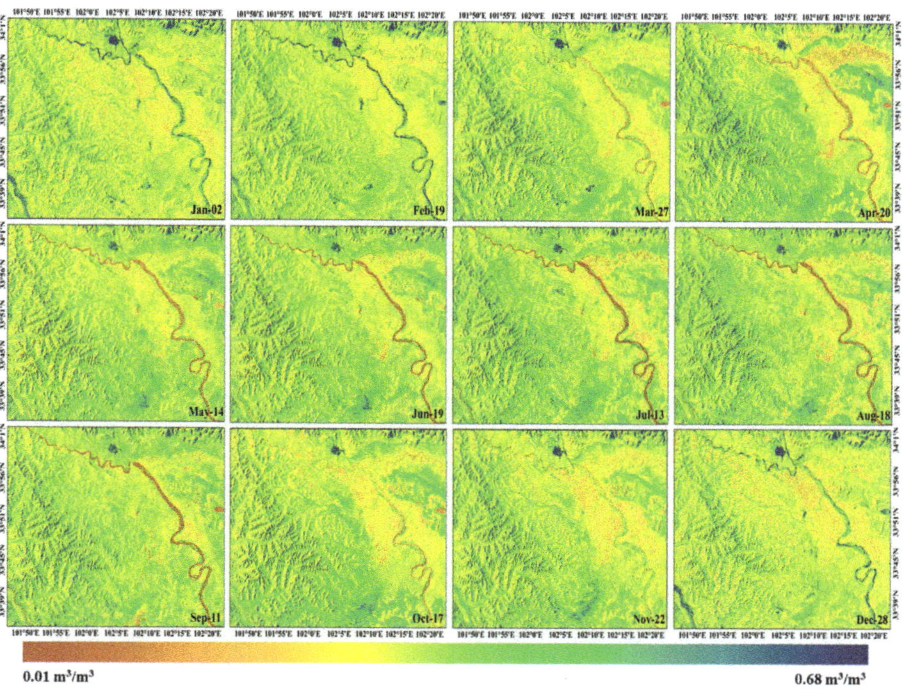

Figure 11. Soil moisture with RF for descending orbit of Maqu network for 2018.

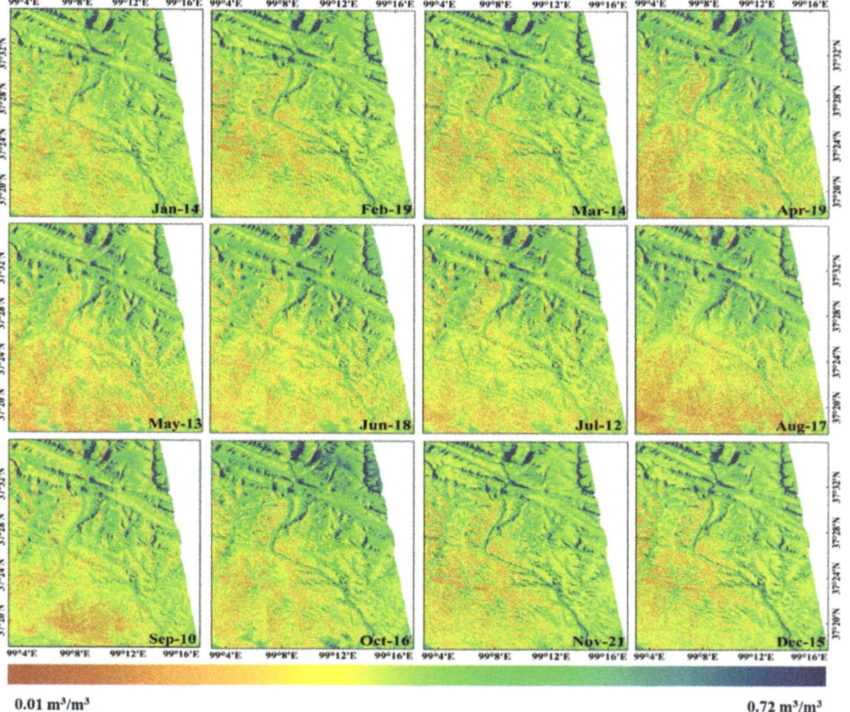

Figure 12. Soil moisture with RF for ascending orbit of Tianjun network for 2020.

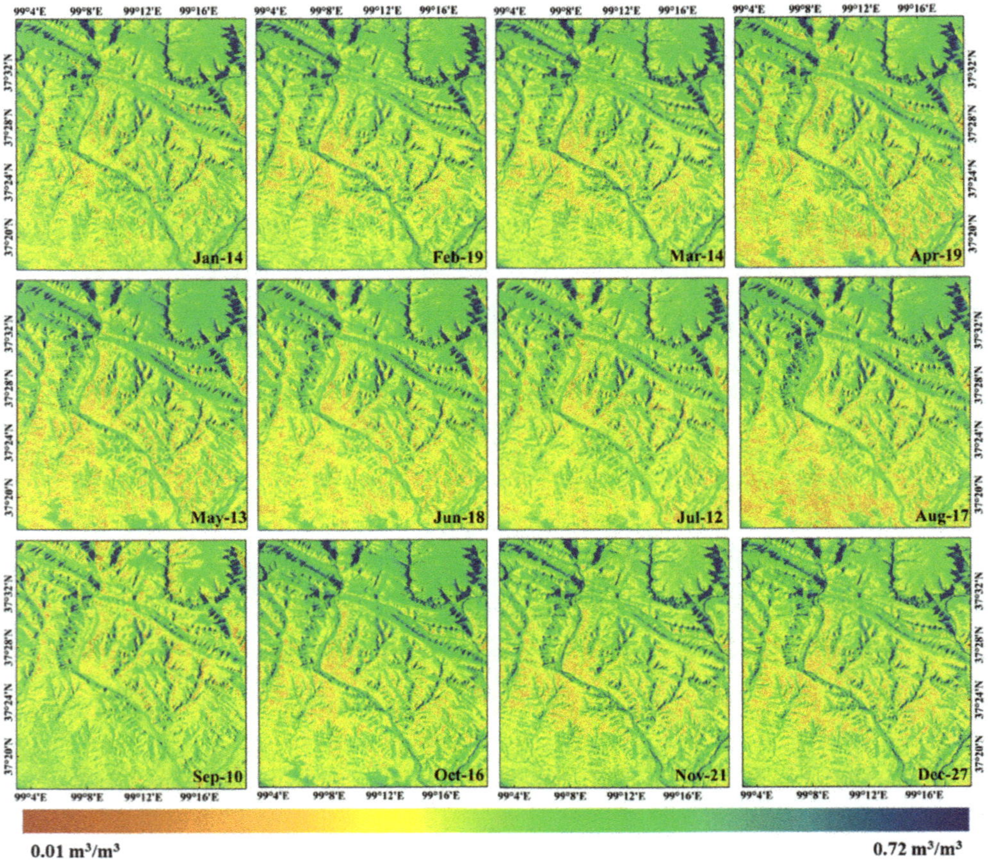

Figure 13. Soil moisture with RF for descending orbit of Tianjun network for 2020.

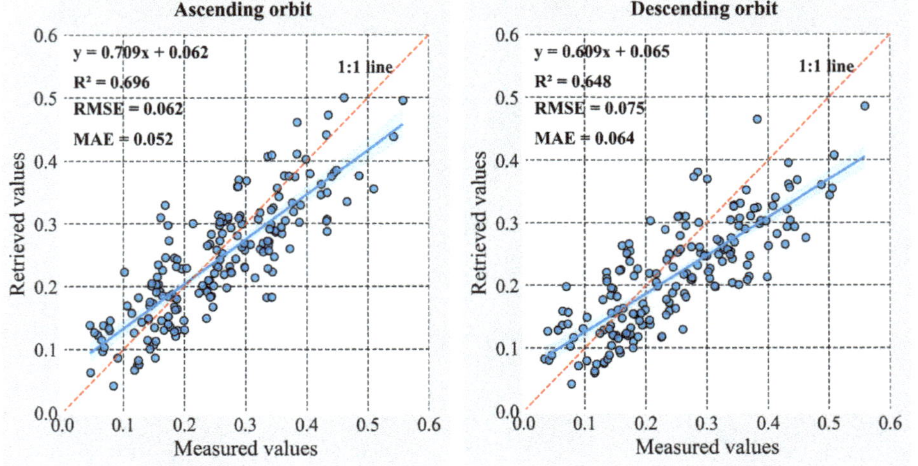

Figure 14. Comparision between measured and retrieved soil moisture values for Maqu network (The blue line is fitted line, and the red line is 1:1 line).

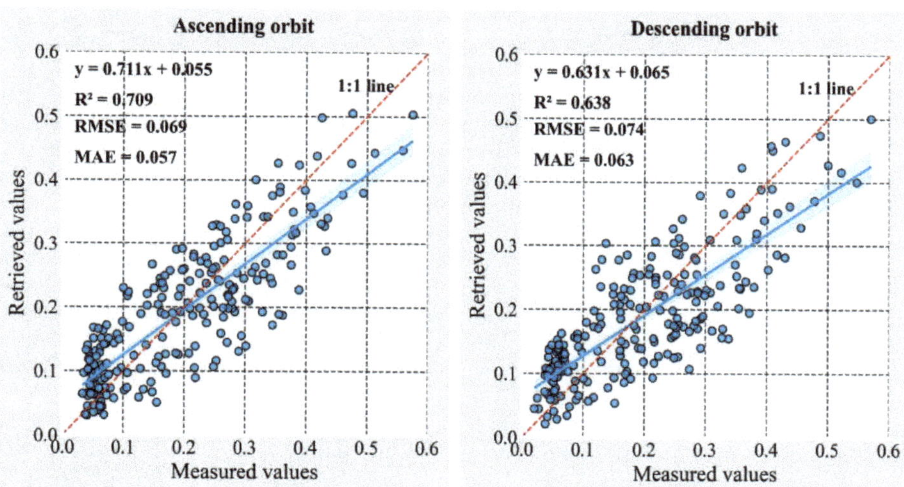

Figure 15. Comparision between measured and retrieved soil moisture values for Tianjun network (The blue line is fitted line, and the red line is 1:1 line).

4. Discussion

The surface roughness is an important parameter in the soil moisture inversion process. The measurement of the surface roughness is difficult in practical experiments for natural and manmade reasons. In addition, the measurement accuracy also substantially differs from that of the actual conditions. Reducing the input of unknown or unobservable parameters is one of the major methods for optimizing the model. Therefore, the surface roughness is replaced by other known parameters, and the empirical models for the ascending and descending orbits are proposed by combining the equations of the VV and VH polarizations based on the AIEM model to decrease the impact of the surface roughness. Four machine learning algorithms (BPNN, SVM, KNN, and RF) are used to further improve the soil moisture retrieval precision in the Naqu network, and these algorithms are commonly applied but have different learning strategies. To further verify the model accuracy, the surface soil moisture for the ascending and descending orbits of the Maqu network for 2018 and Tianjun network for 2020 are retrieved using the RF algorithm, respectively. We found that the retrieval results of these machine learning algorithms are more consistent compared with the empirical model. However, due to the different learning schemes, there are still some minor distinctions in the results of the four algorithms. According to the results of this paper, the RF performance is superior to the other machine learning approaches because the RF model obtains independent regression trees by randomly testing training data [37]. Therefore, the model can overcome noise and avoid the overfitting issue in practical applications. Chen [42] estimated the soil moisture of winter wheat farmlands during the vegetative season based on the machine learning algorithms of support vector regression, random forests (RF), and gradient boosting regression tree, the results also indicated that the performance of the RF algorithm is better than those of the other algorithms. The 12 advanced statistical and machine learning algorithms were used to estimate soil moisture using the Sentinel-1 data [43], and the result indicated that the RF algorithm has satisfactory performance compared with those of the other models.

The AIEM is a forward model that is used to calculate the backscattering coefficient of the bare ground with high estimated precision and low predicted consumption. In this study, we only considered the single scattering situation and ignored multiple scattering ones, which is one of the main reasons for the errors in the AIEM simulation process. Zeng [44] presents the scattering results between the numerical simulations and experimental measurements with the AIEM, which also indicated that multiple scattering has a certain effect on backscattering, and that influence on the HH polarization is higher than that on the VV polarization. Although the penetrability of the C band does not lead to intensive volume scattering, its influence on the actual scattering process is not neglected.

The vegetation water content is a substantial parameter that affects the soil moisture retrieval accuracy. In this study, we selected the Naqu, Maqu, and Tianjun soil moisture monitoring networks as the research areas to achieve surface soil moisture inversion using Sentinel-1 data. The Sentinel-1 data is a C-band (5.405 GHz) synthetic aperture radar that provides dual-polarization, and it also has the ability to penetrate the sparse and low vegetation on the ground surface. Alpine meadow is the main land cover type in the research areas due to the climate, and the soil moisture inversion result is less affected by the surface vegetation. However, there is still vegetation water content interference. Moreover, plant growth is a dynamic process, and its structure and morphology will change significantly over time; however, we could not use the empirical constants to reveal the dynamic changes in the vegetation information, which led to some uncertainty regarding the estimated results.

Although the influence of the surface roughness in this study is reduced by combining the empirical equations of the soil moisture and VV, and VH polarizations, the surface roughness is still a key factor in the soil moisture retrieval process. The issue of surface roughness has received broad attention in recent years, and researchers have proposed relevant models [38–40]. However, uncertainties still exist in the research on the roughness parameterization scheme. The main reason is that the different models are usually developed by using different experimental data; in other words, the soil type, soil texture, and moisture content parameters, and the rough conditions in each model, are different, as are the hypothesized conditions of the model development (for example, the calculated method selection of the soil dielectric constant and soil effective temperature). Overall, every roughness model has its comparative advantages and constrained conditions, and no model can perform well in all circumstances. Further research on the roughness parameterization schemes that can be applied to the complex soil conditions of different soil roughnesses, moistures, soil types, and correlation functions is essential.

The radar response to the soil moisture content is closely related to critical parameters, such as surface roughness, microwave frequency, and incident angle. Ulaby [45] found that the radar response seems to be linear within a range of 15–30% moisture content for all angles, frequencies, polarizations, and surface conditions. Theoretically, the Sentinel-1 images could be employed to inverse soil moisture content well over the range of 15–30% moisture content. When the soil moisture content is higher than 0.3 m^3/m^3, the empirical model markedly underestimates the soil moisture compared to the observation data. Bruckler [46] also confirms this result. Although machine learning algorithms can improve the inversion results, how to further improve the inversion accuracy of the soil moisture with high water content is the next issue to be explored.

5. Conclusions

We select Naqu, Maqu, and Tianjun soil moisture monitoring networks on the QTP as the research areas. The database of the surface microwave scattering is obtained using the AIEM to analyze the response of the surface parameters and radar signal. The soil moisture retrieval models of the empirical and machine learning algorithms for the ascending and descending orbits are proposed by using the Sentinel-1 and soil moisture measurements. Finally, the soil moisture retrieval accuracies of the different models are validated in these research areas.

The major conclusions of this study are abstracted as follows:

(1) The empirical models for the ascending and descending orbits can estimate the surface soil moisture in the Naqu network, but the soil moisture content is markedly underestimated in empirical models when the soil moisture is high. The simulation results of the ascending orbit are better than those of the descending orbit.

(2) The combination of the AIEM and machine learning algorithms can further enhance soil moisture inversion precision. The performances of the machine learning algorithms are substantially superior to that of the empirical model, and the accuracy of the RF model is higher than those of the BPNN, SVM, and KNN models. The inversion accuracies of the soil moisture with the different machine learning algorithms in the ascending orbit are also better than those in the descending orbit.

(3) The RF algorithm achieves a satisfactory performance for the ascending and descending orbits for both the Maqu and Tianjun networks. The rationality and accuracy of the RF algorithm at different locations and times on the QTP are further verified.

Author Contributions: The specific contributions of each author in the paper are listed. L.D. processed data and wrote the manuscript; W.W., R.J. and F.X. contributed important ideas and considerations; R.J. and Y.Z. carried out the soil moisture observation experiment. All authors have read and agreed to the published version of the manuscript.

Funding: This work was funded by the National Nature Science Foundation of China (No. 42130113 and No. 42101411); the National Science and Technology Major Project of China's High Resolution Earth Observation System (Project No. 21-Y20B01-9001-19/22); the Basic Research Innovative Groups of Gansu province, China (Grant No. 21JR7RA068).

Data Availability Statement: Not applicable.

Acknowledgments: Soil moisture measurements are provided by the National Tibetan Plateau Data Center (https://datatodcac.cn/).

Conflicts of Interest: The authors declare no conflict of interest.

References

1. Koster, R.D.; Dirmeyer, P.A.; Guo, Z.; Bonan, G.; Chan, E.; Cox, P.; Gordon, C.T.; Kanae, S.; Kowalczyk, E.; Lawrence, D.; et al. Regions of strong coupling between soil moisture and precipitation. *Science* **2004**, *305*, 1138–1140. [CrossRef] [PubMed]
2. Nguyen, H.H.; Cho, S.; Jeong, J.; Choi, M. A D-vine copula quantile regression approach for soil moisture retrieval from dual polarimetric SAR Sentinel-1 over vegetated terrains. *Remote Sens. Environ.* **2021**, *255*, 112283. [CrossRef]
3. Brocca, L.; Moramarco, T.; Melone, F.; Wagner, W.; Hasenauer, S.; Hahn, S. Assimilation of surface- and root-zone ASCAT soil moisture products into rainfall-runoff modeling. *IEEE Trans. Geosci. Remote* **2012**, *50*, 2542–2555. [CrossRef]
4. Dharssi, I.; Bovis, K.J.; Macpherson, B.; Jones, d.C.P. Operational assimilation of ASCAT surface soil wetness at the Met Office. *Hydrol. Earth Syst. Sci.* **2011**, *15*, 2729–2746. [CrossRef]
5. Brocca, L.; Melone, F.; Moramarco, T.; Wagner, W.; Naeimi, V.; Bartalis, Z.; Hasenauer, S. Improving runoff prediction through the assimilation of the ASCAT soil moisture product. *Hydrol. Earth Syst. Sci.* **2010**, *14*, 1881–1893. [CrossRef]
6. Dobson, M.C.; Ulaby, F.T.; Hallikainen, M.T.; Elrayes, M.A. Microwave dielectric behavior of wet soil-part II_ dielectric-mixing models. *IEEE Trans. Geosci. Remote* **1985**, *23*, 35–46. [CrossRef]
7. Santi, E.; Paloscia, S.; Pettinato, S.; Brocca, L.; Ciabatta, L.; Entekhabi, D. Integration of microwave data from SMAP and AMSR2 for soil moisture monitoring in Italy. *Remote Sens. Environ.* **2018**, *212*, 21–30. [CrossRef]
8. Zeng, J.; Li, Z.; Chen, Q.; Bi, H.; Qiu, J.; Zou, P. Evaluation of remotely sensed and reanalysis soil moisture products over the Tibetan Plateau using in-situ observations. *Remote Sens. Environ.* **2015**, *163*, 91–110. [CrossRef]
9. Ouaadi, N.; Jarlan, L.; Ezzahar, J.; Zribi, M.; Khabba, S.; Bouras, E.; Bousbih, S.; Frison, P.L. Monitoring of wheat crops using the backscattering coefficient and the interferometric coherence derived from Sentinel-1 in semi-arid areas. *Remote Sens. Environ.* **2020**, *251*, 112050. [CrossRef]
10. Zeyliger, A.M.; Muzalevskiy, K.V.; Zinchenko, E.V.; Ermolaeva, O.S. Field test of the surface soil moisture mapping using Sentinel-1 radar data. *Sci. Total Environ.* **2022**, *807*, 151121. [CrossRef]
11. Ma, C.; Li, X.; Wang, J.; Wang, C.; Duan, Q.; Wang, W. A comprehensive evaluation of microwave emissivity and brightness temperature sensitivities to soil parameters using qualitative and quantitative sensitivity analyses. *IEEE Trans. Geosci. Remote* **2017**, *55*, 1025–1038. [CrossRef]
12. Oh, Y. Quantitative retrieval of soil moisture content and surface roughness from multipolarized radar observations of bare soil surfaces. *IEEE Trans. Geosci. Remote* **2004**, *42*, 596–601. [CrossRef]
13. Dubois, P.C.; Zyl, J.v.; Engman, T. Measuring soil moisture with imaging radars. *IEEE Trans. Geosci. Remote* **1995**, *33*, 915–926. [CrossRef]
14. Shi, J.; Jiang, L.; Zhang, L.; Chen, K.; Wigneron, J.P.; Chanzy, A. A parameterized multifrequency-polarization surface emission model. *IEEE Trans. Geosci. Remote* **2005**, *43*, 2831–2841. [CrossRef]
15. Fung, A.K. *Microwave Scattering and Emission Models for Users*; Artech House Inc.: Boston, MA, USA; Artech House Inc.: London, UK, 2010.

16. Fung, A.K. *Microwave Scattering and Emission Model and Their Applications*; Artech House Inc.: Boston, MA, USA; Artech House Inc.: London, UK, 1994.
17. Ulaby, F.T.; Moore, R.K.; Fung, A.K. *Microwave Remote Sensing: Active and Passive, Vol. III from Theory to Applications*; Artech House: Boston, MA, USA, 1986.
18. Fung, A.K.; Li, Z.; Chen, K. Backscattering from a randomly rough dielectric surface. *IEEE Trans. Geosci. Remote* **1992**, *30*, 356–369. [CrossRef]
19. Chen, K.S.; Wu, T.-D.; Tsay, M.-K.; Fung, A.K. A note on the multiple scattering in an IEM model. *IEEE Trans. Geosci. Remote* **2000**, *38*, 249–256. [CrossRef]
20. Baghdadi, N.; Gherboudj, I.; Zribi, M.; Sahebi, M.; King, C.; Bonn, F. Semi-empirical calibration of the IEM backscattering model using radar images and moisture and roughness field measurements. *Int. J. Remote Sens.* **2004**, *25*, 3593–3623. [CrossRef]
21. Dong, L.; Baghdadi, N.; Ludwig, R. Validation of the AIEM through correlation length parameterization at field scale using radar imagery in a semi-arid environment. *IEEE Geosci. Remote Sens.* **2013**, *10*, 461–465. [CrossRef]
22. Zribi, M.; Ciarletti, V.; Taconet, O. Validation of a rough surface model based on Fractional Brownian Geometry with SIRC and ERASME radar data over Orgeval. *Remote Sens. Environ.* **2000**, *73*, 65–72. [CrossRef]
23. Lievens, H.; Vernieuwe, H.; Alvarez-Mozos, J.; Baets, B.D.; Verhoest, N.E. Error in radar-derived soil moisture due to roughness parameterization: An analysis based on synthetical surface profiles. *Sensors* **2009**, *9*, 1067–1093. [CrossRef]
24. Paloscia, S.; Pettinato, S.; Santi, E.; Notarnicola, C.; Pasolli, L.; Reppucci, A. Soil moisture mapping using Sentinel-1 images: Algorithm and preliminary validation. *Remote Sens. Environ.* **2013**, *134*, 234–248. [CrossRef]
25. Kolassa, J.; Reichle, R.H.; Liu, Q.; Alemohammad, S.H.; Gentine, P.; Aida, K.; Asanuma, J.; Bircher, S.; Caldwell, T.; Colliander, A.; et al. Estimating surface soil moisture from SMAP observations using a Neural Network technique. *Remote Sens. Environ.* **2018**, *204*, 43–59. [CrossRef]
26. Yan, R.; Bai, J. A new approach for soil moisture downscaling in the presence of seasonal difference. *Remote Sens.* **2020**, *12*, 2818. [CrossRef]
27. Abowarda, A.S.; Bai, L.; Zhang, C.; Long, D.; Li, X.; Huang, Q. Generating surface soil moisture at 30 m spatial resolution using both data fusion and machine learning toward better water resources management at the field scale. *Remote Sens. Environ.* **2021**, *255*, 112301. [CrossRef]
28. Nguyen, T.T.; Ngo, H.H.; Guo, W.; Chang, S.W.; Nguyen, D.D.; Nguyen, C.T.; Zhang, J.; Liang, S.; Bui, X.T.; Hoang, N.B. A low-cost approach for soil moisture prediction using multi-sensor data and machine learning algorithm. *Sci. Total Environ.* **2022**, *833*, 155066. [CrossRef] [PubMed]
29. Chen, Y.; Yang, K.; Qin, J.; Cui, Q.; Lu, H.; La, Z.; Han, M.; Tang, W. Evaluation of SMAP, SMOS, and AMSR2 soil moisture retrievals against observations from two networks on the Tibetan Plateau. *J. Geophys. Res. Atmos.* **2017**, *122*, 5780–5792. [CrossRef]
30. Yang, K.; Ye, B.; Zhou, D.; Wu, B.; Foken, T.; Qin, J.; Zhou, Z. Response of hydrological cycle to recent climate changes in the Tibetan Plateau. *Clim. Chang.* **2011**, *109*, 517–534. [CrossRef]
31. Yang, K.; Qin, J.; Zhao, L.; Chen, Y.; Tang, W.; Han, M.; Zhu, L.; Chen, Z.; Lv, N.; Ding, B.; et al. A multiscale soil moisture and freeze-thaw monitoring network on the Third Pole. *Bull. Am. Meteorol. Soc.* **2013**, *94*, 1907–1916. [CrossRef]
32. Su, Z.; Wen, J.; Dente, L.; van der Velde, R.; Wang, L.; Ma, Y.; Yang, K.; Hu, Z. The Tibetan Plateau observatory of plateau scale soil moisture and soil temperature (Tibet-Obs) for quantifying uncertainties in coarse resolution satellite and model products. *Hydrol. Earth Syst. Sci.* **2011**, *15*, 2303–2316. [CrossRef]
33. Rumelhart, D.E.; Hinton, G.E.; Williams, R.J. Learning representations by back propagating errors. *Nature* **1986**, *323*, 533–536. [CrossRef]
34. Vapnik, V.N. *The Nature of Statistical Learning Theory*; Springer: New York, NY, USA, 1995; pp. 988–999.
35. Dibike, Y.B.; Velickov, S.; Solomatine, D.; Abbott, M.B. Model induction with support vector machines: Introduction and applications. *J. Comput. Civ. Eng.* **2001**, *15*, 208–216. [CrossRef]
36. Devroye, L.; Gyorfy, L.; Krzyzak, A.; Lugosi, G. On the strong universal consistency of nearest neighbor regression function estimates. *Ann. Stat.* **1994**, *22*, 1371–1385. [CrossRef]
37. Breiman, L. Random forests. *Mach. Learn.* **2001**, *45*, 5–32. [CrossRef]
38. Oh, Y.; Sarabandi, K.; Ulaby, F.T. Semi-empirical model of the ensemble-averaged differential mueller matrix for microwave 569 backscattering from bare soil surfaces. *IEEE Trans. Geosci. Remote* **2002**, *40*, 1348–1355. [CrossRef]
39. Shi, J.; Wang, J.; Hsu, A.Y.; O'Neill, P.E.; Engman, E.T. Estimation of bare surface soil moisture and surface roughness parameter using L-band SAR image data. *IEEE Trans. Geosci. Remote* **1997**, *35*, 1254–1266. [CrossRef]
40. Zribi, M.; Dechambre, M. A new empirical model to retrieve soil moisture and roughness from C-band radar data. *Remote Sens. Environ.* **2002**, *84*, 42–52. [CrossRef]
41. Dabrowska-Zielinska, K.; Musial, J.; Malinska, A.; Budzynska, M.; Gurdak, R.; Kiryla, W.; Bartold, M.; Grzybowski, P. Soil 634 moisture in the Biebrza Wetlands retrieved from Sentinel-1 imagery. *Remote Sens.* **2018**, *10*, 1979. [CrossRef]
42. Chen, L.; Xing, M.; He, B.; Wang, J.; Shang, J.; Huang, X.; Xu, M. Estimating soil moisture over winter wheat fields during growing season using machine-learning methods. *IEEE J. Stars* **2021**, *14*, 3706–3718. [CrossRef]
43. Chaudhary, S.K.; Srivastava, P.K.; Gupta, D.K.; Kumar, P.; Prasad, R.; Pandey, D.K.; Das, A.K.; Gupta, M. Machine learning algorithms for soil moisture estimation using Sentinel-1: Model development and implementation. *Adv. Space Res.* **2022**, *4*, 69. [CrossRef]

44. Zeng, J.; Chen, K.; Bi, H.; Zhao, T.; Yang, X. A comprehensive analysis of rough soil surface scattering and emission predicted by AIEM with comparison to numerical simulations and experimental measurements. *IEEE Trans Geosci. Remote* **2017**, *55*, 1696–1708. [CrossRef]
45. Ulaby, F.T. Radar measurement of soil moisture content. *IEEE Trans. Antennas Propag.* **1974**, *22*, 257–265. [CrossRef]
46. Bruckler, L.; Witono, H.; Stengel, P. Near surface soil moisture estimation from microwave measurements. *Remote Sens. Environ.* **1988**, *26*, 101–121. [CrossRef]

 remote sensing

Article

A Novel Freeze-Thaw State Detection Algorithm Based on L-Band Passive Microwave Remote Sensing

Shaoning Lv [1,2,3], Jun Wen [4,*], Clemens Simmer [3,5], Yijian Zeng [6], Yuanyuan Guo [1] and Zhongbo Su [6]

1 Department of Atmospheric and Oceanic Sciences, Institute of Atmospheric Sciences, Fudan University, Shanghai 200438, China
2 Zhuhai Fudan Innovation Research Institute, Zhuhai 519000, China
3 Institute for Geosciences-Meteorology, University of Bonn, Auf dem Huegel 20, 53121 Bonn, Germany
4 The Plateau Atmosphere and Environment Key Laboratory of Sichuan Province, Chengdu University of Information Technology, Chengdu 610225, China
5 Cloud and Precipitation Exploration Laboratory (CPEX-Lab) of Geoverbund ABC/J, Auf dem Huegel 20, 53121 Bonn, Germany
6 Department of Water Resources, Faculty of Geo-Information Science and Earth Observation (ITC), University of Twente, 7500AE Enschede, The Netherlands
* Correspondence: jwen@cuit.edu.cn

Abstract: Knowing the freeze-thaw (FT) state of the land surface is essential for many aspects of weather forecasting, climate, hydrology, and agriculture. Microwave L-band emission contains rather direct information about the FT-state because of its impact on the soil dielectric constant, which determines microwave emissivity and the optical depth profile. However, current L-band-based FT algorithms need reference values to distinguish between frozen and thawed soil, which are often not well known. We present a new FT-state-detection algorithm based on the daily variation of the H-polarized brightness temperature of the SMAP L3c FT global product for the northern hemisphere, which is available from 2015 to 2021. Exploiting the daily variation signal allows for a more reliable state detection, particularly during the transition periods, when the near-surface soil layer may freeze and thaw on sub-daily time scales. The new algorithm requires no reference values; its results agree with the SMAP FT state product by up to 98% in summer and up to 75% in winter. Compared to the FT state inferred indirectly from the 2-m air temperature and collocated soil temperature at 0–7 cm of the ERA5-land reanalysis, the new FT algorithm has a similar performance to the SMAP FT product. The most significant differences occur over the midlatitudes, including the Tibetan plateau and its downstream area. Here, daytime surface heating may lead to daily FT transitions, which are not considered by the SMAP FT state product but are correctly identified by the new algorithm. The new FT algorithm suggests a 15 days earlier start of the frozen-soil period than the ERA5-land's estimate. This study is expected to extend the L-band microwave remote sensing data for improved FT detection.

Keywords: frozen-soil state estimation; passive microwaves; remote sensing; SMAP FT state product

Citation: Lv, S.; Wen, J.; Simmer, C.; Zeng, Y.; Guo, Y.; Su, Z. A Novel Freeze-Thaw State Detection Algorithm Based on L-Band Passive Microwave Remote Sensing. *Remote Sens.* **2022**, *14*, 4747. https://doi.org/10.3390/rs14194747

Academic Editor: Luca Brocca

Received: 27 July 2022
Accepted: 17 September 2022
Published: 22 September 2022

1. Introduction

Spatial patterns and the timing of freeze–thaw (FT) state transitions over land are highly variable; they strongly impact land–atmosphere interactions and thus the weather; the climate; and hydrological, ecological, and biogeochemical processes [1–4]. In particular, FT state transition leads to differences in hydrological and thermal conductivities/diffusivities, the albedo for shortwave and emissivity for longwave radiation, and latent/sensible heat fluxes [5–7]. The albedo is, e.g., higher for frozen than for unfrozen soil, and the water and energy exchange between the land surface and the atmosphere is reduced because of weaker radiation heating and evaporation while frozen. Changes in the FT state dynamics can also signal climate change [8,9] and invoke permafrost carbon feedback [10]. Moreover, ecosystem responses to seasonal FT-state changes are rapid via significant changes in evapotranspiration, soil respiration, plant photosynthetic activity, liquid water availability, vegetation net primary production, and net ecosystem CO_2 exchange (NEE) with the atmosphere [11–15]. Thus, the knowledge of the FT state is required for modeling work in the above subjects, which invoke different parametrizations for frozen and unfrozen soil [16–20]. However, FT state estimations from in situ temperature observations are limited in scale, and it is not straightforward to deduce the state from soil, skin, or near-surface air temperature.

In contrast, more direct state information results from the very different microwave dielectric constant for frozen and unfrozen soil [21–23]. Accordingly, microwave brightness temperatures (*TBs*) change sharply during FT state transitions. For instance, NASA's MEaSUREs (Making Earth System Data Records for Use in Research Environments) program provides two global daily products for the land FT state based on a single-channel algorithm [24]. One covers the years 1979 to 2017 and exploits the 37 GHz channels of three satellite-based passive microwave sensors by exploring the TB values under respective landscape frozen and thawed reference states [25]: the scanning multichannel microwave radiometer (SMMR), the special sensor microwave/imager (SSM/I), and the special sensor microwave imager/sounder (SSMIS). For MEaSUREs, 37 GHz is selected because of its high correlation with the near-surface air temperature. Moreover, lower-frequency L-band sensors on SMOS (Soil Moisture and Sea Salinity) and SMAP (Soil Moisture Active and Passive) are more suitable for FT-change detection because of their deeper penetration depth and their sensitivity to soil moisture [26,27]. For the L band, the difference leads to emissivities of ~0.6 for unfrozen and ~0.9 for frozen soil, with a much deeper penetration depth for the latter [28–30]. The cross-polarized gradient ratio (XPGR) at the L band between H and V polarization is used to analyze the SMAP observations. Similar to the higher-frequency single-channel algorithms, the use of XPGR needs reference values for thawed and frozen-soil states and a threshold value for discrimination between both [23,31]. The baseline F/T detection algorithm of SMAP [32] requires at least 20 days to find reference values for the frozen state, which can be challenging for short interim frozen periods induced, e.g., by synoptic-scale cold waves.

Since the XPGR method or the baseline, the F/T detection algorithm of SMAP first needs to identify the frozen/thawed TB reference values, and these results rely on how to construct the reference. Instead, we propose a new FT algorithm that builds its parameters on the TB signal characters. The new FT algorithm has a similar basis to the durinal amplitude variation (DAV) approach [33] applied to higher frequency passive microwave measurements for snow and ice sheet applications [34,35]. It has been proved that the DAV of passive microwave signals are sensitive to FT state changes [33], which are dynamic and complex and vary continuously in space and time. Estimating the FT state changes from the DAV signal is functional because the conditions driving FT changes, e.g., radiation balance and air temperature, change on broad time scales, spanning sub-daily, daily, synoptic, seasonal, and annually interdecadal [36]. Especially in cold arid regions, which are prone to experience FT state transitions, soil moisture fluctuations due to evaporation and precipitation, and their L-band signals, are comparatively low on daily and synoptic scales.

In this study, we use the daily *TB* cycle and its connection with changing penetration depths during FT state changes to develop a new FT state-detection algorithm, which exploits variance-based filtering on DAV signals ($|\Delta TB|$) at the L band between 6 a.m. and 6 p.m. (local time)—the overpassing times for the SMAP and SMOS satellites. Wherever data overlap occurs, as is typical at high latitudes, data that were acquired closest to 6 a.m. and 6 p.m. local solar times are chosen as stated in the SMAP FT product handbook [26]. The method is based on microwave transfer theory and does not need reference values. The structure of the paper is as follows: Section 2.1 describes the data used and the study area, including the SMAP FT product and the ERA5-land reanalysis. Section 2.2.1 details the SMAP FT product, followed by our new method in Section 2.2.2. The statistics required for implementing the new method are explained in Section 2.3. The results are found in

Section 3.1 (example at a single site) and 3.2 (the north hemisphere). Section 3.3 evaluates the new FT algorithm by comparing its result with the current SMAP FT product and using the categorical triple collocation (CTC) method. Section 3.4 quantifies the uncertainties of the new method. Conclusions are in Section 4, with a discussion on the relation between the 2-m air temperature and the soil FT states presented in Section 5.

2. Methodology

2.1. Study Area and Data

We use the following three data sets in this study:

(1) SMAP *TB* observations and the derived FT-state indicator [26] with 36 km × 36 km spatial resolution at 6 a.m. and 6 p.m. local time. The SMAP L3 product includes an FT state indicator besides H and V polarized TB observations at the L band (1.41 GHz). The data is available starting on 30 March 2015 [26]. The derived FT-state indicator (SMAP L3c FT product) is available globally from 85.044°S to 85.044°N, and we focus on the domain from 20°N to 85.044°N. We use the SMAP *TBs* for the new algorithm and the binary FT-state indicator for frozen (1) and thawed soil (0), including the transition direction for its evaluation. Moreover, SMAP also provides a 9-km spatial resolution FT product. As noted on https://nsidc.org/data/SPL3FTP_E/versions/3 (accessed on 1 April 2021), the 9-km product is derived from SMAP-enhanced Level-1C brightness temperatures (SPL1CTB_E). For SPL1CTB_E, Backus–Gilbert optimal interpolation techniques are used to extract enhanced information from SMAP antenna temperatures before they are converted to brightness temperatures. Since the Backus–Gilbert optimal interpolation techniques added more noise, we prefer 36 km × 36 km spatial resolution in this study [37]. Only H-polarization is used in this study because the DAV signals between H/V polarizations are small.

(2) Hourly 2m-air (T_{2m}), skin (T_{skin}), and soil temperatures from the ERA5-land reanalysis available on https://cds.climate.copernicus.eu/cdsapp#!/dataset/reanalysis-era5-land?tab=overview (accessed on 1 April 2021) [38]. ERA5-land provides a consistent representation of the evolution of land state variables over several decades at a higher resolution (0.1° × 0.1°) than ERA5 (0.25° × 0.25°). ERA5-land has been produced by replaying the land component of the ECMWF ERA5 climate reanalysis at an enhanced resolution. ERA5-land also provides soil profile information that is vital to the analysis of L-band *TBs*.

Since L-band observations may have—depending on the FT-state—deep penetration depths, neither particular variable, such as 2m-air, skin, or soil temperatures with diurnal changes in ERA5-land, is suitable for comparison with the daily FT indicator at a daily scale. In SMAP FT calibration/validation, the average of the air temperature and soil temperature at 5 cm is used to infer that the FT state corresponds to the L-band signal. We took the same scheme as the average of daily 2 m-air temperature (T_{2m}) and collocated 0–7 cm soil temperature (ERA-land-assessment data hereafter), which are used as an FT state reference to evaluate the existing and the new FT algorithms. According to longitude, the hourly data are interpolated to 6 a.m. and 6 p.m. local time. Considering the detectable range of the L band, the FT state reference can be inferred from 2-m air/skin/5 cm soil/10 cm soil temperatures. However, the inferred FT state from these temperatures may contradict each other at the moment, and it is hard to judge which one represents the signal detected by the L band.

In SMAP FT Cal/Val, the in situ 2m-air temperature and the soil temperature at 5 cm are used to validate and calibrate SMAP's FT state indicator [32]. The in situ soil moisture at 10 cm and the skin temperature are taken as the FT state reference, not as the ground truth in the SMAP's FT Cal/Val. Although T_{2m} is often used to estimate soil FT states [39–41], it shall be noted that ERA-land-assessment data are not the condition for judging frozen/thawed soil from the reanalysis but an indicator of the thermal conditions near the surface regarding the land–atmosphere interaction. In this study, all variables from ERA5-land have been interpolated with the nearest method to match the 36 km resolution of SMAP products.

To better demonstrate how the new FT algorithm works, we selected the location of the Xilinhot site (115.93°E, 42.04°N) [42] already used in other microwave remote sensing studies [43–45] to illustrate the functioning of the new FT detection algorithm because of its meteorological conditions, which are typical of regions experiencing FT-state changes. Xilinhot site grows crops, corn, oat, and buckwheat in summer, and this landscape represents one of the main surface types on the globe; 31–43% of the land cover in the northern hemisphere belongs to this climate type [46]. From the ERA5-land reanalysis data, we use the grid data covering Xilinhot to represent the site's land surface and meteorological conditions. From the data for 6 a.m. and 6 p.m. local time (UTC + 08), the daily differences of T_{2m}, T_{skin}, and soil temperature at 0–7 cm (*stl*1), 7–28 cm (*stl*2), 28–100 cm (*stl*3), and 100–189 cm (*stl*4) are computed and used for interpreting the satellite-observed *TB* signals for the location of the Xilinhot site.

2.2. Methodology

2.2.1. The SMAP F/T Algorithm

The SMAP FT algorithm [26] is based on the so-called relative frost factor FF_{rel},

$$FF_{rel} = \frac{FF_{NPR} - FF_{fr}}{FF_{th} - FF_{fr}} \tag{1}$$

where FF_{NPR} is the frost factor defined as the normalized polarization ratio,

$$FF_{NPR} = \frac{TB_v - TB_h}{TB_v + TB_h} \tag{2}$$

and FF_{fr}/FF_{th} is the reference frost factor for the frozen/thawed state, respectively. FF_{fr} is the average FF_{NPR} for January and February (winter), and FF_{th} is each year's average FFNPR for July and August (summer).

The SMAP FT status (FT_{SMAP}) is derived from FF_{rel} for each location via

$$FT_{SMAP} = \begin{cases} thaw, & if \ \ FF_{rel} > threshold \\ frozen, & if \ FF_{rel} \le threshold \end{cases} \tag{3}$$

where a threshold of 0.5 was used globally.

The algorithm relies on the quality of the two reference values FF_{fr} and FF_{th}. Their estimation requires at least 20 days of relatively stable frozen (or unfrozen) conditions [47]. FF_{th} is hard to identify at higher latitudes and altitudes where the ground is frozen throughout the year, while FF_{fr} is hard to determine for the midlatitudes where the soil is not completely frozen from the surface down to the L-band penetration depth. According to the SMAP FT handbook [32], FF_{NPR} needs to be larger than an arbitrary value of 0.1, which excludes relatively dry areas that undergo minor dielectric constant changes during FT transitions. FT-SCV(Freeze/Thaw algorithm using Single Channel TBV) is used as an extended algorithm to overcome this defect in FF_{NPR}. FT-SCV does not reply to the freeze/thaw reference derived from the winter/summer period, but it correlates with surface air temperature from global weather stations [32]. The SMAP freeze/thaw products contain both FT-SCV and NPR algorithms. When there is enough difference between freeze and thaw references, the NPR threshold method is applied, and when the reference difference is too small, a single channel algorithm is adopted.

2.2.2. The New FT Algorithm

The new FT algorithm uses the strong *TB* variations over the day caused by freezing in the night and thawing over the day, which happens over a period of days at the beginning and end of the totally frozen period. For the L band, which is longer than previous DAV applications, the signal can penetrate ice and snow over the soil surface and be related to the FT state of the soil. To retrieve this signal from the microwave transfer theory, we start with the zeroth-order microwave transfer model given by

$$TB = \varepsilon T_{eff} \tag{4}$$

where ε is the emissivity, which depends on the soil dielectric constant and mainly varies with soil moisture and the FT-state of the soil. T_{eff} is the vertically integrated soil temperature profile weighted with the soil dielectric profile as (Lv et al. 2016a)

$$T_{eff} = T_1(1 - e^{-\tau_1}) + \sum_{i=2}^{n-1} T_i(1 - e^{-\tau_i})\prod_{j=1}^{i-1} e^{-\tau_j} + T_n \prod_{j=1}^{n-1} e^{-\tau_j} \tag{5}$$

where T is soil temperature, τ is soil optical depth, and the subscripts i and j are the layer numbers counting from the top of the soil (1) to the bottom of a layered soil slab (n) influencing *TB*. Because of the much deeper penetration depth of frozen soil, the attenuation of radiation emitted from lower layers is strongly reduced [48], which enlarges the depth down to which the integration for T_{eff} must be performed; thus, the deeper soil layers with their only minor daily and even seasonally varying temperatures dominate the *TBs* of frozen soil (see *TB* variations during winter in Figure 1). Thus, especially in winter, the *TB*s of frozen soil are mostly higher than in summer, containing the influence of soil temperature and soil dielectric constant.

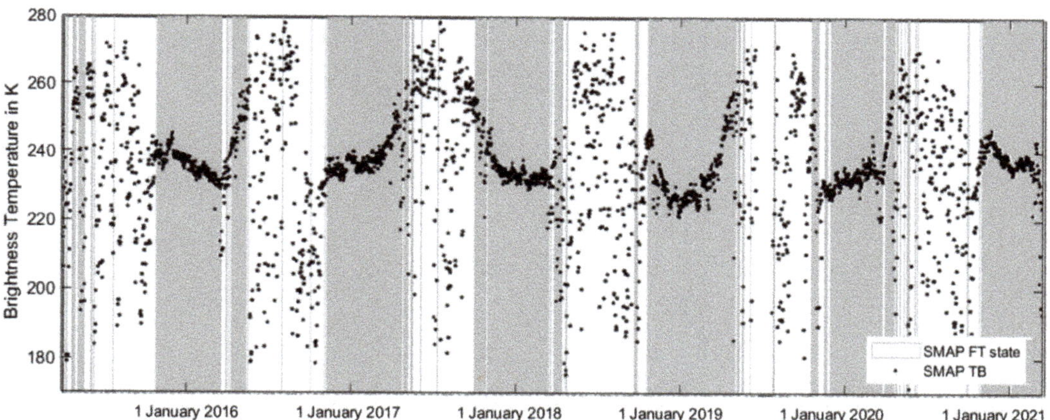

Figure 1. SMAP H-polarization *TB* time series and the derived reference FT state (grey for frozen and white for unfrozen) extracted for the location of the Xilinhot site (43°30′–45°N, 115°–117°E).

When unfrozen, soil moisture variations due to evaporation lead to *TB* increases of only up to 15 K in a day [33]. An exception to significant daily *TB* changes for unfrozen soil is precipitation, which can reduce *TB*s by tens of K. *TB* can change in the same range due to daily soil temperature variations via T_{eff} ((Equations (4) and (5)). *TB* changes during FT transitions are in the range and larger than the precipitation signal because of the huge ε difference between frozen and unfrozen soil. When frozen, emissivity—and thus *TB* variations—are very small and only slightly depend on soil composition, such as the clay/sand fraction and organic matter, which also affect the emissivity of unfrozen soil. Thus, daily *TB* changes for unfrozen soil—except for precipitation—are much smaller than those caused by freezing and thawing. Any FT transition typically begins and ends at the surface. Thus, L-band radiometers can sense the start and end of FT transitions. The new FT algorithm exploits the daily *TB* difference caused by FT state transitions, and we assume the day without enough SMAP *TB* (i.e., an absence of *TB* at 6 a.m., 6 p.m., or both) interpolated with the nearest FT result.

We use the following formalism for FT-state detection. Let

$$\begin{cases} TB_{ih\_6am} = \varepsilon_{6am} T_{eff\_6am} \\ TB_{ih\_6pm} = \varepsilon_{6pm} T_{eff\_6pm} \end{cases} \tag{6}$$

TB_{ih_6am}/TB_{ih_6pm} are the *TB*s observed by SMAP at 6 a.m. and 6 p.m. local time on day *i* in H-polarization (h) with $\varepsilon_{6am}/\varepsilon_{6pm}$ the respective soil emissivities and $T_{eff\_6am/pm}$ the respective T_{eff}. The DAV signals between H and V polarizations have few differences [33] when the soil surface is frozen both at 6 a.m. and 6 p.m., and neglecting the impact of soil temperature changes on the dielectric constant, i.e., $\varepsilon_{6am} = \varepsilon_{6pm} = \varepsilon$. On a daily scale, this is reasonable because other factors, such as the sub-grid open water fraction, terrain heterogeneity, and tree cover, will not have diurnal changes. Precipitation will be excluded by air temperature > 0 °C, and snowmelt will lead to large DAV. The *TB* difference between both is using Equation (6) given by

$$\begin{aligned} \Delta TB_i &= TB_{ih\_6pm} - TB_{ih\_6am} \\ &= \varepsilon\left(T_{eff\_6pm} - T_{eff\_6am}\right) \\ &= \varepsilon\left[\Delta T_1(1-e^{-\tau_1}) + \sum_{i=2}^{n-1}\Delta T_i(1-e^{-\tau_i})\prod_{j=1}^{i-1} e^{-\tau_j} + \Delta T_n \prod_{j=1}^{n-1} e^{-\tau_j}\right] \end{aligned} \tag{7}$$

At 6 a.m./pm, soil temperature and moisture profile gradients are less sharp than at noon, and ΔTB_i will be much smaller than the temperature differences (ΔT_i) in any layer, since $\varepsilon < 1$, $(1-e^{-\tau_1}) < 1$, $(1-e^{-\tau_i})\prod_{j=1}^{i-1} e^{-\tau_j} < 1$, and $\prod_{j=1}^{n-1} e^{-\tau_j} < 1$, i.e.,

$$|\Delta TB_i| < \max(|\Delta T_i|) \tag{8}$$

Equation (8) takes the daily scale as the diurnal definition DAV approaches. However, Equation (8) is not valid for unfrozen soil because ε will change with soil moisture over the day due to evaporation and precipitation, which will dominate ΔTB_i. However, ΔTB_i will also be small when no precipitation happens between both times and when the sky is cloudy, and low winds reduce evaporation. Thus, ΔTB_i is not enough to infer the FT state. A sudden heat/cold wave can interrupt a daily FT state transition, which may induce a large ΔTB_i with soil staying frozen or unfrozen throughout the event. Such synoptical scale heat/cold waves make identifying the beginning/end of the yearly freezing difficult. To avoid this problem, as well as the absence of *TB* in the low latitudes due to the revisit period of SMAP, we interpolate the DAV absences with the nearest valid values.

Thus, to filter out the influence of synoptic variations and cloudy and/or low wind days [31], we use, in addition, the ΔTB_i variance over β days

$$\mathrm{var}(\Delta TB)_\beta = \frac{1}{\beta}\sum_{i=-(\beta-1)/2}^{i=(\beta-1)/2}[\Delta TB_i - E(\Delta TB_i)]^2 \tag{9}$$

$\mathrm{var}(\Delta TB)_\beta$ is not a new parameter but to keep $|\Delta TB_i|$ filtering out the synoptic weather interference. The selection β = 7 filters out the impact of atmospheric Rossby waves in the midlatitudes (3–5 days) at locations experiencing annual FT cycling in the mid-latitudes [49]. This averaging will filter out the impact of days with low ΔTB_i caused by cloudy days or synoptic weather systems. The days after or before β days will also be checked by Equation (10) below. In this case, if the freezing or thawing state transition, e.g., due to synoptic weather systems, lasts for more than five days, we can still find the annual begins/ends of an FT cycle. Therefore, the influence of synoptic events is excluded.

Then the new algorithm is

$$FT_{new} = \begin{cases} thaw, & if \ \ \mathrm{var}(\Delta TB)_\beta \geq \gamma^2 \ \ or \ |\Delta TB_i| \geq \gamma \\ frozen, & if \ \ \mathrm{var}(\Delta TB)_\beta < \gamma^2 \ \ \ and \ |\Delta TB_i| < \gamma \end{cases} \tag{10}$$

with γ a threshold brightness temperature square in terms of both $|\Delta TB_i|$ (for instantaneous) and $\mathrm{var}(\Delta TB)_\beta$ (for the synoptic weather scale). By Equations (9) and (10), the new FT algorithm contains a synoptic time scales background to daily values as variance-based filtering. For example, sunny days will lead to $\Delta TB_i \geq \gamma$ [33]; cloudy/slow winds days will be filtered out by $\mathrm{var}(\Delta TB)_\beta \geq \gamma^2$ because these days do not last longer than the period of an atmospheric Rossby wave period in the middle latitudes. We calculate all $|\Delta TB_i|$ from the grid inferred by SMAP FT products as the freeze state (Figure 2) and obtain $\gamma = 8K$ by statistically computing $|\Delta TB_i|$ and $\mathrm{var}(\Delta TB)_\beta$ over the northern hemisphere to keep 95% confidence for cases where T_{2m}< 0 °C. The bias of ERA-land–air temperature and collocated 0–7 cm soil temperature is about 1 K, which provides 95% ± 3% uncertainty [50]. Any day that can obtain $|\Delta TB_i|$ from SMAP will be checked by Equation (10). For a day that $|\Delta TB_i|$ is not available, it will be filled with an FT value depending on the nearest FT_{new}. In arid regions, $\Delta TB_i \geq \gamma$ would always work because the heat capacity of dry soil is much smaller than that of wet soil. Equation (10) will treat the arid region as a thawed state. For wet snow, if there is no more water melted in the daytime, then TB will not be affected too much. If water is melting, Equation (10) will be treated as a thawed state, and the wet snow-covered ground will still be part of the land surface FT state.

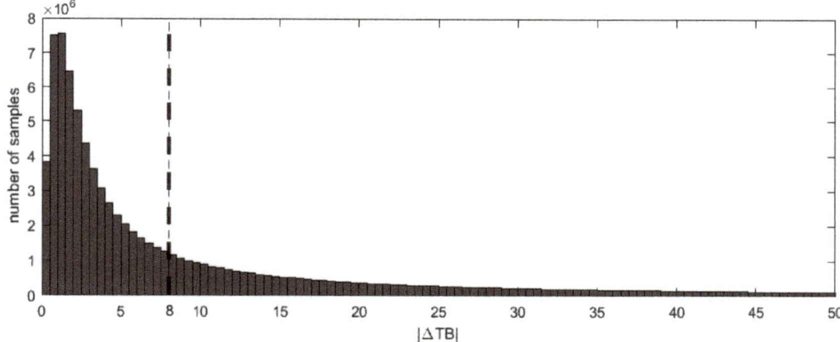

Figure 2. $|\Delta TB|$ with β = 7 by SMAP TB data contained in SMAP L3 radiometer global daily 36 km EASE-grid freeze/thaw state; data over the northern hemisphere where 95% of samples are within 8 K.

2.3. Evaluation of the New FT-State Detection Algorithm

By Equation (10), one can compute the starting and ending times of the frozen-soil period in winter, i.e., the first/last freeze state in an annual FT cycle. Applying Equation (9), it requires at least one FT value per day which affects accuracy in the low latitudes.

Before a comparison with the half-daily SMAP FT products, we have to scale it to daily resolution by

$$FT_{SMAP-daily} = \begin{cases} thaw, & if\, FT\, status_{6am} = 0 \quad or \quad FT\, status_{6pm} = 0 \\ frozen, & if\, FT\, status_{6am} = FT\, status_{6pm} = 1 \end{cases} \tag{11}$$

Equation (11) produces a bias towards thawed states but matches the concept of the new FT algorithm because $|\Delta TB_i|$ would be smaller when the states at 6 a.m. and 6 p.m. are consistent. Hence, the agreement is defined as the fraction of days in which the new method and SMAP FT have the exact daily FT state inference against the total number of days (see Equation (11)). Specifically, both the SMAP's FT product and the new algorithm contain 964 × 203 grids in the latitude between the equator to 83.6320°N and 2148 days after 31 March 2015. Instead, "agreement" computes the percentage of pairs (one from the new FT and the other from SMAP) that are consistent along longitude/latitude/time. SMAP also adopts a similar FT product-accuracy assessment method to Equation (11) [32]. The difference is that SMAP needs to compare at 6 a.m./pm instead of the daily scale.

With the above steps, the categorical triple collocation (CTC) method [51,52] is used for validating the binary FT state. Triple collocation (TC) is a method to verify the accuracy of three sets of observations without surface measurements. The method assumes the following relationship between the observed quantity (M) and the true value (X):

$$M_i = A_i + B_i X + \varepsilon_i \tag{12}$$

This method assumes that each group of observations is independent of the other and that observations and errors are separate. However, for binary variables (such as the freeze–thaw state of soil), Equation (12) should be rewritten as

$$M_i = X + \varepsilon_i \tag{13}$$

CTC is based on triple collocation (TC) [53,54] by relaxing the assumptions of TC sufficiently to allow its application to binary and categorical variables. CTC provides relative performance rankings but not absolute values of performance metrics.

The CTC method evaluates the estimation accuracy of binary variables by introducing equilibrium accuracy (α)

$$\alpha_i = \frac{1}{2}(\psi_i + \eta_i) \tag{14}$$

In the formula, ψ is the probability of being correctly estimated as thawed, and η is the probability of being correctly estimated as frozen. The following relationship exists between the covariance matrix (Q) of the static variable and the equilibrium accuracy α,

$$Q_{ij} = Cov(M_i, M_j) = \begin{cases} 1 - E(E(M_i))^2, for \;\; i = j \\ Var(X)(2\alpha_i - 1)(2\alpha_j - 1), for \;\; i \neq j \end{cases} \tag{15}$$

This relationship is extended to non-static variables with noticeable seasonal changes (such as a freeze–thaw state):

$$Q_{ij} = Cov(M_i, M_j) = \begin{cases} 1 - E(E(M_i/t))^2, for \;\; i = j \\ 4E(p(t))(1 - E(p(t)))(2\alpha_i - 1)(2\alpha_j - 1), for \;\; i \neq j \end{cases} \tag{16}$$

In the formula $p(t) \equiv pM\ (M = 1 \mid t)$, t is time (unit is day). The CTC algorithm requires that the random errors between each group of observation systems are conditionally independent if the relative accuracy W_i is defined as

$$W_i = 2(2\alpha_i - 1)\sqrt{E(p(t))(1 - E(p(t)))} \tag{17}$$

Three groups of different observation systems in Equation (17) can form three equations, as follows:

$$W = \begin{bmatrix} \sqrt{\frac{Q_{12}Q_{13}}{Q_{23}}} \\ \sqrt{\frac{Q_{12}Q_{23}}{Q_{13}}} \\ \sqrt{\frac{Q_{23}Q_{13}}{Q_{12}}} \end{bmatrix} \tag{18}$$

Since W_i is a monotonically increasing function with αi as the independent variable, the ordering of W represents the ordering of the observation system errors from small to large. The relative error between different observation systems can be obtained by arranging the W vector calculated by Equation (18) in descending order [54].

3. Results

3.1. The Demonstration of the New Method at the Xilinhot Site

Figure 3a shows the in situ air temperatures and the SMAP *TB* observed at or near the Xilinhot site from March 2015 to March 2021. The green lines are the beginnings and ends of the annual freezing cycles inferred from the new FT algorithm. The intervals between green lines in summer indicate the thawed state, and in winter for the frozen state. *TB* ranges between 240 K and 280 K in the thawed soil state. The seasonal *TB* amplitude more or less follows the variation of air, skin, and upper soil temperatures without phase delay. Under frozen conditions, *TB* does not exhibit a clear seasonal variation because of a more stable soil emissivity. Figure 3b proves the inference from Equation (8): $|\Delta T_{skin}|$ is the maximum difference between the 6 p.m. and 6 a.m. calculated from skin temperature (T_{skin}) in the ERA5-land (Figure 3a). $|\Delta T_{skin}|$ is relatively stable, about 10–15 K through the years, while $|\Delta TB_i|$ in winter is two times smaller than in summer.

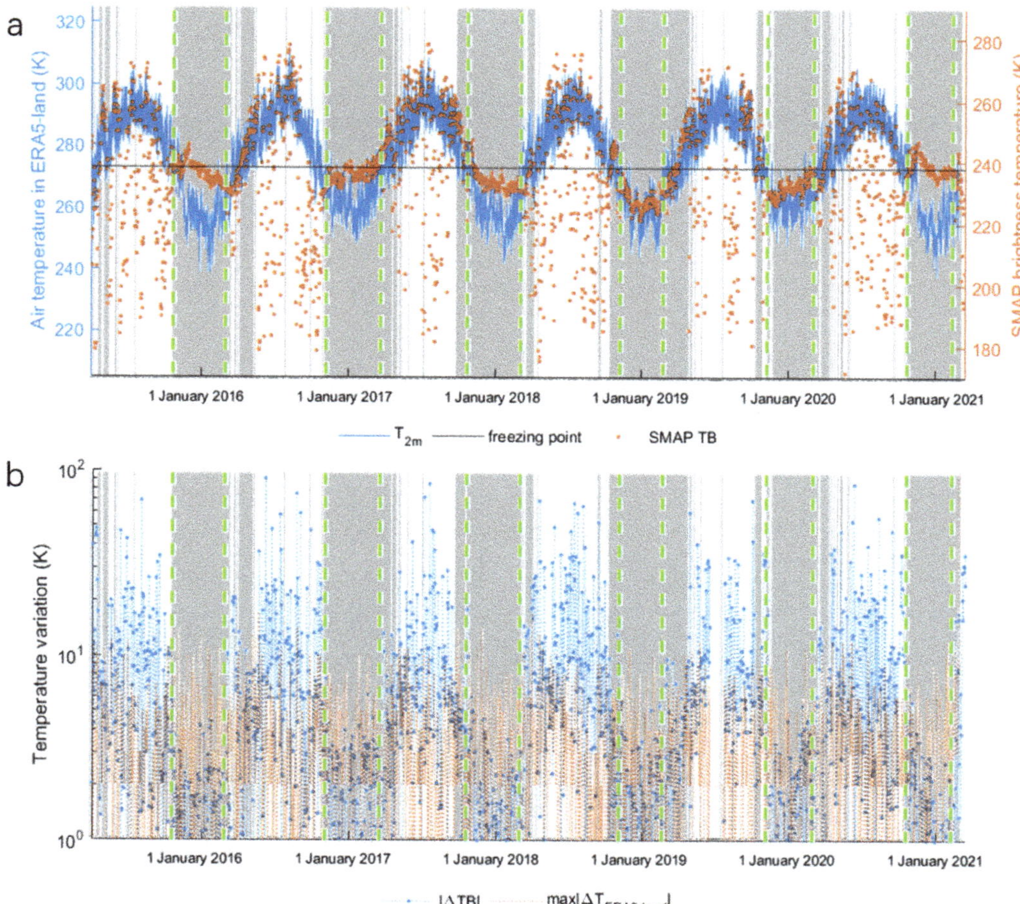

Figure 3. (**a**) Time-series of 2-meter air temperature T_{2m}, and SMAP *TB* at the Xilinhot site; (**b**) time series of $|\Delta TB|$ computed from the SMAP *TB*s and $\max|(\Delta T_{ERA5-land})|$. The vertical green dashed lines indicate the beginning and ending day of the soil frozen state as inferred from the new FT algorithm. The gray shades the frozen state inferred from SMAP FT product.

The ΔTB_i time series in (Figure 4) shows mostly a substantial intra-annual variation in summer from −40 K to 40 K due to soil moisture variations, which is mainly influenced by the precipitation connected to the East Asia Monsoon. In winter, ΔTB_i varies only from −8 K to 8 K. However, there are $|\Delta TB_i| \leq 8$ K is summer as well. To filter out these isolated small $|\Delta TB_i|$ cases in summer, $\mathrm{var}(\Delta TB)_\beta$ is constructed as in Equation (9). In Figure 4, $\mathrm{var}(\Delta TB)_\beta$ is close to zero but ranges from 0 to 200 K in summer. By comparison with ΔTB_i, $\mathrm{var}(\Delta TB)_{\beta=7}$(the red line in Figure 4) successfully filters out the small ΔTB_i of summer cases. Figure 4 also shows the sudden changes of both ΔTB_i and $\mathrm{var}(\Delta TB)_{\beta=7}$ in the beginning and ending time of the frozen-soil period according to the new algorithm.

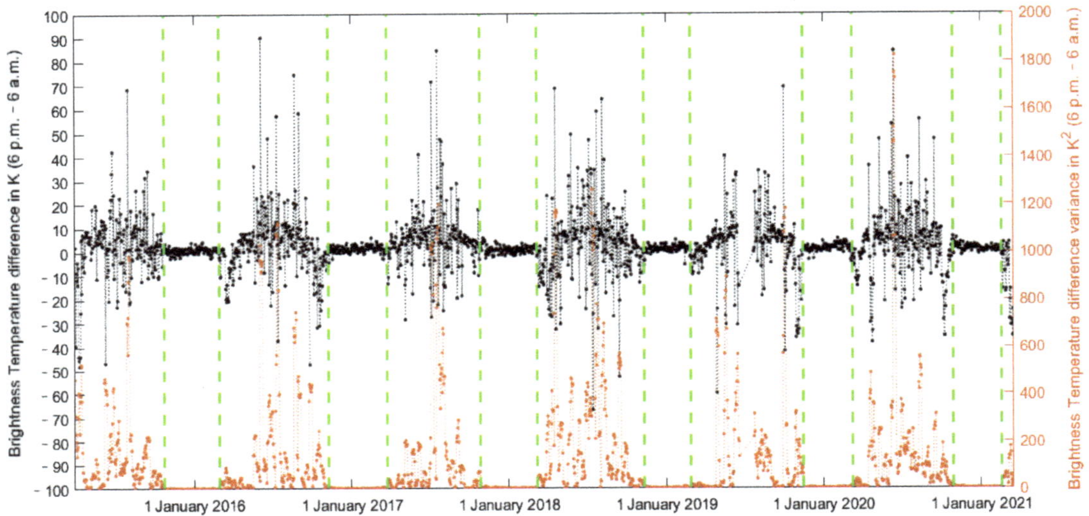

Figure 4. An illustration of diagnosis example at the Xilinhot site. The time series of ΔTB (Equation (7)) is represented by the black line from April 2015 to March 2021. The red line represents its seven-day moving variance as in Equation (9). The green dashed lines represent the beginning/ending of the soil frozen state inferred from the new FT algorithm.

At the beginning and after the end of an annual frozen-soil period (Figure 4), TB_{ih_6pm} is often less than TB_{ih_6am}(valley in the time series), while the soil and air temperatures have opposite behavior due to solar heating during daylight hours. Days with $TB_{ih\_6pm} < TB_{ih\_6am}$ can be explained by daytime melting of the uppermost few centimeters of the soil, which reduces the topsoil emissivity. According to [55], intra-daily FT state transitions between the completely thawed and frozen-soil state may last for tens of days at the Maqu site on the Tibetan plateau [55]. The new FT detection algorithm takes this peculiarity during the transition phase into account and identifies the FT state of the bulk soil rather than of a thin surface layer as detected by the existing TB-based algorithms. Moreover, the new FT detection algorithm requires no local reference values.

3.2. *Result over the Northern Hemisphere*

Figures 5 and 6 show, respectively, the beginnings between August and December and the endings between January and May of the frozen-soil periods from 2015–2021. For most of the northern hemisphere, there are hardly any frozen-soil states in the northern hemisphere during June and July.

Siberia and Tibet experience the earliest soil freezing. In East Asia, the beginning of soil freezing follows the latitude and the East Asia Summer Monsoon [56] propagation in a meridional direction. Regions with an earlier retreat of the East Asia Summer Monsoon also freeze earlier, i.e., the region reaching from Mongolia to southern China shows a northwest–southeast freezing beginning gradient in November [57]. Over Europe, the gradient direction of the start times exhibits a southwest–northeast pattern towards Siberia. The freezing begins about two months later than in other regions with the same latitude (40°N~60°N) [58]. In North America, the freezing starts in October in the north and invades the south in December. Winter 2015–2016 and winter 2017–2018 show more severe soil freezing than other winters. Deep blue colors along 120°W mark the Rocky Mountains.

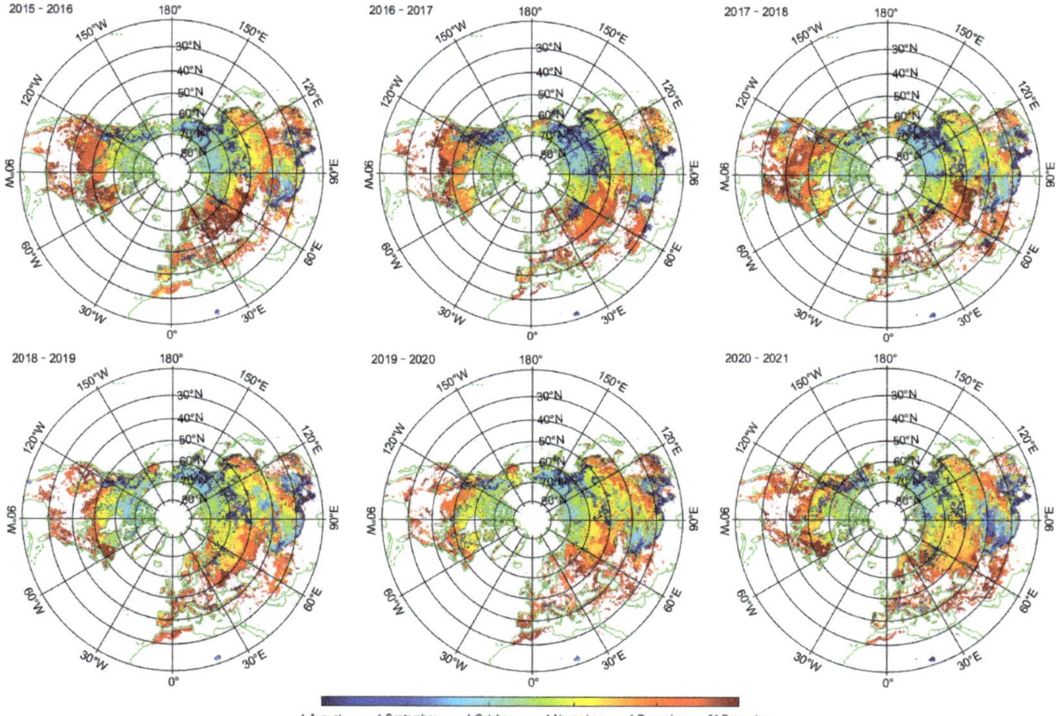

Figure 5. The beginning of the frozen-soil periods of the six winters as inferred by the new FT algorithm over the northern hemisphere (latitude > 20°N).

The pattern of the frozen-soil ending dates (Figure 6) is similar to the one of the starting dates. The frozen-soil period lasted the longest in Siberia, Tibet, and the Rocky Mountains. Winter 2017–2018 has the latest ending time of the frozen-soil period in North America and Europe. The contours of the frozen-soil period's beginning vary by tens of days from year to year. Parts of Siberia and Tibet never have unfrozen soil (deep blue colors); the soils are also frozen during June and July, which are not shown in the figures. Color transitions in Figure 6 are smoother and more continuous than in Figure 5 and show a gentler date gradient. For instance, the gradient over Eurasia is spotted spatially in Figure 5. An air temperature drop may explain the jagged color contours in Figure 5. These delicate patterns reflecting topography, sea-land distribution, and climatological types are not observed in Figure 6, implying a more gradual and slower heating process in spring.

By combining Figures 5 and 6, we obtain the duration of the annual frozen cycle as in Figure 7. The most extended frozen-soil period (>200 days) is located near the North Pole and over some regions in Tibet (Figure 7). The sharpest gradient in the length of the frozen period stretches from the eastern part of Siberia to the northeast part of China following the latitudes; here, the freezing duration decreases from more than 300 days to 90 days in about ten days per latitude degree. The freezing duration does not change much between the years and exhibits similar patterns.

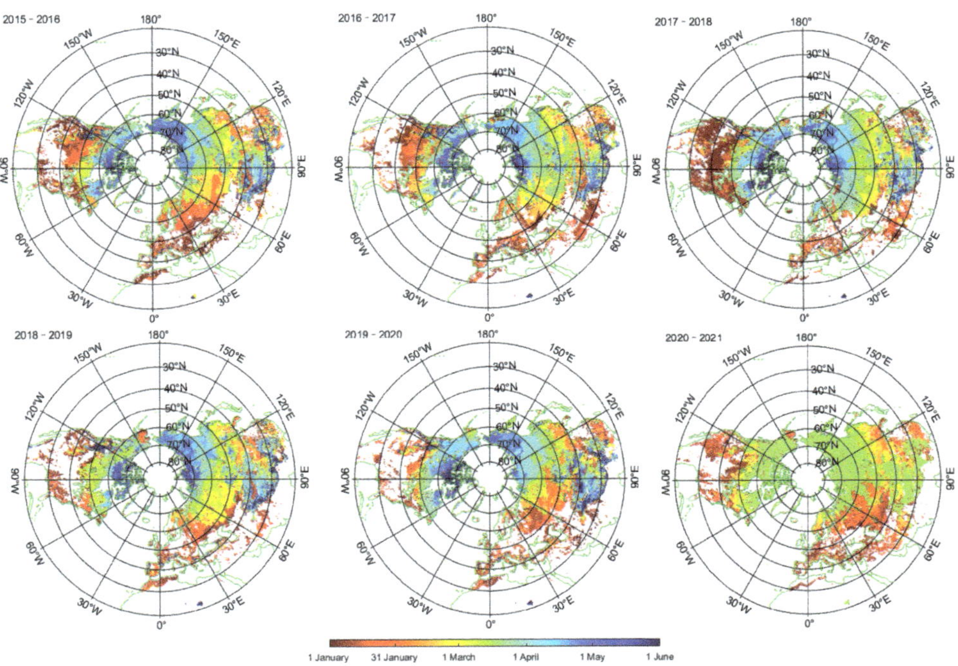

Figure 6. The ending of the frozen-soil periods of the six winters as inferred by the new FT algorithm.

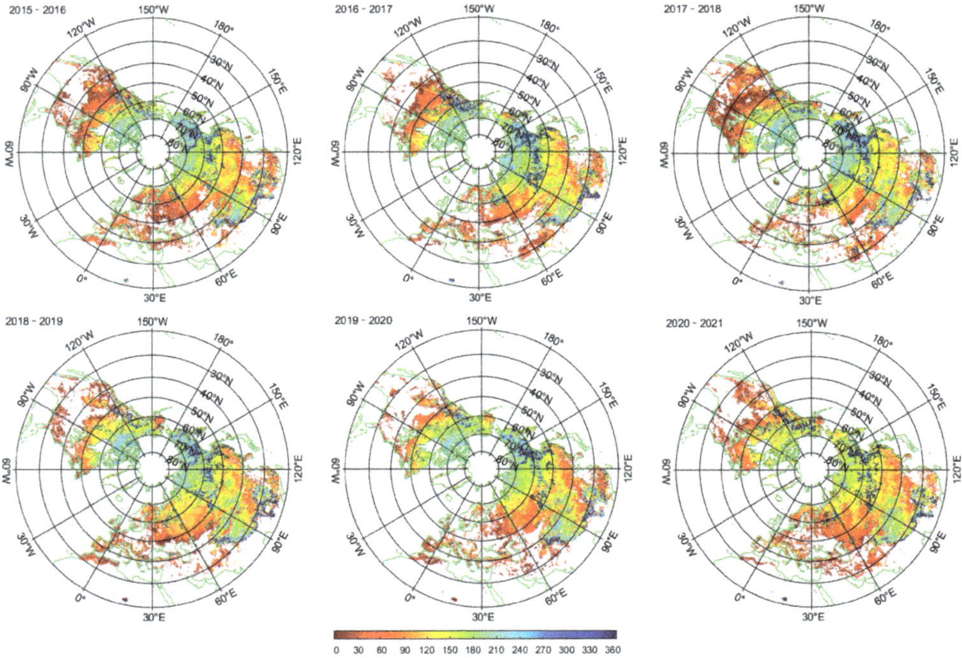

Figure 7. The duration of the frozen period inferred from the SMAP L3 global H-polarization brightness temperature product.

3.3. The Comparison with the SMAP FT Products

In this section, we will compare the results from the new FT algorithm with SMAP's regarding time variation and spatial distribution. An average of 2-m air temperature and collocated 0–7 cm soil temperature will also be used as a reference and compared with both FT results.

Figure 8 shows the zonal average agreement evolution along the annual FT transition zones between 30°N and 80°N. Although the SMAP FT product contains the data in 0°N–30°N, the zones that are supposed to be freeze-free or permanently frozen (>83°N) are not included in Figure 8. The new FT detection algorithm results mostly agree well with the SMAP FT product (Figure 8). In a transition zone, the overall agreement is only 0.6–0.7. The agreement of the detected beginning and ending times in the 30°N–40°N latitude belt is lower than the zones above 40°N because of the Tibet Plateau (25°N–40°N, see Figure 9a). The agreement between the two algorithms can be below 0.5 in winter (the blue area in Figure 8, especially ≥80°N).

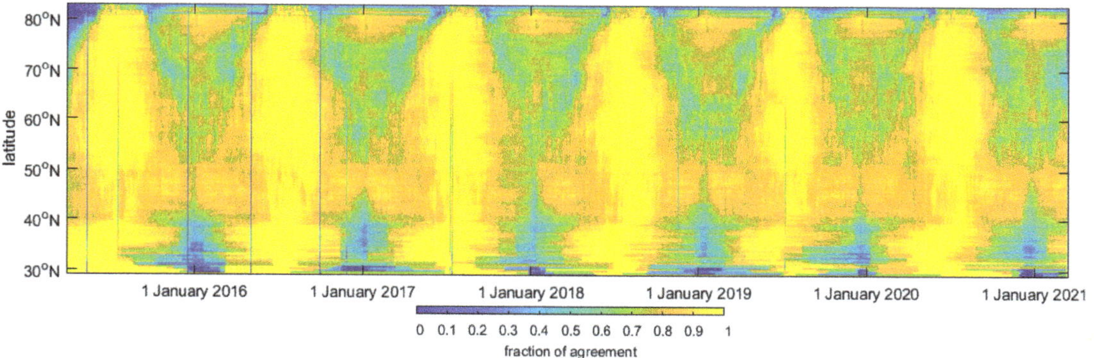

Figure 8. The fraction of agreement time series along the latitudes.

In summer, both agree up to 0.98 and 0.70–0.75 in the wintertime. However, the lowest agreement does not happen in the deep winter but some time ahead and after January by showing the two valleys as seen for each year in Figure 8. These valleys are the freezing/thawing transition period in the northern hemisphere that obtains the lowest agreement.

The difference in Figure 8 is due to the hypothesis in both algorithms. The SMAP FT algorithm requires FF_{fr} and FF_{th}; their estimate needs sufficiently long periods of wet and frozen soil, which becomes increasingly difficult with decreasing latitude. Thus, FF_{fr} and FF_{th} can be unreliable for the transition zones where the frozen emission character is not typical and may have interannual changes. For instance, the FF_{fr} and FF_{th} change variation can reach 4% and 20%, respectively, and strongly depends on the samples. Table 1 illustrates FF_{fr} and FF_{th} variation at the Xilinhot site by adopting different samples every year. Another reason for the difference in Figure 8 is the assumption adopted by the new algorithm. For one side, the new algorithm cannot identify a frozen-soil period shorter than seven days (as shown in Equation (9)). This leads to the thawed/frozen error in the new FT algorithm where the 3–5 days heat/cold events may result in different FT states from the two algorithms.

Figure 9a compares the SMAP FT state product and the new algorithm regarding the spatial distribution of the agreement. The Tibetan Plateau shows the lowest agreement value in Asia, especially its southern margin close to the Himalayas Mountains. While for most of the plateau, the agreement stays above 0.7, it is below 0.5 at some points organized in a belt in an east–west direction, probably due to the complex terrain that affects the ERA5-land quality [59]. The area downstream of the plateau, including the center parts of China, also shows strong disagreement between both estimates. A zone with a low agreement

(0.7–0.8) is found between 50 and 60°N, especially in the Lake Baikal region, northern Europe, and the east coast of Canada and Alaska. A zone with high agreement is found in eastern China, central Asia, southwestern Europe, and the southern U.S. Figure 9b,c shows the same agreement map between the new/SMAP's FT algorithm and the FT stated inferred from ERA-land-assessment data by a binary judgment of the freezing point (273.15 K), i.e., $T < 273.15$ K is frozen, and $T > 273.15$ K is thawed. Figure 9b,c shows that despite some mismatch between the results of the new algorithm and the SMAP FT, both overall agree by more than 70%. The new FT state detection better agrees with T_{2m} in the mid-latitudes than the SMAP FT algorithm but is worse in latitudes above 60°N and low latitudes below 30°N. Figure 9c also shows the influence of SMAP's NPR threshold method and the single channel algorithm with a clear boundary between high/low latitudes.

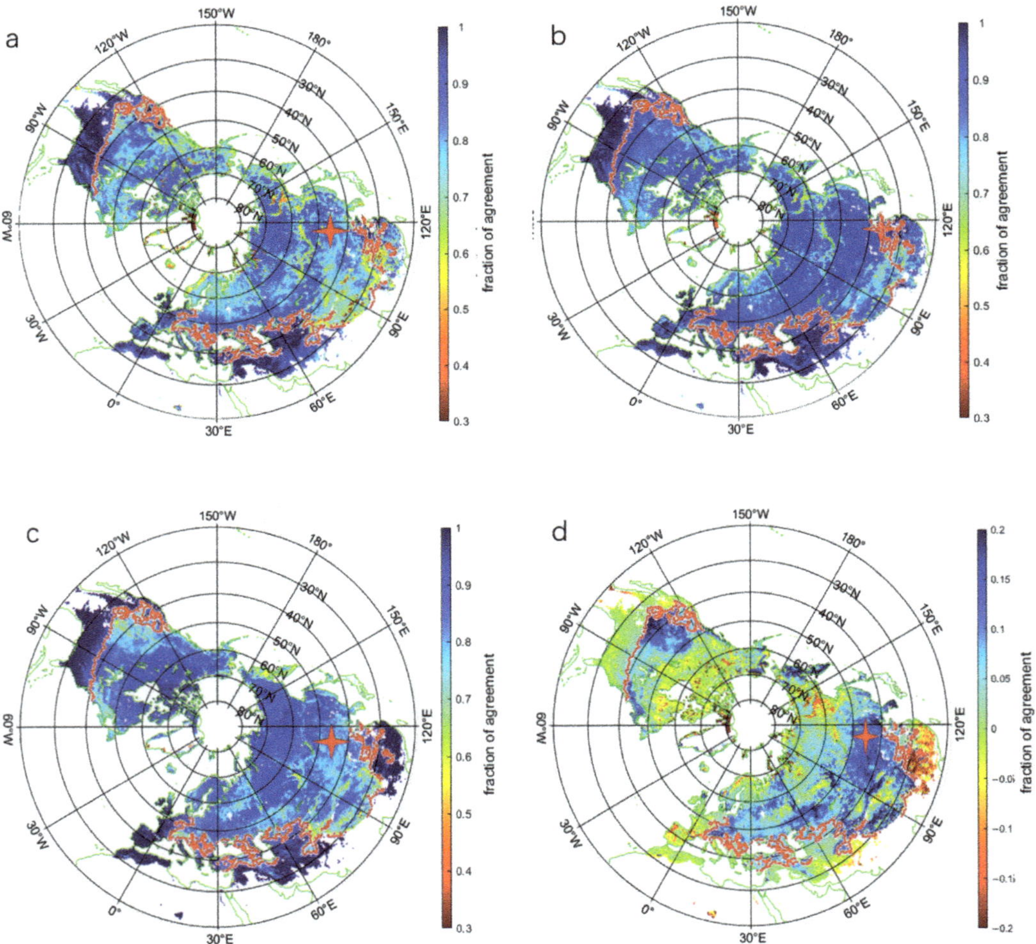

Figure 9. The spatial pattern of the fraction of agreement in the northern hemisphere between (**a**) the new method and SMAP L3 FT state products; (**b**) the new method and ERA-land-assessment data; (**c**) the SMAP L3 FT state products and ERA-land-assessment data; and (**d**) the difference by b minus c. The red cross marks the location of Xilinhot, and the red line is the boundary between SMAP's NPR threshold method and the single channel algorithm.

Table 1. FF_{fr} and FF_{th} from SMAP's algorithm at Xilinhot site.

Summer	2015	2016	2017	2018	2019
FF_{th}	0.1085	0.1118	0.1078	0.1095	0.1074
winter	2015–2016	2015–2016	2015–2016	2015–2016	2015–2016
FF_{fr}	0.0251	0.0277	0.0302	0.0274	0.0295

Figure 10 shows the CTC result, i.e., the new method F/T, ERA-land F/T, and SMAP F/T. SMAP F/T is confidently ranked the highest at high latitudes (≥60N°, Figure 10a). There are no dominant products that represent the FT state truth in the northern hemisphere in Rand Second (Figure 10c). However, for the mid-latitudes in East Asia, which includes the XInlinhot site, the new FT is more confident than SMAP's FT product (Figure 10c).

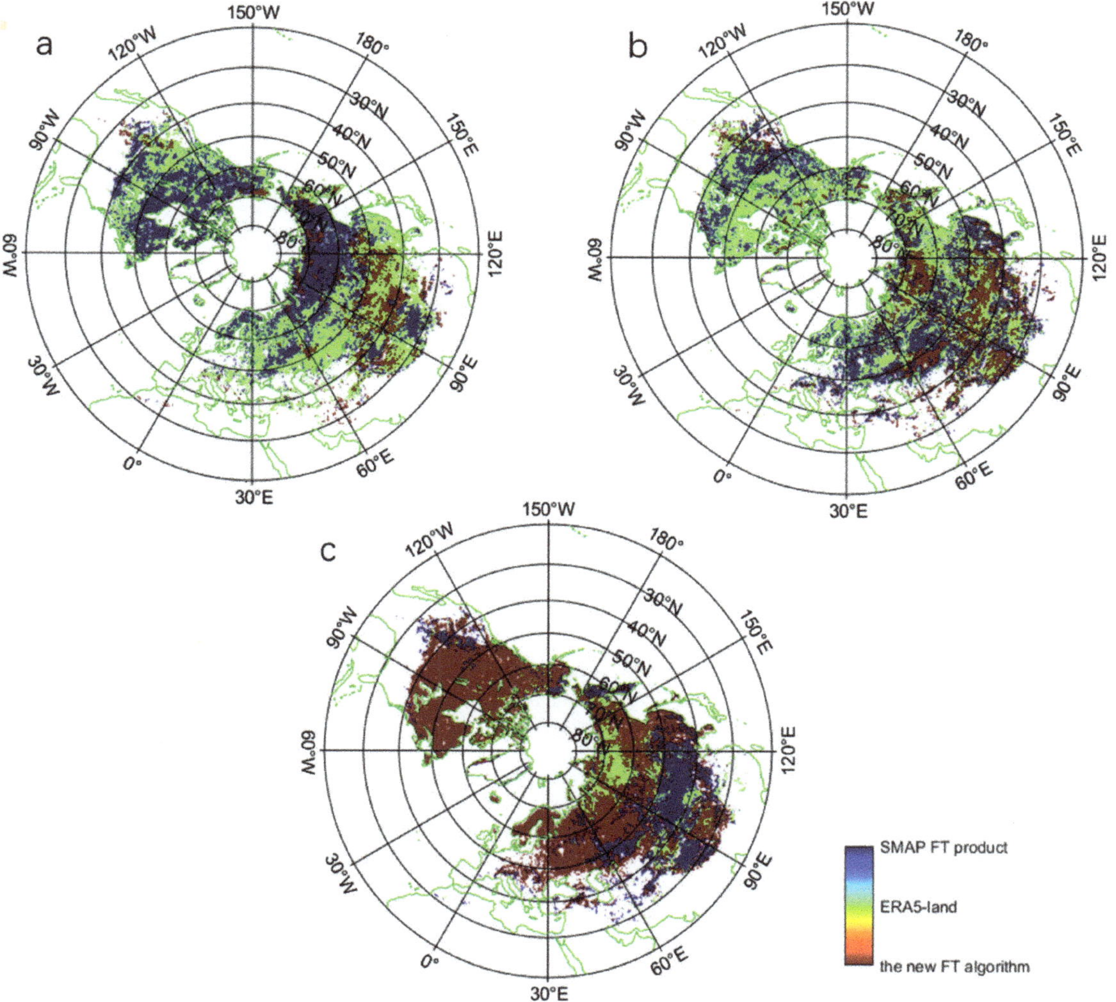

Figure 10. Map of the measurement system ranked first (**a**), second (**b**), and third (**c**) in the northern hemisphere (≥20N°) using CTC.

3.4. Sensitivity Test

Parameters β and γ in the new FT state detection have been selected based on the typical length of synoptic weather systems and the 95% confidence level outcome. β (days) is the window length (days) over which the variance in Equation (9) is estimated and used as a decision criterium in Equation (11), which filters out sporadic changes by weather events. γ is a threshold temperature to judge if the observed SMAP TB signal is related to frozen soil in Equation (10). The values for these parameters have been set in an ad hoc fashion; here, we analyze the sensitivity of the results against the soil freeze/thaw state inferred from ERA-land-assessment data to the variation of these parameters by varying β from 5 to 11 K (Figures 11a and 12a,b) and γ from 3 to 11 days (Figures 11b and 12c,d).

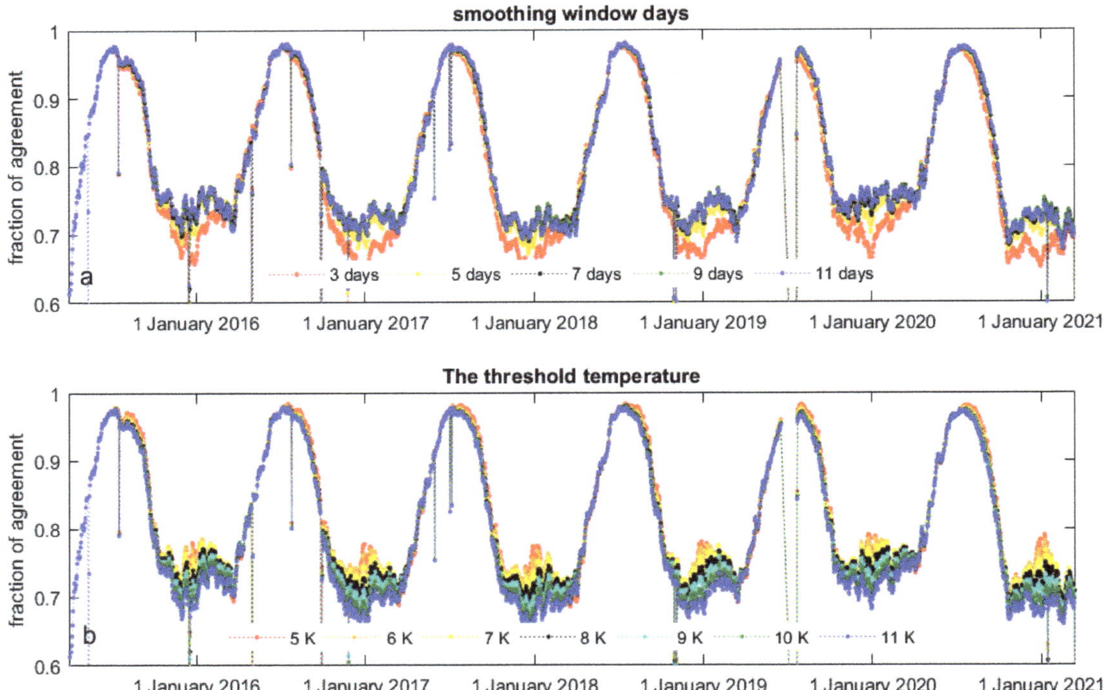

Figure 11. The time-series of the fraction of agreement between the new method and the FT state inferred from ERA-land-assessment data in the northern hemisphere regarding (**a**) β, the window length for the variance; and (**b**) γ, the threshold temperature difference.

When only changing one parameter, the degree of agreement between the new FT estimates and T_{2m} changes significantly only in winter (Figure 11). The impact of changing only β is also larger in winter than in summer (Figure 11a). A smaller β reduces the agreement from above 0.73 (at β = 7) to 0.65 when β = 3, while a larger β only barely increases the agreement between both estimates. A smaller γ improves the agreement in winter (Figure 11b) by more than 0.1. Overall, the variation of the two parameters leads only to significant changes in the agreement between both estimates in the thawing period. The spatial distribution of the change of agreement between both FT state estimates relative to the default values for the two parameters is shown in Figure 12. The Tibetan Plateau and some Northern Europe areas behave opposite to Siberia, the western coast of North America, and the region around latitude 30°N.

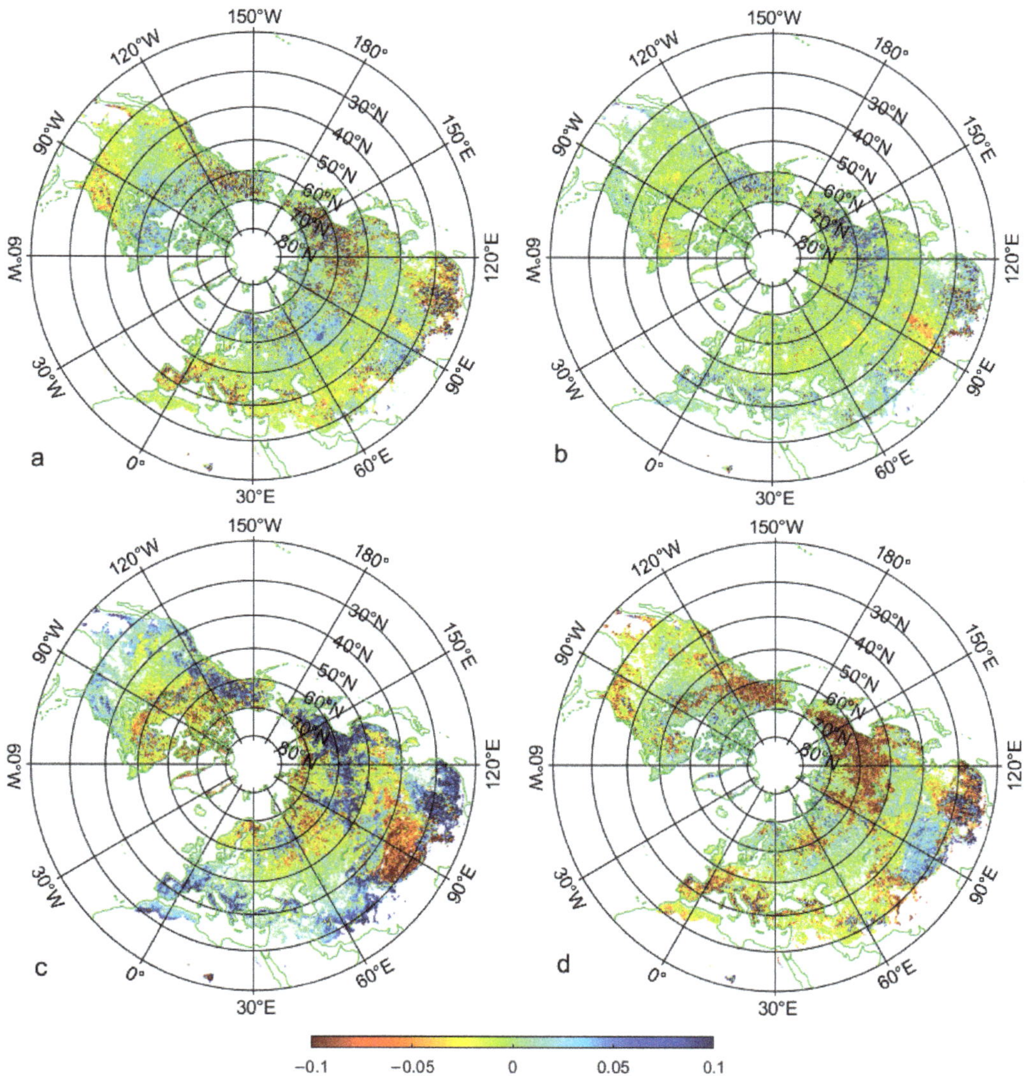

Figure 12. The spatial pattern of the fraction of agreement difference compared to Figure 9b for β = 3 (**a**), β = 11 (**b**), γ = 5 (**c**), and γ = 11 (**d**).

4. Discussion

Although the result reveals the potential for retrieving the FT state from the DAV signals at the L band, several issues need to be addressed for further development as fellows,

The primary issue of the FT product-accuracy assessment of passive microwave remote sensing FT products is how to measure and define the FT on the ground, especially for a deeper penetrated band such as the L band. We lack a precise "ground truth" of the soil's freeze/thaw state. The SMAP FT team uses WMO's air temperature, and WMO's air temperature is also vulnerable in the agreement assessment. For instance, what is the best way to deal with the scale mismatch between the weather station and SMAP's footprint? How do we account for sub-grid open water fraction, terrain heterogeneity, tree cover, precipitation, snowmelt, and so on with the weather station data? These problems can be avoided by taking an average of ERA5-landland–air temperature and collocated 0–7 cm soil temperature in the evaluation, and we are aware that the ERA5-landland–air temperature and collocated 0–7 cm soil temperature are not appropriate for validation which needs FT ground profile truth for sure. ERA-land-assessment data FT inference is used in this study, and SMAP uses a more complicated scheme by considering T_{2m}, T_{skin}, T_{5cm}, and T_{10cm}. Some studies use soil temperature to evaluate SMAP FT products, and with 0–5 cm, the soil temperature at SMAP grids containing CVS stations is about 70% [60]. Since in situ T_{2m} is usually to either infer the ground FT states or those used in the FT products accuracy assessment, it is interesting to know the relationship between air temperature and FT state on a global scale. However, when it comes to soil temperatures such as T_{skin}, T_{5cm},, T_{10cm}, or more profound layers, it is hard to say which layer can correspond to the *TB* signal best. Another option is in situ soil temperature at 5 cm, as in SMAP's FT product-accuracy assessment. However, either the detectable depth or the footprint of SMAP's radiometer does not match ERA5-land. Indeed, the soil temperature from a particular layer cannot compare with SMAP's FT product because the penetration depth of the L-band signal is dynamic [61,62], especially for frozen soil that may range from a few centimeters to meters. Thus, the comparison in this study, such as Figures 9–12, is hard to consider as an evaluation of the new FT algorithm or SMAP's. Still, the comparison shows that the new FT algorithm can capture FT signals as well as the SMAP's official one.

Since ERA-land-assessment data is adopted in this study, Figure 13 shows the frozen land cover and annual accumulations inferred from the new FT state algorithm and the thermal conditions near the surface from $T_{2m} < 273.15$ K in the northern hemisphere. Both are very similar, while the maxima reached for both data sets are different due to their different grid sizes; SMAP data are given in area-equal grids 36 km in diameter (about 1300 km^2), while ERA-land-assessment data used a $1° \times 1°$ lat-lon grid (0 to 1000 km^2). However, the frozen land area detected by the new FT algorithm is about 15 days in advance of T_{2m}, especially in spring (the shift of black lines). It coincides with the challenges in using ERA-land-assessment data as an accuracy assessment for L-band-derived estimates of FT since the radiometer measurements are sensitive to the near-surface soil layer. The time lag in springs is reasonable because soil absorbs the solar radiation and then heats air temperature. The accumulated frozen land areas are also different. Usually, the air temperature will lead to more frozen soil flags, except for the year 2019. Figure 13 shows that T_{2m} is also more appropriate to be the "ground truth" in the beginning than the ending of an annual FT cycle. A similar situation exists for T_{skin}, T_{5cm},, T_{10cm}, or more profound layers. The penetration depth of the L-band detection varies severely during the FT transitions; thus, it would not be suitable to evaluate the FT algorithm with a static depth. Moreover, the penetration depth and sensing depth issues for soil moisture retrieving from L band are still unclear [28,62–65]. The problem is more complex for frozen soil because the dielectric profile is not continuous if freeze–thaw transitions happen in a profile's middle layers, and a complete frozen soil profile shall theoretically have much deeper penetration. Thus, we cannot even get a precise penetration depth for the case of a freeze–thaw transition. By using the air temperature, we avoid this complicated situation. Otherwise, selecting which layer and according to what standard compared with the SMAP FT products would be vulnerable. In this study, we use the same method as the SMAP handbook, which is considered the state-of-art in this topic, to obtain the agreement.

Thus, clarifying the "ground truth" is critical to developing the FT remote sensing algorithm at the L band. Future work needs more in situ accuracy assessment activities in the field experiment, not only for the satellites but from tower-based or airborne platforms. The L-band TB detected by the satellites covers the FT states in tens of kilometers, and most importantly, the FT has a vertical structure. The field experiment is expected to identify the sensing depth, which is also a vague concept for soil moisture remote sensing at the L band during the FT transition periods. Instead, the ERA-land-assessment data makes the temporal and spatial scales more comparable with the FT result in this study. Specifically, (1) the vertical scale: the air temperature is a composite indicator of the FT state for a certain layer; (2) the temporal scale: the resolution of the DAV values is daily. It is reliable to use daily air temperature and collocated 0–7 cm soil temperature, not instant soil temperature, in the assessment.

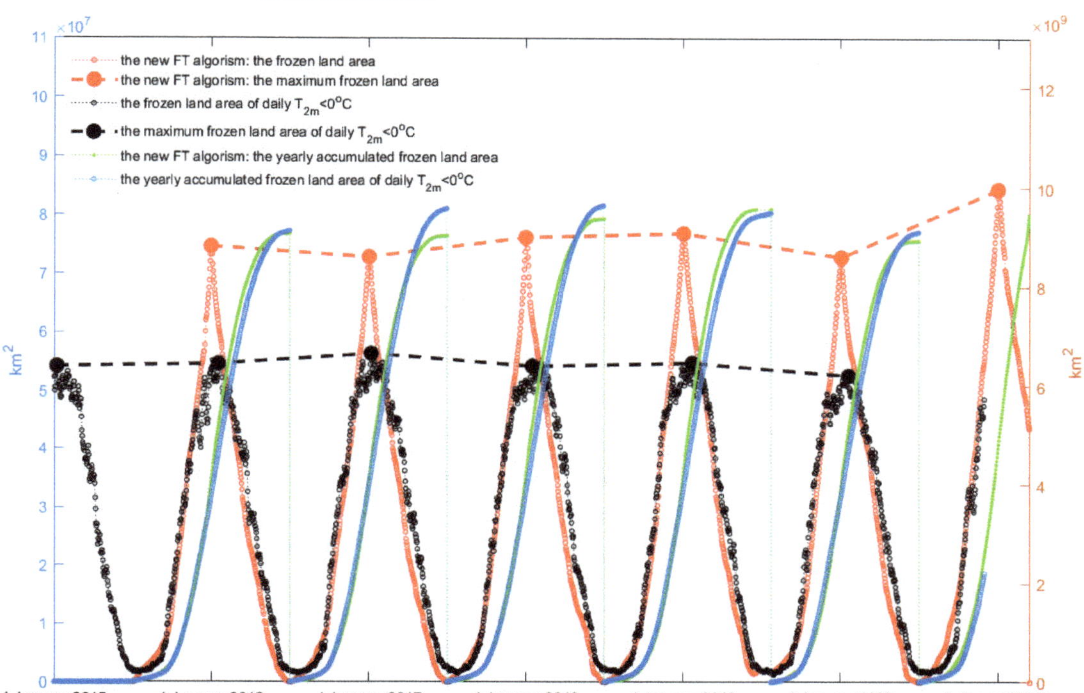

Figure 13. The cover of the frozen land detected by the new FT algorithm and $T_{2m} < 273.15$ K.

Besides lacking appropriate ground truth, another limitation of the new algorithm is the absence of the DAV data in the low latitudes (≤45°N). The orbits of SMAP and SMOS make the revisit period over in the low latitudes more frequent than two times a day but only one value in 3 days. The new FT algorithm is based on a DAV signal as in Equation (8) and further extends to the synoptic scale by the parameter β. Thus, a lot of data are absent in the low latitudes area. To overcome this drawback, we have to define the annual FT scale's beginning and ending times and fill the DAV absence with the nearest interpolation. This produces dummy signals that affect the accuracy in low latitudes. In all cases, the daily L3_FT products incorporate (AM and PM satellite overpasses) data for the current day, as well as past days' information (to a maximum of 3 days, necessary only near the southern margin of the FT domain) to ensure complete coverage of the FT domain in each day's product.

5. Conclusions

We developed a new FT state-detection algorithm based on the difference in the microwave brightness temperature between 6 a.m. descending and 6 p.m. ascending half-orbit passes that are relatively small over frozen soil due to the large penetration depth leading to an effective temperature dominated by stable deeper soil temperatures. The new FT state detection agrees well with SMAP's FT state detection, with a minimum of above 0.72 in winter. The new algorithm can reach a comparable agreement regarding the spatial distribution as the SMAP FT product does against ERA-land-assessment data FT inference in the northern hemisphere. The algorithm is rather stable to variations of its ad hoc set parameters. The limitation is that this algorithm is only applicable to the L band, considering its penetration depth and corresponding soil effective temperatures; it cannot be applied to other bands. The new algorithm presented in this study is expected to extend the application of L-band passive microwave remote sensing data in freezing-thawing conditions.

Since the sensitivity of the L-band signal to the F/T state varies in terms of land cover type and climate regions, the new algorithm shall be validated with in-site FT observations at various sites. For instance, the sub-grid open-water fraction, tree cover, precipitation, and snowmelt critical to TB signals at the L band shall be checked. These dynamics will enlarge/reduce the DAV signals and further the efficiency of parameters β and γ. On the other hand, the terrain is a significant factor because we can see Tibet and the Rocky Mountains in Figure 9. Although the new algorithm shows comparable agreement with SMAP's FT products, further optimization, and validation work, should be carried out in the future.

Author Contributions: Data curation, Y.Z., Y.G. and Z.S.; Formal analysis, J.W.; Funding acquisition, C.S.; Writing—original draft, S.L. All authors have read and agreed to the published version of the manuscript.

Funding: This research was funded by the National Natural Science Foundation of China (Grant 42075150); by the Natural Science Foundation of Shanghai (No. 21ZR1405500); and by the Deutsche Forschungsgemeinschaft (DFG) via the research group FOR2131 on "Data Assimilation for Improved Characterization of Fluxes across Compartmental Interfaces", subproject P2.

Data Availability Statement: SMAP data is available at: https://smap.jpl.nasa.gov/data/. ERA5-Land data is available at: https://cds.climate.copernicus.eu/cdsapp#!/search?type=dataset&text=era5-land.

Conflicts of Interest: The authors declare no conflict of interest.

References

1. Walvoord, M.A.; Kurylyk, B.L. Hydrologic Impacts of Thawing Permafrost-A Review. *Vadose Zone J.* **2016**, *15*, 1–20. [CrossRef]
2. Schuur, E.A.G.; McGuire, A.D.; Schädel, C.; Grosse, G.; Harden, J.W.; Hayes, D.J.; Hugelius, G.; Koven, C.D.; Kuhry, P.; Lawrence, D.M.; et al. Climate change and the permafrost carbon feedback. *Nature* **2015**, *520*, 171–179. [CrossRef]
3. Zeng, Y.; Su, Z.; Barmpadimos, I.; Perrels, A.; Poli, P.; Boersma, K.F.; Frey, A.; Ma, X.; de Bruin, K.; Goosen, H.; et al. Towards a Traceable Climate Service: Assessment of Quality and Usability of Essential Climate Variables. *Remote Sens.* **2019**, *11*, 1186. [CrossRef]
4. Yu, L.; Fatichi, S.; Zeng, Y.; Su, Z. The role of vadose zone physics in the ecohydrological response of a Tibetan meadow to freeze-thaw cycles. *Cryosphere* **2020**, *14*, 4653–4673. [CrossRef]
5. Hu, G.; Zhao, L.; Wu, X.; Li, R.; Wu, T.; Xie, C.; Pang, Q.; Zou, D. Comparison of the thermal conductivity parameterizations for a freeze-thaw algorithm with a multi-layered soil in permafrost regions. *CATENA* **2017**, *156*, 244–251. [CrossRef]
6. Gao, J.Q.; Xie, Z.H.; Wang, A.W.; Luo, Z.D. Numerical simulation based on two-directional freeze and thaw algorithm for thermal diffusion model. *Appl. Math. Mech.-Engl.* **2016**, *37*, 1467–1478. [CrossRef]
7. Zhao, T.; Shi, J.; Hu, T.; Zhao, L.; Zou, D.; Wang, T.; Ji, D.; Li, R.; Wang, P. Estimation of high-resolution near-surface freeze/thaw state by the integration of microwave and thermal infrared remote sensing data on the Tibetan Plateau. *Earth Space Sci.* **2017**, *4*, 472–484. [CrossRef]
8. Koven, C.D.; Ringeval, B.; Friedlingstein, P.; Ciais, P.; Cadule, P.; Khvorostyanov, D.; Krinner, G.; Tarnocai, C. Permafrost carbon-climate feedbacks accelerate global warming. *Proc. Natl. Acad. Sci. USA* **2011**, *108*, 14769–14774. [CrossRef]
9. Koven, C.D.; Riley, W.J.; Stern, A. Analysis of Permafrost Thermal Dynamics and Response to Climate Change in the CMIP5 Earth System Models. *J. Clim.* **2013**, *26*, 1877–1900. [CrossRef]
10. Zhao, P.; Xu, X.; Chen, F.; Guo, X.; Zheng, X.; Liu, L.; Hong, Y.; Li, Y.; La, Z.; Peng, H.; et al. The Third Atmospheric Scientific Experiment for Understanding the Earth-Atmosphere Coupled System over the Tibetan Plateau and Its Effects. *Bull. Am. Meteorol. Soc.* **2018**, *99*, 757–776. [CrossRef]
11. Kimball, J.S.; White, M.A.; Running, S.W. BIOME-BGC simulations of stand hydrologic processes for BOREAS. *J. Geophys. Res. -Atmos.* **1997**, *102*, 29043–29051. [CrossRef]
12. Li, F.Y.; Newton, P.C.D.; Lieffering, M. Testing simulations of intra- and inter-annual variation in the plant production response to elevated CO2 against measurements from an 11-year FACE experiment on grazed pasture. *Glob. Chang. Biol.* **2014**, *20*, 228–239. [CrossRef] [PubMed]
13. Cramer, W.; Kicklighter, D.W.; Bondeau, A.; Iii, B.M.; Churkina, G.; Nemry, B.; Ruimy, A.; Schloss, A.L.; Participants Potsdam, N.P.P.M.I. Comparing global models of terrestrial net primary productivity (NPP): Overview and key results. *Glob. Chang. Biol.* **1999**, *5*, 1–15. [CrossRef]

14. Matzner, E.; Borken, W. Do freeze-thaw events enhance C and N losses from soils of different ecosystems? A review. *Eur. J. Soil Sci.* **2008**, *59*, 274–284. [CrossRef]
15. Wang, S.; Zhang, Y.; Lu, S.; Su, P.; Shang, L.; Li, Z. Biophysical regulation of carbon fluxes over an alpine meadow ecosystem in the eastern Tibetan Plateau. *Int. J. Biometeorol.* **2016**, *60*, 801–812. [CrossRef] [PubMed]
16. Xie, Z.H.; Liu, S.; Zeng, Y.J.; Gao, J.Q.; Qin, P.H.; Jia, B.H.; Xie, J.B.; Liu, B.; Li, R.C.; Wang, Y.; et al. A High-Resolution Land Model With Groundwater Lateral Flow, Water Use, and Soil Freeze-Thaw Front Dynamics and its Application in an Endorheic Basin. *J. Geophys. Res. -Atmos.* **2018**, *123*, 7204–7222. [CrossRef]
17. Swenson, S.C.; Lawrence, D.; Lee, H. Improved simulation of the terrestrial hydrological cycle in permafrost regions by the Community Land Model. *J. Adv. Modeling Earth Syst.* **2012**, *4*, M08002. [CrossRef]
18. Yu, L.; Zeng, Y.; Su, Z. Understanding the mass, momentum, and energy transfer in the frozen soil with three levels of model complexities. *Hydrol. Earth Syst. Sci.* **2020**, *24*, 4813–4830. [CrossRef]
19. Yu, L.Y.; Zeng, Y.J.; Wen, J.; Su, Z.B. Liquid-Vapor-Air Flow in the Frozen Soil. *J. Geophys. Res. -Atmos.* **2018**, *123*, 7393–7415. [CrossRef]
20. Mwangi, S.; Zeng, Y.; Montzka, C.; Yu, L.; Su, Z. Assimilation of Cosmic-Ray Neutron Counts for the Estimation of Soil Ice Content on the Eastern Tibetan Plateau. *J. Geophys. Res. -Atmos.* **2020**, *125*, e2019JD031529. [CrossRef]
21. Yashchenko, A.S.; Bobrov, P.P. Impact of the Soil Moisture Distribution in the Top Layer on the Accuracy Moisture Retrieval by Microwave Radiometer Data. *IEEE Trans. Geosci. Remote Sens.* **2016**, *54*, 5239–5246. [CrossRef]
22. Schwank, M.; Stahli, M.; Wydler, H.; Leuenberger, J.; Matzler, C.; Fluhler, H. Microwave L-band emission of freezing soil. *IEEE Trans. Geosci. Remote Sens.* **2004**, *42*, 1252–1261. [CrossRef]
23. Rautiainen, K.; Lemmetyinen, J.; Schwank, M.; Kontu, A.; Ménard, C.B.; Mätzler, C.; Drusch, M.; Wiesmann, A.; Ikonen, J.; Pulliainen, J. Detection of soil freezing from L-band passive microwave observations. *Remote Sens. Environ.* **2014**, *147*, 206–218. [CrossRef]
24. Kim, Y.; Kimball, J.S.; Glassy, J.; McDonald, K.C. MEaSUREs Global Record of Daily Landscape Freeze/Thaw Status, Version 4 [dataset]. 2017. [CrossRef]
25. Kim, Y.; Kimball, J.S.; McDonald, K.C.; Glassy, J. Developing a Global Data Record of Daily Landscape Freeze/Thaw Status Using Satellite Passive Microwave Remote Sensing. *IEEE Trans. Geosci. Remote Sens.* **2011**, *49*, 949–960. [CrossRef]
26. Xu, X.; Dunbar, R.; Derksen, C.; Colliander, A.; Kim, Y.; Kimball, J. SMAP L3 Radiometer Global and Northern Hemisphere Daily 36 km EASE-Grid Freeze/Thaw State, Version 3, Thaw State, Version, 3. 2020. Available online: https://nsidc.org/data/spl3ftp/versions/2 (accessed on 1 May 2020).
27. Rautiainen, K.; Parkkinen, T.; Lemmetyinen, J.; Schwank, M.; Wiesmann, A.; Ikonen, J.; Derksen, C.; Davydov, S.; Davydova, A.; Boike, J.; et al. SMOS prototype algorithm for detecting autumn soil freezing. *Remote Sens. Environ.* **2016**, *180*, 346–360. [CrossRef]
28. Lv, S.; Zeng, Y.; Su, Z.; Wen, J. A Closed-Form Expression of Soil Temperature Sensing Depth at L-Band. *IEEE Trans. Geosci. Remote Sens.* **2019**, *57*, 4889–4897. [CrossRef]
29. Zhao, T.; Shi, J.; Zhao, S.; Chen, K.; Wang, P.; Li, S.; Xiong, C.; Xiao, Q. Measurement and Modeling of Multi-Frequency Microwave Emission of Soil Freezing and Thawing Processes. In Proceedings of the 2018 Progress in Electromagnetics Research Symposium (PIERS-Toyama), Toyama, Japan, 1–4 August 2018; pp. 31–36. [CrossRef]
30. Lv, S.; Zeng, Y.; Wen, J.; Zhao, H.; Su, Z. Estimation of Penetration Depth from Soil Effective Temperature in Microwave Radiometry. *Remote Sens.* **2018**, *10*, 519. [CrossRef]
31. Kim, Y.; Kimball, J.S.; Glassy, J.; Du, J.Y. An extended global Earth system data record on daily landscape freeze-thaw status determined from satellite passive microwave remote sensing. *Earth Syst. Sci. Data* **2017**, *9*, 133–147. [CrossRef]
32. Dunbar, S.; Xu, X.; Colliander, A.; Derksen, C.; Kimball, J.; Kim, Y. Algorithm Theoretical Basis Document (ATBD). SMAP Level 3 Radiometer Freeze/Thaw Data Products. *JPL CIT JPL* **2020**, *500*, 33.
33. Sharifnezhad, Z.; Norouzi, H.; Prakash, S.; Blake, R.; Khanbilvardi, R. Diurnal Cycle of Passive Microwave Brightness Temperatures over Land at a Global Scale. *Remote Sens.* **2021**, *13*, 817. [CrossRef]
34. Kopczynski, S.; Ramage, J.; Lawson, D.; Goetz, S.; Evenson, E.; Denner, J.; Larson, G. Passive microwave (SSM/I) satellite predictions of valley glacier hydrology, Matanuska Glacier, Alaska. *Geophys. Res. Lett.* **2008**, *35*, 2008GL034615. [CrossRef]
35. Tedesco, M. Snowmelt detection over the Greenland ice sheet from SSM/I brightness temperature daily variations. *Geophys. Res. Lett.* **2007**, *34*, 2006GL028466. [CrossRef]
36. Guo, D.; Wang, H. Simulated change in the near-surface soil freeze/thaw cycle on the Tibetan Plateau from 1981 to 2010. *Chin. Sci. Bull.* **2014**, *59*, 2439–2448. [CrossRef]
37. Chaubell, J.; Yueh, S.; Entekhabi, D.; Peng, J. Resolution enhancement of SMAP radiometer data using the Backus Gilbert optimum interpolation technique. In Proceedings of the 2016 IEEE International Geoscience and Remote Sensing Symposium (IGARSS), Beijing, China, 10–15 July 2016; pp. 284–287. [CrossRef]
38. Muñoz-Sabater, J.; Dutra, E.; Agustí-Panareda, A.; Albergel, C.; Arduini, G.; Balsamo, G.; Boussetta, S.; Choulga, M.; Harrigan, S.; Hersbach, H.; et al. ERA5-Land: A state-of-the-art global reanalysis dataset for land applications. *Earth Syst. Sci. Data Discuss* **2021**, *13*, 4349–4383. [CrossRef]
39. Baker, D.G.; Ruschy, D.L. Calculated and Measured Air and Soil Freeze-Thaw Frequencies. *J. Appl. Meteorol. Climatol.* **1995**, *34*, 2197–2205. [CrossRef]

40. Fortin, G. Variability and frequency of the freeze thaw cycles in Quebec region, 1977-2006. *Can. Geogr. -Geogr. Can.* **2010**, *54*, 196–208. [CrossRef]
41. Ho, E.; Gough, W.A. Freeze thaw cycles in Toronto, Canada in a changing climate. *Theor. Appl. Climatol.* **2006**, *83*, 203–210. [CrossRef]
42. Zhao, T.; Shi, J.; Entekhabi, D.; Jackson, T.J.; Hu, L.; Peng, Z.; Yao, P.; Li, S.; Kang, C.S. Retrievals of soil moisture and vegetation optical depth using a multi-channel collaborative algorithm. *Remote Sens. Environ.* **2021**, *257*, 112321. [CrossRef]
43. Yamano, H.; Chen, J.; Tamura, M. Hyperspectral identification of grassland vegetation in Xilinhot, Inner Mongolia, China. *Int. J. Remote Sens.* **2003**, *24*, 3171–3178. [CrossRef]
44. Shi, C.X.; Xie, Z.H.; Qian, H.; Liang, M.L.; Yang, X.C. China land soil moisture EnKF data assimilation based on satellite remote sensing data. *Sci. China-Earth Sci.* **2011**, *54*, 1430–1440. [CrossRef]
45. Wu, S.; Chen, J. Validation of AMSR-E soil moisture products in Xilinhot grassland. In Proceedings of the Remote Sensing for Agriculture, Ecosystems, and Hydrology XIV, Edinburgh, UK, 24–26 September 2012; p. 85311J.
46. Chapin, F.S.; Sala, O.E.; Huber-Sannwald, E. *Global Biodiversity in a Changing Environment: Scenarios for the 21st Century*; Springer Science & Business Media: Berlin/Heidelberg, Germany, 2013.
47. Derksen, C.; Xu, X.L.; Dunbar, R.S.; Colliander, A.; Kim, Y.; Kimball, J.S.; Black, T.A.; Euskirchen, E.; Langlois, A.; Loranty, M.M.; et al. Retrieving landscape freeze/thaw state from Soil Moisture Active Passive (SNAP) radar and radiometer measurements. *Remote Sens. Environ.* **2017**, *194*, 48–62. [CrossRef]
48. Lv, S.; Zeng, Y.; Wen, J.; Zheng, D.; Su, Z. Determination of the Optimal Mounting Depth for Calculating Effective Soil Temperature at L-Band: Maqu Case. *Remote Sens.* **2016**, *8*, 476. [CrossRef]
49. Blackmon, M.L. A Climatological Spectral Study of the 500 mb Geopotential Height of the Northern Hemisphere. *J. Atmos. Sci.* **1976**, *33*, 1607–1623. [CrossRef]
50. Cao, B.; Gruber, S.; Zheng, D.; Li, X. The ERA5-Land soil temperature bias in permafrost regions. *Cryosphere* **2020**, *14*, 2581–2595. [CrossRef]
51. McColl, K.A.; Roy, A.; Derksen, C.; Konings, A.; Alemohammad, S.H.; Entekhabi, D. Triple collocation for binary and categorical variables: Application to validating landscape freeze/thaw retrievals. *Remote Sens. Environ.* **2016**, *176*, 31–42. [CrossRef]
52. Scott, K.A. Assessment of Categorical Triple Collocation for Sea Ice/Open Water Observations: Application to the Gulf of Saint Lawrence. *IEEE Trans. Geosci. Remote Sens.* **2019**, *57*, 9659–9673. [CrossRef]
53. Stoffelen, A. Toward the true near-surface wind speed: Error modeling and calibration using triple collocation. *J. Geophys. Res. Ocean.* **1998**, *103*, 7755–7766. [CrossRef]
54. Li, H.; Chai, L.; Crow, W.; Dong, J.; Liu, S.; Zhao, S. The reliability of categorical triple collocation for evaluating soil freeze/thaw datasets. *Remote Sens. Environ.* **2022**, *281*, 113240. [CrossRef]
55. Su, Z.; Wen, J.; Zeng, Y.; Zhao, H.; Lv, S.; Van Der Velde, R.; Zheng, D.; Wang, X.; Wang, Z.; Schwank, M.; et al. Multiyear in-situ L-band microwave radiometry of land surface processes on the Tibetan Plateau. *Sci. Data* **2020**, *7*, 317. [CrossRef]
56. Wang, B.; Ho, L.; Zhang, Y.; Lu, M.-M. Definition of South China Sea monsoon onset and commencement of the East Asia summer monsoon. *J. Clim.* **2004**, *17*, 699–710. [CrossRef]
57. Cai, M.; Ren, R.C. Meridional and downward propagation of atmospheric circulation anomalies. Part I: Northern Hemisphere cold season variability. *J. Atmos Sci.* **2007**, *64*, 1880–1901. [CrossRef]
58. Harris, C.; Arenson, L.U.; Christiansen, H.H.; Etzemuller, B.; Frauenfelder, R.; Gruber, S.; Haeberli, W.; Hauck, C.; Holzle, M.; Humlum, O.; et al. Permafrost and climate in Europe: Monitoring and modelling thermal, geomorphological and geotechnical responses. *Earth-Sci. Rev.* **2009**, *92*, 117–171. [CrossRef]
59. Huang, X.; Han, S.; Shi, C. Multiscale Assessments of Three Reanalysis Temperature Data Systems over China. *Agriculture* **2021**, *11*, 1292. [CrossRef]
60. Kraatz, S.; Jacobs, J.M.; Schröder, R.; Cho, E.; Cosh, M.; Seyfried, M.; Prueger, J.; Livingston, S. Evaluation of SMAP Freeze/Thaw Retrieval Accuracy at Core Validation Sites in the Contiguous United States. *Remote Sens.* **2018**, *10*, 1483. [CrossRef]
61. Zheng, D.; Li, X.; Zhao, T.; Wen, J.; van der Velde, R.; Schwank, M.; Wang, X.; Wang, Z.; Su, Z. Impact of Soil Permittivity and Temperature Profile on L-Band Microwave Emission of Frozen Soil. *IEEE Trans. Geosci. Remote Sens.* **2020**, *59*, 4080–4093. [CrossRef]
62. Zheng, D.; Li, X.; Wang, X.; Wang, Z.; Wen, J.; van der Velde, R.; Schwank, M.; Su, Z. Sampling depth of L-band radiometer measurements of soil moisture and freeze-thaw dynamics on the Tibetan Plateau. *Remote Sens. Environ.* **2019**, *226*, 16–25. [CrossRef]
63. Roy, A.; Toose, P.; Williamson, M.; Rowlandson, T.; Derksen, C.; Royer, A.; Berg, A.A.; Lemmetyinen, J.; Arnold, L. Response of L-Band brightness temperatures to freeze/thaw and snow dynamics in a prairie environment from ground-based radiometer measurements. *Remote Sens. Environ.* **2017**, *191*, 67–80. [CrossRef]
64. Rowlandson, T.L.; Berg, A.A.; Roy, A.; Kim, E.; Lara, R.P.; Powers, J.; Lewis, K.; Houser, P.; McDonald, K.; Toose, P.; et al. Capturing agricultural soil freeze/thaw state through remote sensing and ground observations: A soil freeze/thaw validation campaign. *Remote Sens. Environ.* **2018**, *211*, 59–70. [CrossRef]
65. Wang, J.; Jiang, L.; Cui, H.; Wang, G.; Yang, J.; Liu, X.; Su, X. Evaluation and analysis of SMAP, AMSR2 and MEaSUREs freeze/thaw products in China. *Remote Sens. Environ.* **2020**, *242*, 111734. [CrossRef]

Article

Using of Remote Sensing-Based Auxiliary Variables for Soil Moisture Scaling and Mapping

Zebin Zhao [1], Rui Jin [1,2,*], Jian Kang [1], Chunfeng Ma [1] and Weizhen Wang [1]

1 Heihe Remote Sensing Experimental Research Station, Key Laboratory of Remote Sensing of Gansu Province, Northwest Institute of Eco-Environment and Resources, Chinese Academy of Sciences, Lanzhou 730000, China; zhaozebin@lzb.ac.cn (Z.Z.); kangjian@lzb.ac.cn (J.K.); machf@lzb.ac.cn (C.M.); weizhen@lzb.ac.cn (W.W.)

2 CAS Center for Excellence in Tibetan Plateau Earth Sciences, Chinese Academy of Sciences, Beijing 100101, China

* Correspondence: jinrui@lzb.ac.cn; Tel.: +86-0931-4967965

Abstract: Soil moisture is one of the core hydrological and climate variables that crucially influences water and energy budgets. The spatial resolution of available soil moisture products is generally coarser than 25 km, which limits their hydro-meteorological and eco-hydrological applications and the management of water resources at watershed and agricultural scales. A feasible solution to overcome these limitations is to downscale coarse soil moisture products with the support of higher-resolution spatial information. Although many auxiliary variables have been used for this purpose, few studies have analyzed their applicability and effectiveness in arid regions. To this end, we comprehensively evaluated four commonly used auxiliary variables, including NDVI (Normalized Difference Vegetation Index), LST (Land Surface Temperature), TVDI (Temperature Vegetation Dryness Index), and SEE (Soil Evaporative Efficiency), against ground-based soil moisture observations during the vegetation growing season in the Heihe River Basin, China. Performance metrics indicated that SEE is most sensitive ($R^2 \geq 0.67$) to soil moisture because it is controlled by soil evaporation limited by the available soil moisture. The similarity of spatial patterns also showed that SEE best captures soil moisture changes, with the STD (standard deviation) of the HD (Hausdorff Distance) less than 0.058 when compared with PLMR (Polarimetric L-band Multi-beam Radiometer) soil moisture products. In addition, soil moisture was mapped by RF (Random Forests) using both single auxiliary variables and 11 types of multiple auxiliary variable combinations. SEE was found to be the best auxiliary variable for scaling and mapping soil moisture with accuracy of 0.035 cm^3/cm^3. Among the multiple auxiliary variables, the combination of LST, NDVI, and SEE was found to best enhance the scaling and mapping accuracy of soil moisture with 0.034 cm^3/cm^3.

Keywords: soil moisture; auxiliary variable; Hausdorff Distance; Random Forests; scaling; mapping

Citation: Zhao, Z.; Jin, R.; Kang, J.; Ma, C.; Wang, W. Using of Remote Sensing-Based Auxiliary Variables for Soil Moisture Scaling and Mapping. *Remote Sens.* **2022**, *14*, 3373. https://doi.org/10.3390/rs14143373

Academic Editor: Nicolas Baghdadi

Received: 20 May 2022
Accepted: 11 July 2022
Published: 13 July 2022

1. Introduction

Soil moisture plays a key role in the water cycle and heat exchange of surface–vegetation–atmosphere columns [1–3]. Information about the spatiotemporal distribution of soil moisture is essential for drought monitoring [4], evapotranspiration forecasting [3,5], water resource management [6–8], and crop yield estimation [9]. In recent years, advances in active and passive microwave remote sensing have made this the main technique for measuring soil moisture distribution at regional and global scales [10,11].

Over the past years, several global soil moisture products based on satellite-based passive microwave sensors have become available [12]. Among these, the Advanced Microwave Scanning Radiometer Earth Observing System (AMSR-E) has a finer spatial resolution of 25 km [13–15], while Soil Moisture and Ocean Salinity (SMOS) [16], launched in 2009, has a resolution of 40 km, and the Soil Moisture Active Passive (SMAP) [17,18], launched in 2015, has a resolution of 36 km. These passive microwave remote sensing products aim to map global soil moisture within an accuracy of 0.04 cm^3/cm^3 to satisfy hydro-climatological and hydro-meteorological requirements and the management of water resources at watershed and agricultural scales.

For practical applications, however, soil moisture maps with spatial resolutions higher than 10 km are urgently required at watershed scales because soil texture and structure, vegetation coverage, and topography lead to high soil moisture spatial variability [19]. For example, Pan and Wood (2010) found that a higher spatial resolution of soil moisture can improve hydrology assimilations [20]. Agricultural applications often require a finer spatial resolution than 1 km [21], and weather forecasting and hydrological applications also need a higher spatial resolution than 10 km. In recent years, the downscaling of soil moisture has thus received wide attention to satisfy watershed-scale research and application needs [22–25]. To obtain a finer distribution of soil moisture, auxiliary variables at a higher spatial resolution are indispensable for capturing the spatial pattern of soil moisture in remote sensing pixels. For example, a combination of high-resolution Normalized Difference Vegetation Index (NDVI), albedo, and Land Surface Temperature (LST) data can be used to downscale soil moisture from 25 km to 1 km [26].

Auxiliary variables are also beneficial for the upscaling of multi-point ground-based soil moisture observations. Generally, the footprint of a remote sensing pixel is much larger than the representative area of in situ soil moisture measurement, which leads to a huge scale gap when validating soil moisture remote sensing products, especially when the spatial heterogeneity of soil moisture is increased by the compound influences of precipitation, soil texture, vegetation cover, and topography. At present, Wireless Sensor Network (WSN) [27] and COsmic-ray Soil Moisture Observing System (COSMOS) [28] are two promising methods capable of overcoming the scale mismatch between point observations and remote sensing pixels. Various upscaling methods have been developed to upscale the multi-points soil moisture observations, including Bayesian linear regression [29], ridge regression [30], and Kriging [30–32]. Introducing auxiliary variables related to soil moisture into the upscaling process does not only compensate the weak spatial representation of the sparse in situ soil moisture measurements, but also takes into account the trend of change of soil moisture with time, improving upscaling accuracy [11].

Until now, multispectral remote sensing data ranging from visible, near-infrared, thermal-infrared to microwave bands have been used directly or indirectly as auxiliary variables relative to soil moisture. Because of the different remote sensing principles (reflection, radiation, and scattering properties) in different wavelengths, each type of auxiliary variable reflects soil moisture at different depths and under different vegetation cover conditions. Optical remote sensing indirectly takes advantage of the strong absorption in the visible bands and strong reflection in the near-infrared bands of vegetation; the relationship between soil moisture and spectral reflectance in the visible/near-infrared bands can be used to determine the soil moisture at the surface or in the top millimeters of soil. In general, the absorption in the visible bands increases with soil moisture. Recently, a series of vegetation indices based on optical remote sensing was developed to scale soil moisture via vegetation health conditions and water stress. During the entire vegetation growing season, there is an obvious positive correlation between NDVI [33] and soil moisture. Based on NDVI, the Vegetation Condition Index (VCI) [34,35], Anomaly Vegetation Index (AVI) [36], Temperature Condition Index (TCI) [35], and Temperature Vegetation Index (TVI) [37] were developed to comprehensively reflect soil water deficiencies in different years. However, soil moisture retrieval is limited in these bands due to their limited capability to penetrate clouds and vegetation, and the careful correction required to eliminate atmospheric effects [38].

Monitoring soil moisture with LST observed by a thermal-infrared remote sensor can be traced back to the 1970s. LST was utilized as a feasible proxy to infer soil moisture in drought-affected areas [39] thanks to soil thermal properties. Researchers demonstrated that, in cloudless conditions, LST reflects soil moisture when the land surface is bare [40]. Both the Crop Water Stress Index (CWSI) [6] and the Apparent Thermal Inertia (ATI) [41] are based on land surface emissivities and LSTs for estimating the variability of soil moisture under bare soil and low vegetation coverage conditions. The Moderate Resolution Imaging Spectroradiometer (MODIS)-derived ATI has been successfully used to map soil moisture (1 km) with an accuracy of 0.031 cm^3/cm^3 in the Babao River Basin, China, where the land cover type is primarily alpine meadow [31]. Because of the limited sensitivity of LST to soil moisture over vegetated areas, auxiliary variables such as the Temperature Vegetation Dryness Index (TVDI) [42] and the Vegetation Temperature Condition Index (VTCI) [43] were proposed to estimate soil moisture and monitor drought. For example, Kang et al. (2015) found that the Advanced Spaceborne Thermal Emission and Reflection Radiometer (ASTER) TVDI describes the heterogeneity of soil moisture within a microwave pixel [44]. Peng et al. (2015) utilized the VTCI calculated from MODIS to downscale the Climate Change Initiative (CCI) soil moisture from 0.25° to 5.6 km [45], while the result showed reasonable agreement with soil moisture observations. However, building the TVDI or VTCI feature space requires not only a large number of samples to cover the extremely dry to humid conditions, but also stable and uniform meteorological conditions in the study area. Additionally, there is uncertainty in the determination of wet and dry edges. Due to the close coupling relationship between soil moisture and evapotranspiration in a water limited region, the Soil Evaporative Efficiency (SEE) [46], defined as the ratio of the actual to potential soil evaporation, was employed to reflect soil moisture in last decades. Merlin et al. (2009) utilized the SEE to downscale the SMOS soil moisture product from a 40 km resolution to 10 km/4 km resolution [22]. However, calculation of soil evaporation directly from remote sensing is difficult, and its uncertainty increases with precipitation and irrigation.

Microwave remote sensing, which utilizes radiative and scattering signals to monitor hydrological variables, maps soil moisture based on the microwave dielectric sensitivity. The combination of active and passive L-band microwave remote sensing holds strong potential for improving the spatial resolution of soil moisture. Active microwave remote sensing can provide much higher spatial resolution, and passive sensors can provide frequent observations, albeit with coarser spatial resolution. With the support of high-resolution microwave backscattering as an auxiliary variable, higher resolution microwave brightness temperature data can be easily obtained by decomposing the passive microwave brightness temperature of the L-band; thus, higher resolution soil moisture can be retrieved [47]. However, both soil surface roughness and soil structure are the main limitations for soil moisture estimation [48] with these methods.

Recently, researchers combined land surface variables estimated from optical, thermal-infrared, and microwave sensors to scale soil moisture with satisfactory accuracy and resolution [49–52]. For example, many researchers have combined land surface albedo and brightness temperature to retrieve soil moisture with a high accuracy (RMSE $\leq$ 0.048 cm^3/cm^3) [53,54]. Knipper et al. (2017) compared the roles of albedo and onboard brightness temperature in improving the resolution of soil moisture products (1 km) and found that brightness temperature provides optimal precision for the spatial variability of soil moisture [55].

A variety of auxiliary variables used to scale soil moisture have been explored during the past decades. However, a systematic evaluation of their applicability and contribution to the scaling transformation of soil moisture, especially in croplands, is still missing. Therefore, in this study, four auxiliary variables quantified from remote sensing images, including NDVI, LST, TVDI, and SEE, were analyzed to show their applicable conditions, the sensitivity to soil moisture, correlations, and sensing depth to soil moisture, and to provide useful information for the scale conversion of soil moisture. Random Forest (RF) was then applied to identify the auxiliary variable that is most suitable for mapping soil moisture, eventually trying to use RF to solve the problems of nonlinearity between soil moisture and auxiliary variables, and mutual correlation between variables, and hoping to use optimal auxiliary variable to guide RF to obtain high-precision of soil moisture.

The paper is organized as follows. Section 2 briefly describes the study area and the key datasets supporting the analysis, including remote sensing images and the Hydro-meteorological Observation Network and Ecohydrological Wireless Sensor Network (EHWSN). Section 3 describes in detail four auxiliary variables closely related to soil moisture and evaluates their performance with in situ soil moisture, the spatial consistency evaluation method between the auxiliary variables and the Polarimetric L-band Multi-beam Radiometer (PLMR)-retrieved soil moisture products are introduced to thoroughly demonstrate the selected optimal auxiliary variable, and Random Forests (RF) is introduced to map soil moisture. Section 4 presents the results and discussions, and is followed by conclusions in Section 5.

2. Study Area and Materials

2.1. Study Area

The middle reach of the Heihe River Basin (HRB) in the arid region of northwest China has been selected as the experimental study area thanks to its rich observation infrastructure maintained by the Heihe Watershed Allied Telemetry Experimental Research (HiWATER) project, which includes 21 Automatic Meteorological Stations (AMSs) and 50 WATERNET nodes located in an observation matrix (5.5 km × 5.5 km) [56]. Additionally, HiWATER provides synchronous space–sky–ground observation datasets that satisfy the experimental requirements.

The extent of the study area ranges from latitude 38.75°N to 39.00°N and from longitude 100.20°E to 100.55°E with elevations of 850–2000 m (Figure 1). The land cover types mainly include farmland, orchard, village, road, irrigation channel, bare land, and desert. The major crops are seed maize, wheat, and vegetables [56]. The climate of the study region is dry with long and cold winters, while the summer is hot and short; occasional sandstorms occur in the period from March to May. The annual mean air temperature is 6 °C, and the mean maximum and minimum temperatures generally occur in July and January, respectively. The annual precipitation is about 100–250 mm, and 70% occurs from June to September. The potential evapotranspiration is as high as 1200–1800 mm per year. Within the research area, the irrigation districts of Daman and Yingke have a complete irrigation infrastructure with densely distributed main canals, branch canals, lateral canals, field ditches, and sublateral canals. The crops are irrigated approximately once every twenty days during the growing season with water from rivers and supplemented by wells. The significant spatiotemporal variations of soil moisture and evapotranspiration in this region result in both the fractured landscape and a rotational irrigation system.

2.2. Materials

All data in this study were provided freely by the HiWATER project (http://westdc.westgis.ac.cn/hiwater, accessed on 1 June 2017), which is a multi-scale integrated observation experiment relying on satellites, aerial sensors, and ground observations started in the HRB in 2012 [56]. The data used in this study include remote sensing images and the Hydro-meteorological Observation Network data and Ecohydrological Wireless Sensor Network (EHWSN) data.

2.2.1. Remote Sensing Images

ASTER reflectivity and its LST products [57,58] were utilized to estimate the auxiliary variables. Nine ASTER L1B images (15 June, 24 June, 10 July, 2 August, 11 August, 18 August, 27 August, 3 September, and 12 September of 2012) were applied after a geometric and radiative calibration. The visible and near-infrared bands (band 2 and band 3) of ASTER were used to calculate the Vegetation Index (VI). Other auxiliary indices, such as TVDI and SEE, were derived based on VI and ASTER LST products.

The soil moisture products (700 m) of PLMR [59] were used as a reference to evaluate the spatial distribution patterns of the auxiliary variables. The PLMR soil moisture products on 30 June, 7 July, 10 July, 26 July, and 2 August of 2012 were retrieved from six L-band microwave channels (three incident angles at 7°, 21.5°, 38.5°, and dual-polarizations) by the Levenberg–Marquardt (LM) optimal algorithm. The product's accuracy is better than

0.04 cm^3/cm^3 compared with WATERNET observations, which is suitable for the study of soil moisture scale conversion and validation of remote sensing products.

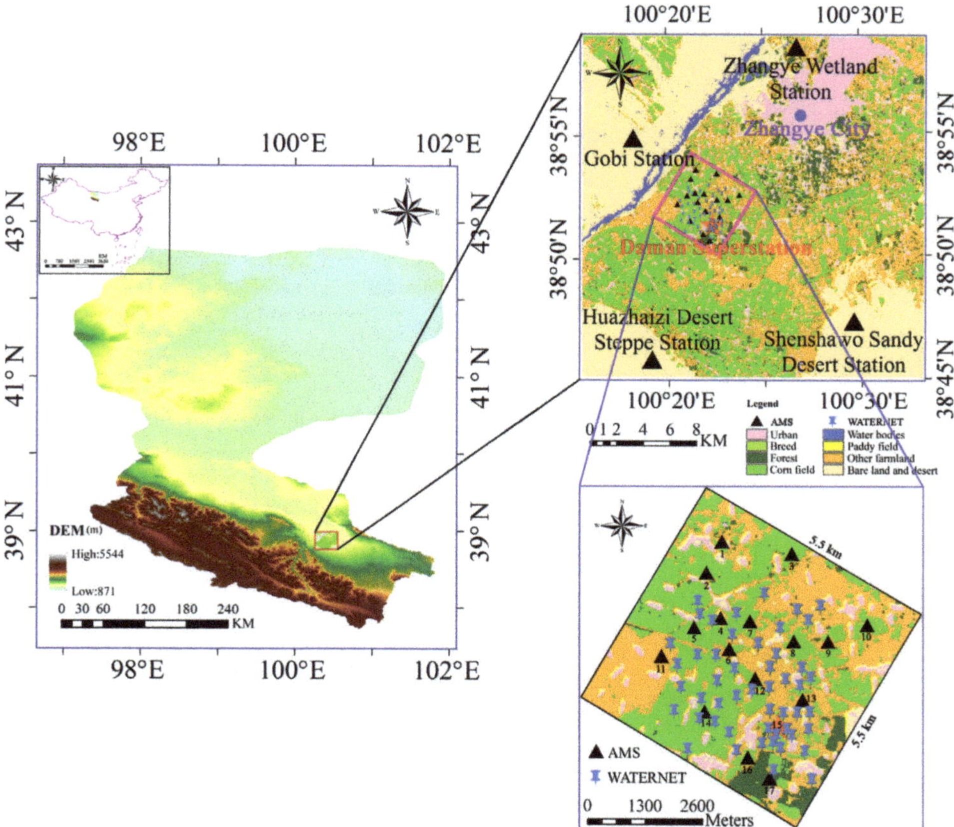

Figure 1. The location of the study area and the distribution of observation instruments (left is the HRB DEM, and right is the land use/land cover of 10 July 2012).

2.2.2. Hydro-Meteorological Observation Network

The hydro-meteorological network in the middle reach of HRB, located in the southwest of the Zhangye City, includes a Daman superstation (AMS_15) and 20 ordinary AMSs (Figure 1). The Daman superstation includes a multiscale observation system for surface fluxes and soil moisture profiles equipped with a 40 m boundary layer tower, one lysimeter, two Eddy Covariance (EC) systems (4.5 m and 17 m), four groups of Large Aperture Scintillometer (LAS), one cosmic-ray probe, and one stable isotopic observation system. The difference is that the ordinary AMSs lack a layer of EC (one EC system) compared with the Daman superstation [60].

2.2.3. Ecohydrological Wireless Sensor Network (EHWSN)

The EHWSN was installed in the intensive observation matrix of 5.5 km × 5.5 km extent (Figure 1), located in the hydro-meteorological network [27]. The EHWSN includes 50 soil moisture/temperature WATERNET nodes. Their distributions were spatially optimized with the stratified non-homogeneity method [61,62] to capture the spatiotemporal variations of soil moisture, soil temperature, and land surface temperature in the heterogeneous surface.

The soil moisture at the depths of 4 cm and 10 cm measured by each WATERNET node and AMS (Daman superstation and 20 ordinary AMSs) were combined to evaluate the applicability of remote sensing auxiliary variables to construct the RF training samples and to verify the RF mapped soil moisture. The 0 cm soil temperature observations were utilized to verify the component temperature of ASTER LST.

3. Methodology

The ASTER images were first introduced to calculate four frequently used auxiliary variables, including NDVI, LST, TVDI, and SEE. The performance metrics between time series of these auxiliary variables and soil moisture observations at 4 cm and 10 cm were evaluated to quantitatively determine their representativeness. The Hausdorff Distance (HD) method was introduced to evaluate the spatial consistency between these auxiliary variables and PLMR-retrieved soil moisture products to validate the feasibility of the auxiliary variables at large scales. Eventually, the machine learning method RF was used to map soil moisture using both single auxiliary variables and multiple auxiliary variables combinations, respectively.

3.1. Auxiliary Spatial Variables

3.1.1. Normalized Difference Vegetation Index (NDVI)

NDVI synthetically reflects the differences in chlorophyll absorption features between visible and near-infrared bands [33]. This auxiliary variable is sensitive to changes in chlorophyll in plant leaves, and is an indirect indicator of soil water content. Due to the influence of vegetation, soil moisture, and atmosphere, NDVI shows large seasonal and regional variations.

3.1.2. Land Surface Temperature (LST)

Due to its key role in surface–atmosphere interactions, LST is one of the essential variables in hydrology, meteorology, climatology and ecology [39]. Generally, there is a negative correlation between LST and soil moisture due to the soil heat capacity changing with soil moisture. In this study, the LST remote sensing products (15 m) were from the dataset named "HiWATER: ASTER LST dataset in 2012 in the middle reaches of the HRB". The products were obtained by the temperature and emissivity separation algorithm, after the atmospheric calibration of ASTER L1B data using the MODIS atmospheric profile product (MOD07) and the radiative transfer model MODTRAN (MODerate spectral resolution atmospheric TRANsmittance algorithm and computer model). These products demonstrated a reasonable accuracy, with an average bias of 0.5 K and average RMSE less than 2 K [57].

3.1.3. Temperature Vegetation Dryness Index (TVDI)

The estimation of TVDI [40,42,63] is based on a two-dimensional feature space constructed by NDVI and LST (Figure 2). The construction of the feature space requires that the study area is large enough to cover canopy coverages ranging from bare soil to dense vegetation, and surface soil moisture varying from dry to saturated. The scatter plots of NDVI and LST generally form a triangle or trapezoid space (Figure 3) in which the location of a pixel represents a certain soil moisture and evapotranspiration. Therefore, TVDI is more suitable for the estimation of soil moisture in arid regions. The Formula (1) to calculate TVDI is:

$$TVDI = \frac{LST_{NDVI} - LST_{NDVI,min}}{LST_{NDVI,max} - LST_{NDVI,min}} \quad (1)$$

where LST_{NDVI} (K) is the land surface temperature at a pixel; $LST_{NDVI,max}$ and $LST_{NDVI,min}$ are the maximum and minimum LST (2) in the study area, respectively, under the same NDVI conditions of LST_{NDVI}. $LST_{NDVI,max}$ under differing vegetation cover is assumed to represent pixels with unavailable soil moisture content, and forms a "dry edge" at the upper boundary of the feature space. $LST_{NDVI,min}$ slowly increases with vegetation cover and describes pixels close to potential evapotranspiration, which form a lower, nearly horizontal, "wet edge" [64–67]. Generally, linear regression is performed to fit the dry and wet edge:

$$\begin{cases} LST_{NDVI,max} = a_1 + b_1 \times NDVI \\ LST_{NDVI,min} = a_2 + b_2 \times NDVI \end{cases} \tag{2}$$

where a_1 and b_1 are the intercept and slope for the wet edge, while a_2 and b_2 are those for the dry edge, respectively. At the dry edge (*TVDI* = 1), soil moisture is close to the wilting point; latent heat flux is assumed to be 0 W/m^2 and the sensible heat flux reaches its maximum. The wet edge is usually simplified as a line parallel to the NDVI axis (Figure 2) when the slope of the wet edge is close to 0, and the soil water content is near to field capacity. For a pixel in Figure 2, the ratio of section A and B is defined as TVDI, where A is the vertical distance between the pixel and the wet edge, whereas B is the vertical distance between the dry and wet edge. If the pixel (NDVI, LST) is closer to the dry edge, the TVDI is higher and the soil moisture is lower. Otherwise, the situation is reversed.

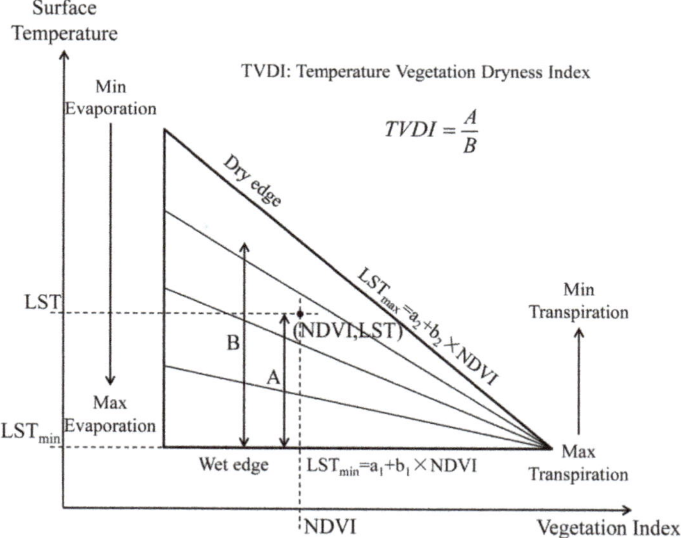

Figure 2. Definition of the TVDI [42].

The algorithm described in this section was applied to all nine ASTER data acquired over our study area. Figure 3 shows the plots of LST against NDVI in a two-dimensional space for the data acquired on 15 June, 24 June, 10 July, 2 August, 11 August, 18 August, 27 August, 3 September, and 12 September of 2012, and the corresponding dry and wet edges determined automatically. Each feature space constructed by the LST changed under the NDVI with a step of 0.005. Furthermore, the dry and wet edges were fitted by maximum and minimum LST under corresponding NDVI. The maximum LST gradually decreased, whereas the minimum LST slowly increased with the NDVI. This figure confirms that the pixels in the study area form a triangle or trapezoid space in the two-dimensional feature space NDVI/LST, and the dry and wet edges are determined on the basis of the triangle or trapezoid space. The negative slope associated with the wet edge might result from the high rate of evapotranspiration from canopies compared with bare soil surfaces under a non-limiting water situation.

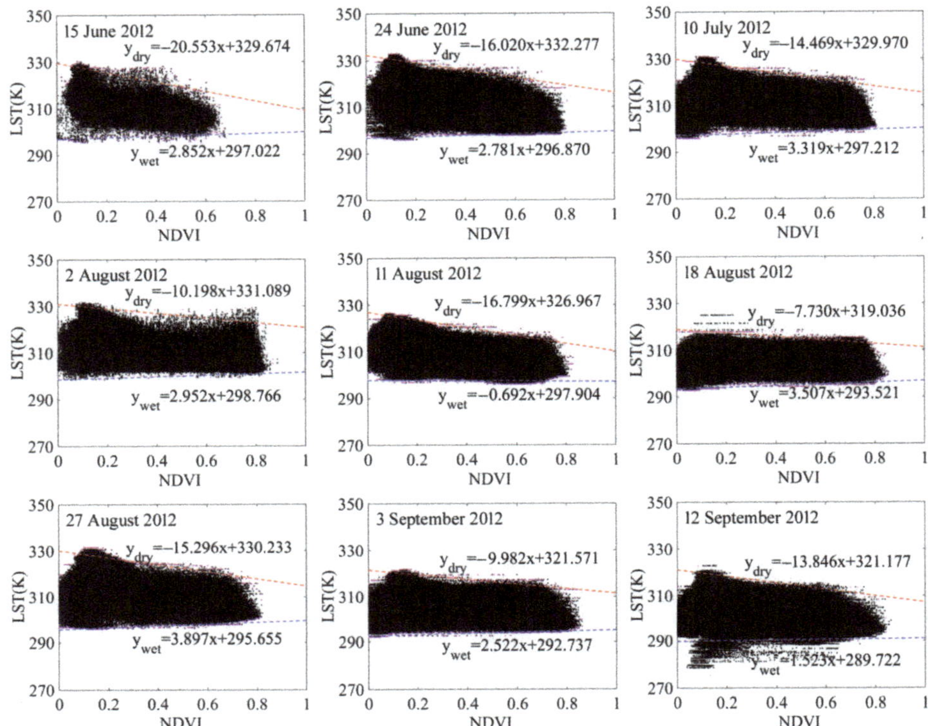

Figure 3. NDVI/LST triangle or trapezoid scatter plots derived by ASTER LST and NDVI at the middle reach of HRB.

3.1.4. Soil Evaporation Efficiency (SEE)

SEE is defined as the ratio of actual to potential soil evaporation. The rationale for choosing SEE as fine-scale information is based on its strong correlation with surface soil moisture [22]. However, there are large difficulties and uncertainties in estimating soil evaporation directly from remote sensing due to the influence of atmospheric conditions, surface soil moisture, vegetation coverage, and soil texture. Therefore, Nishida et al. (2003) put forward a method to indirectly estimate SEE using surface soil temperature and vegetation fraction over heterogeneous surfaces [46]. SEE is a function of surface soil temperature (3):

$$SEE = \frac{T_{s,max} - T_s}{T_{s,max} - T_{s,min}} \tag{3}$$

where $T_{s,max}$ and $T_{s,min}$ are the maximum and minimum surface soil temperature, respectively. T_s is the surface soil temperature obtained by decomposing the ASTER LST based on NDVI [68] as following:

(I) When NDVI < 0.2, the pixel mainly consists of bare soil and the ASTER LST is considered as the surface soil temperature (4):

$$T_s = LST \tag{4}$$

(II) When $0.2 \leq$ NDVI < 0.5, the land surface is regarded as a compound of bare soil and vegetation, where the LST is a combination of vegetation canopy temperature and surface soil temperature. The surface soil temperature T_s (5) was obtained based on the vegetation fraction and vegetation canopy temperature:

$$T_s = \left[\frac{\varepsilon \times LST^4 - \varepsilon_c \times (T_c)^4 \times f_c}{\varepsilon_s \times (1 - f_c)}\right]^{\frac{1}{4}} \quad (5)$$

$$T_c = \varepsilon \times LST \quad (6)$$

$$f_c = 1 - \left(\frac{NDVI - NDVI_{min}}{NDVI_{max} - NDVI_{min}}\right)^K \quad (7)$$

$$\varepsilon = \varepsilon_c \times f_c + \varepsilon_s \times (1 - f_c) \quad (8)$$

where T_c is the vegetation canopy temperature (6), f_c is the vegetation fraction (7), ε is the surface emissivity (8), ε_s and ε_c are the emissivities of the bare soil and full vegetation covered surface, and $NDVI_{min}$ and $NDVI_{max}$ are the minimum and maximum $NDVI$, respectively.

(III) When NDVI $\geq$ 0.5, the pixels are regarded as fully vegetated, the canopy temperature is substituted by LST (9), and the surface soil temperature is obtained with Equation (6).

$$T_c = LST \quad (9)$$

The components temperature decomposed from ASTER LST are verified with the observed 0 cm soil temperature by the Hydro-meteorological Network. The result demonstrates that there is a good consistency (R^2 = 0.91, RMSE $\leq$ 3.10 K) between the decomposed surface soil temperature and observed 0 cm soil temperature (Figure 4), which is superior to the retrieved surface soil temperature with accuracy of 4 K [69,70]. Kustas and Norman (1999) developed a simple two-source model for modeling surface soil temperature with 4 K lower than the radiometric temperature observations [69]. Zhao et al. (2014) utilized the radiative transfer theory to estimate the RMSEs within 4 K between the retrieved and observed surface soil temperature [70]. Therefore, it is considered that the component temperature decomposition method in this paper is reliable and the decomposed 0 cm soil temperature can be used for SEE calculation.

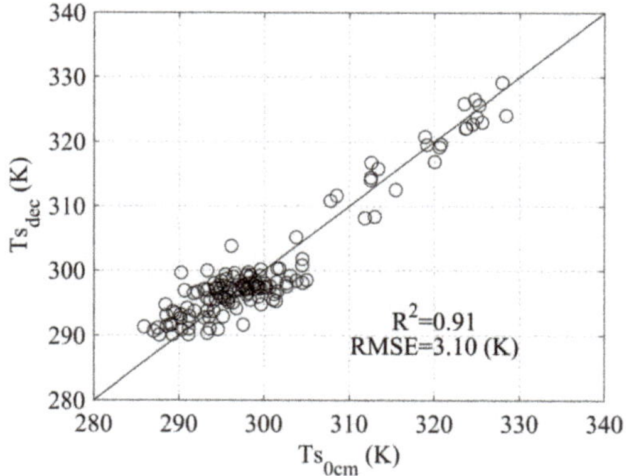

Figure 4. Scatterplot of the decomposed (Ts_{dec}) and observed (Ts_{0cm}) 0 cm soil temperature.

3.2. Performance Metrics

Correlation analysis was employed to compare the sensibilities of the auxiliary variables to soil moisture at depths of 4 cm and 10 cm during the entire vegetation growing season from 15 June to 12 September, 2012. The coefficient of determination (R^2) was used to evaluate the strength of the relationship. According to the overpass time of ASTER, the soil moisture observations at GMT 04:10 were selected as reference to quantitatively analyze the applicability of each auxiliary variable.

3.3. Hausdorff Distance (HD)

The HD method was introduced to evaluate the consistency of the spatial distribution between the auxiliary variables and the PLMR-retrieved soil moisture products. The HD is a type of maximum and minimum distance that compares two finite point sets A and B (10). HD has higher tolerance to perturbations in the locations of the points than the binary correlation technique because it measures their proximity rather than their exact correspondence [71]. Unlike most shape comparison methods that require constructing point-to-point correspondences between two images, HD can be calculated without the explicit pairing of points and only requires calculating the maximum distance between two datasets [72,73]. If the values of HD are floating in a small range of a certain median, there is a good spatial consistency between the A and B.

HD is defined as:

$$H(A,B) = max(h(A,B), h(B,A)) \tag{10}$$

$$h(A,B) = \max_{a \in A} \min_{b \in B} \|a - b\| \tag{11}$$

$$h(B,A) = \max_{b \in B} \min_{a \in A} \|b - a\| \tag{12}$$

where $A = \{a_1, a_2, \cdots, a_m\}$ and $B = \{b_1, b_2, \cdots, b_n\}$ are the point datasets of two images, and a_i and b_i are the pixels value of A and B. $h(A,B)$ is the directed HD from A to B (11), and $h(B,A)$ is that from B to A (12). $\|\cdot\|$ is the Euclidean norm of a vector. In this study, A and B are the auxiliary variables and the PLMR soil moisture product, respectively, both covering the extent between latitudes 38.75°N to 39.00°N and longitudes 100.20°E to 100.55°E in the middle reach of HRB. Two representative images (10 July and 2 August of 2012) of PLMR soil moisture products were selected to compare data accessibility and consistency. The auxiliary variables were normalized to the PLMR soil moisture product based on (13) to make their units and range of values comparable:

$$X = \frac{x - x_{min}}{x_{max} - x_{min}} (PLMR_{max} - PLMR_{min}) \tag{13}$$

where X is the normalized value, x is the auxiliary variable's value, x_{max} and x_{min} are the maximum and minimum values of the auxiliary variable, and the $PLMR_{max}$ and $PLMR_{min}$ are the maximum and minimum values of PLMR product, respectively.

3.4. Random Forests (RF)

The RF [74,75] is a machine learning algorithm based on the bagging integrated learning theory [76] and the random subspace method [77]. RF has high prediction accuracy and tolerance for anomalies and noise in the data, with rare over-fitting in practical application. RF has been widely used in various applications, such as data mining [78–80], bioinformatics research [81,82], and information classification [82–84]. RF's application accuracy can be enhanced further by fusing multi-source information.

Similar to other classifier models, such as deep learning and artificial neural networks, RF is based on a dataset of decision trees that is determined by more than one variable. Assuming that there is a multi-decision tree classification model $\{h(X, \Theta_k), k = 1, 2, 3, \cdots\}$, $\{\Theta_k\}$ is an independent identically distributed random vector and k is the count of decision tree models included. Hence, for any given variables X, each tree casts a unit vote for the most popular class and the final RF classifier is determined.

The RF method was introduced to map soil moisture by the recommended auxiliary variable. Multiple auxiliary variables combinations were also used to map soil moisture by RF and compare them with a single auxiliary variable. Four single auxiliary variables and 11 types ($C_4^2 + C_4^3 + C_4^4$) of auxiliary variables combinations A (LST, NDVI), B (LST, TVDI), C (LST, SEE), D (NDVI, TVDI), E (NDVI, SEE), F (TVDI, SEE), G (LST, NDVI, TVDI),

H (LST, NDVI, SEE), I (LST, TVDI, SEE), J (NDVI, TVDI, SEE), and K (LST, NDVI, TVDI, SEE) were selected for mapping. For all single auxiliary variables and their combinations, a 30 times and 30 folds cross-train were used for the training samples.

The procedure of RF in this study was as follows:

(I) The training dataset p was constituted by the 4 cm-depth soil moisture observed by AMS and WATERNET. The bootstrap sampling method was used to extract the samples from the auxiliary variables with the condition that the volume of the samples dataset was same as that of the training dataset.
(II) Decision trees were constructed by the sample dataset and then p types of corresponding classification results were achieved.
(III) Each tree in the RF casts a unit vote for the most popular class. The final soil moisture products were predicted by the mean of ballot vote according to the p types of classification results.
(IV) Correlation analysis and RMSE between the predicted and observed soil moisture were carried out to quantitatively evaluate the RF mapping accuracy.

4. Results and Discussion

4.1. Correlation between Auxiliary Variables and Soil Moisture Observations

Four auxiliary variables (NDVI, LST, TVDI, and SEE) were correlated with observed soil moisture measured over heterogeneous surfaces in the arid area.

Based on the R^2 values, the strengths of the correlations between the auxiliary variables and soil moisture at the superficial zone and the root zone were in the order of SEE > TVDI > LST > NDVI (Figure 5). The auxiliary variable with a higher R^2 value with soil moisture performed better in the scaling transformation of soil moisture. There was an expected but moderate relationship between NDVI and soil moisture during the vegetation growing season (Figure 5a). In contrast, there was a negative but slightly stronger correlation (R^2 = 0.49) between LST and soil moisture (Figure 5b) than with NDVI due to the rapid rise of LST with water stress. There was also a negative correlation between TVDI and soil moisture (Figure 5c), as this auxiliary variable can better reflect the soil moisture status at a depth of 10 cm (R^2 = 0.68) than that at a depth of 4 cm (R^2 = 0.63). The higher the TVDI, the lower the soil moisture, and thus TVDI can indicate a water deficiency in agricultural production. Compared with these three auxiliary variables, the data points of SEE and soil moisture were more concentrated (Figure 5d). The correlations between SEE and soil moisture were better than for other auxiliary variables. The R^2 between the SEE and soil moisture was 0.67 at 4 cm depth, and was 0.74 at 10 cm depth.

In an arid environment, soil moisture links surface phenology with subsurface water storage, and strongly influences the surface water cycle and energy partitioning due to the strong coupling effect between soil moisture and evapotranspiration. Moreover, soil moisture in the root zone also controls vegetation health and percent coverage [85]. Vegetation health is closely related to transpiration, which is limited by soil moisture in the arid region of HRB. As an important characteristic of vegetation health, NDVI reflects vegetation transpiration by reflectivity, and is mainly used to represent the growth conditions of vegetation in the ecosystem. Therefore, NDVI always indirectly represents soil moisture changes. Nevertheless, because of the heterogeneous surface, the distribution of soil moisture and NDVI varied widely in HRB, leading to a weak correlation (R^2 = 0.41) between NDVI and soil moisture (Figure 5a). As a key factor of vegetation growth, stomatal resistance to transpiration is significantly affected by LST and is partly controlled by soil moisture availability. Suitable LST and soil moisture ensure vegetation health transpiration. LST is always influenced by the soil heat capacity and conductivity. These two thermal properties are functions of soil type and change with soil moisture. As a consequence, soil moisture largely controls LST through the energy balance of the land surface. The higher the soil moisture, the less energy is available for sensible heat flux of the surface [42]. Generally, soil heat capacity decreases with soil moisture increase, and this leads to a decrease in LST. Therefore, LST always indirectly reflected the extent to which vegetation absorbs root zone soil moisture (Figure 5b). However, the correlation between LST and soil moisture (R^2 = 0.49) was also influenced by the impact of land surface heterogeneity, soil types, climate conditions, and vegetation coverage.

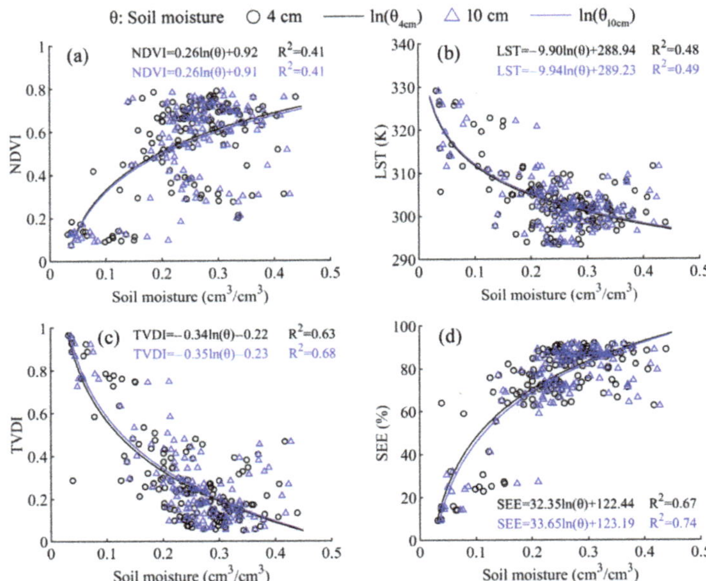

Figure 5. The scatter plots of four auxiliary variables and soil moisture at 4 cm (black) and 10 cm (blue). (**a**–**d**) represent the correlations and fitting relationships between NDVI, LST, TVDI, SEE and soil moisture, respectively.

TVDI, derived based on the two-dimensional feature space formed by LST and NDVI, has the advantages of both NDVI and LST. Therefore, TVDI could reflect soil moisture more accurately ($R^2 \geq 0.63$) than LST (R^2 = 0.41) or NDVI (R^2 = 0.49) (Figure 5c). However, the TVDI method requires a dry limit in the NDVI/LST triangle space [64] and unsaturated soil moisture.

Evaporation is controlled by soil temperature, vapor pressure deficit, soil moisture content, and other meteorological factors. To accurately simulate soil moisture content, it is essential to estimate the evaporation rate from the land surface. In wet conditions, soil evaporation approaches potential evaporation because it occurs at the surface of the soil and the deep moisture rises up through the capillaries and reaches the soil surface. In general, soil moisture content changes below field capacity and a dry layer exists at the soil surface. When soil evaporation occurs, soil moisture diffuses to the surface soil through the soil pores to replenish it. Therefore, soil evaporation is mainly controlled by the soil moisture content. In contrast, the SEE is a direct function of soil evaporation, so it was more sensitive ($R^2 \geq 0.67$) to soil moisture (Figure 5d). In the NDVI/LST triangle space, the surface soil temperature corresponds to soil evaporation whereas the LST corresponds to the land surface evapotranspiration with changes of NDVI. Notably, the LST is closer to the surface soil temperature when the NDVI is smaller. Therefore, the SEE is more sensitive to soil moisture than the TVDI by considering the parameterization of SEE (Equation (3)). Having described all of above, the SEE is presented as a more sensitive agency variable to soil moisture than other three variables (Figure 5).

4.2. Consistency of Spatial Pattern between Auxiliary Variables and PLMR Soil Moisture Products

The consistency in spatial patterns of the auxiliary variables and PLMR soil moisture products was measured with HD. Box and whisker plots (Figure 6) showed that the HDs between four auxiliary variables and PLMR soil moisture products were stable (normally distributed). However, the HD of NDVI and SEE converged more (maximum and minimum intervals) than the LST and TVDI. In contrast, the outliers of SEE were minimal and one-sided by comparison with NDVI. At the same time, the minimum HDs of SEE were less than for the other three auxiliary variables. Besides, the standard deviations (STD) of HD (Table 1) also demonstrated that the HD of SEE to soil moisture were closer (STD $\leq$ 0.058) to the median values than in the NDVI, LST, and TVDI. These features comprehensively indicated that there was a better spatial consistency between the SEE and the PLMR soil moisture than that with other auxiliary variables. The spatial pattern distribution also showed that the variation ranges of SEE were basically consistent with that of PLMR soil moisture (Figure 7). In summary, the SEE is strongly recommended as the most optimal auxiliary variable for representing the distribution of soil moisture based on the HDs (Figure 6), the correlation coefficients (Figure 5), and visual interpretation (Figure 7).

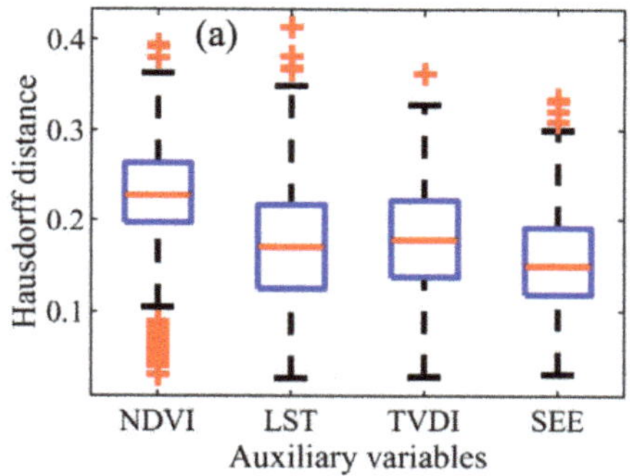

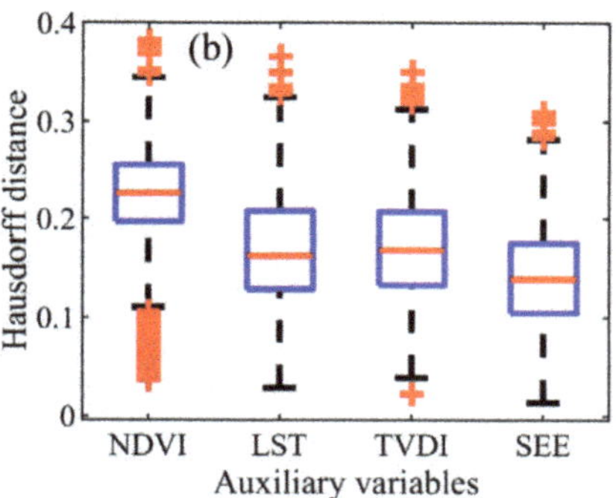

Figure 6. Box and whisker plots of HD between the auxiliary variables and PLMR soil moisture products on the days of 10 July (**a**) and 2 August (**b**) of 2012 (five parameters: median, lower, and upper extremes, and lower and upper quartiles are shown in the figure. The median is plotted as a red line and the box is delimited by the quartiles and the whiskers by the extremes, and the red '+' indicates outliers).

Table 1. The standard deviations (STD) of Hausdorff Distance (HD) on 10 July and 2 August, 2012.

Indices	NDVI	LST	TVDI	SEE
HD STD (20120710)	0.073	0.067	0.066	0.058
HD STD (20120802)	0.073	0.065	0.061	0.056

4.3. Mapping Soil Moisture by Random Forests (RF)

To evaluate the feasibility of using the selected auxiliary variable for soil moisture inversion, we mapped soil moisture by RF with each single auxiliary variable including NDVI, LST, TVDI, and SEE separately (Figure 8). The soil moisture mapping results accuracies were relatively poorer for NDVI (R^2 = 0.58, RMSE = 0.058 cm^3/cm^3), LST (R^2 = 0.72, RMSE = 0.048 cm^3/cm^3), and TVDI (R^2 = 0.77, RMSE = 0.043 cm^3/cm^3). The performance of SEE was optimal (R^2 = 0.86, RMSE = 0.035 cm^3/cm^3) compared with observations. This result indicates that the recommended auxiliary variable SEE, which is controlled by soil surface evaporation and is limited by the available soil moisture, can be used to accurately reflect soil moisture.

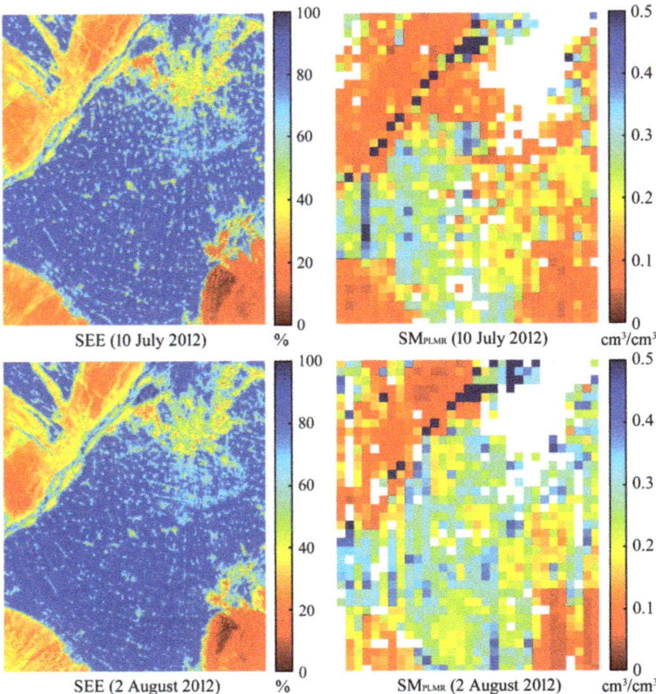

Figure 7. Comparison of the SEE and PLMR soil moisture products (SM_{PLMR}) on 10 July and 2 August 2012.

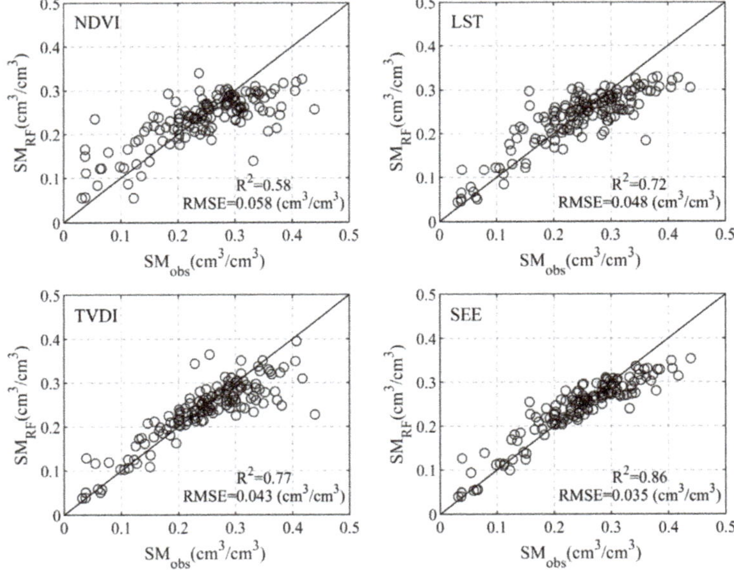

Figure 8. Scatter plots of the observed soil moisture and RF mapped soil moisture with single auxiliary variables.

Eleven types of multiple auxiliary variables combinations were also used to map soil moisture. The Taylor diagram (Figure 9) shows the R, RMSE, and STD of the RF mapping results compared with observed soil moisture at 4 cm by Daman superstation, ordinary AMSs, and WATERNETs. The results showed that multiple auxiliary variable combinations all performed well for mapping soil moisture, with 0.90 < R < 0.93, 0.033 cm^3/cm^3 < RMSE < 0.038 cm^3/cm^3, and 0.068 cm^3/cm^3 < STD < 0.074 cm^3/cm^3 between mapped and observed soil moisture (Table 2). This performance can be explained by the fact that all auxiliary variables combinations contain both NDVI, which reflects the variability of the land cover, and LST, which reflects the complicated soil properties. However, combinations C, H, and K, which contain the auxiliary variable SEE, were better (R^2 = 0.86, RMSE = 0.034 cm^3/cm^3) than other combinations. The main reason is that the SEE is closely related to soil evaporation and better reflects soil moisture. Therefore, considering the limitations of accuracy, sample calculation and complexity of the feature space construction of TVDI, combination H was found to be optimal to map soil moisture by RF. Additionally, comparing the mapping of soil moisture with four single auxiliary variables and their combinations demonstrated that SEE better reflects soil moisture than most multi-variable combinations (e.g., combination A). However, the combination H (R^2 = 0.86, RMSE = 0.034 cm^3/cm^3) performed similarly to SEE (R^2 = 0.86, RMSE = 0.035 cm^3/cm^3). Whether the SEE or the combination H was used for soil moisture mapping, the accuracies were less than the SMAP and SMOS soil moisture products' accuracy of 0.04 cm^3/cm^3. Therefore, we can conclude that SEE is an optimal auxiliary variable for soil moisture mapping and the multiple auxiliary variables combination H (LST, NDVI, and SEE) is expected to enhance mapping accuracy.

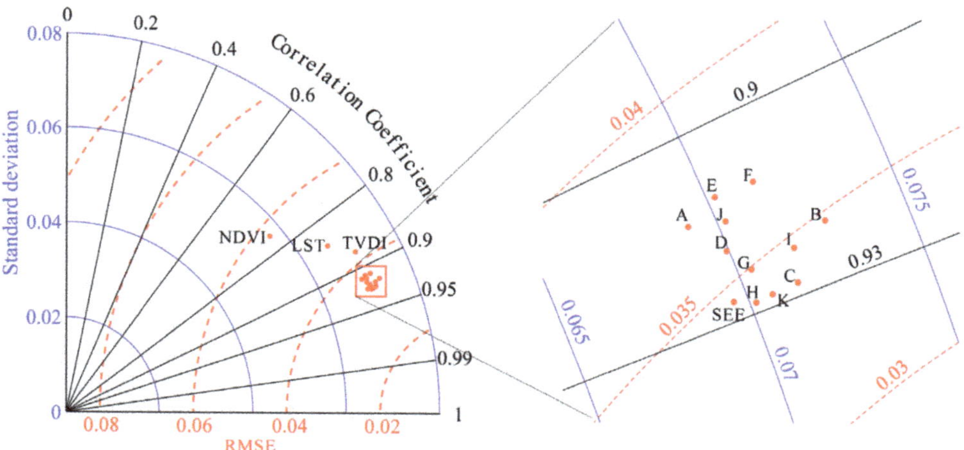

Figure 9. Taylor diagram of the comparison of the RF mapped and observed soil moisture. A (LST, NDVI), B (LST, TVDI), C (LST, SEE), D (NDVI, TVDI), E (NDVI, SEE), F (TVDI, SEE), G (LST, NDVI, TVDI), H (LST, NDVI, SEE), I (LST, TVDI, SEE), J (NDVI, TVDI, SEE), and K (LST, NDVI, TVDI, SEE).

Table 2. The coefficient of determination (R^2), root mean square errors (RMSE), and standard deviations (STD) between RF results by each multiple auxiliary variables combination and observed soil moisture.

Group	A	B	C	D	E	F	G	H	I	J	K
R^2	0.84	0.85	0.86	0.85	0.83	0.83	0.85	0.86	0.85	0.84	0.86
RMSE	0.037	0.035	0.034	0.036	0.037	0.037	0.035	0.034	0.035	0.036	0.034
STD	0.069	0.073	0.071	0.070	0.070	0.071	0.070	0.070	0.072	0.070	0.071

The performance of using the combination H to map soil moisture in the middle reach of HRB was evaluated. The scatter plots of in situ observed (except for 11 August because of lack of observations) and RF-mapped soil moisture (Figure 10) showed that the correlations between the two datasets were $0.80 \leq R^2 \leq 0.91$, $0.021\ cm^3/cm^3 \leq RMSE \leq 0.046\ cm^3/cm^3$. The mean RMSE ($0.034\ cm^3/cm^3$) was smaller than $0.04\ cm^3/cm^3$, which is the SMAP/SMOS accuracy and satisfies the research of drought monitoring and water resource management.

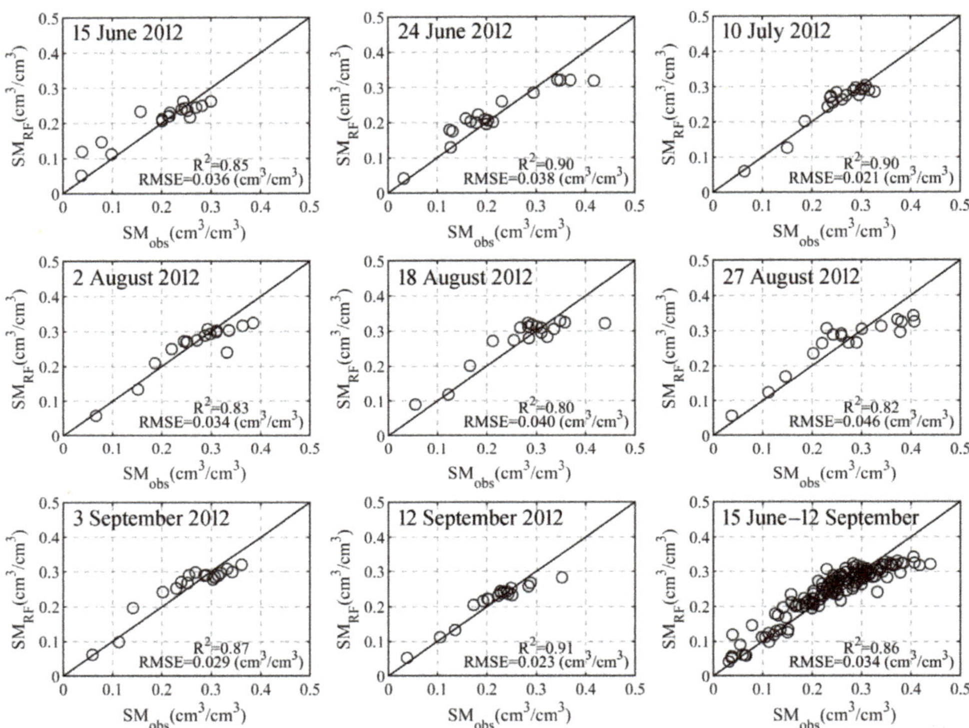

Figure 10. Scatter plots of the observed soil moisture and RF mapped soil moisture with the combination H (LST, NDVI, and SEE).

The RF-mapped soil moisture agrees well with the observations at 4 cm of Daman Superstation (Figure 11). The statistical analysis showed that soil moisture mapped by RF using the combinations C, H, and K compared to the observations with an RMSE of $0.034\ cm^3/cm^3$. The soil moisture mapped by SEE alone had the lowest RMSE of $0.035\ cm^3/cm^3$. Precipitation observations at Daman Superstation recorded significant rainfall on 5 June, 17 June, 27 June, 6 July, 16 July, 20 July, 29 July, 6 August, 12 August, 17 August, 31 August, and 23 September. Irrigation events took place on 7 June, 3 July, 28 July, and 25 August. Considering both the region's size and the uniform crop distribution, we concluded that precipitation or irrigation occurred in the whole study area. RF-mapped soil moisture based on combination H with a resolution of 15 m (Figure 12) was consistent with both precipitation and irrigation events in the study area. For example, soil moisture mapping results indicated that there was higher soil moisture on 2 August due to the residual moisture from the irrigation event of 28 July. Soil moisture on 18 August and 3 September was relatively higher due to the strong rainfall that occurred one or three days before. Whereas the values were comparatively lower on 10 July because land surface evapotranspiration removed most of the water from the irrigation and precipitation that occurred on 3 July and 6 July, respectively. In addition, on 15 June and 12 September, the soil moisture values were lowest due to the lack of precipitation or irrigation. However, due to the uneven distribution of the sampling points in the study area, some biases existed in the RF mapping results. For example, the RF samples lacked points in both the river and desert zones, and thus the RF mapping results do not reflect soil moisture changes in these zones. Therefore, an even distribution of representative in situ observations and multi-source information, such as other high-precision remote sensing data, will be needed for future RF samples.

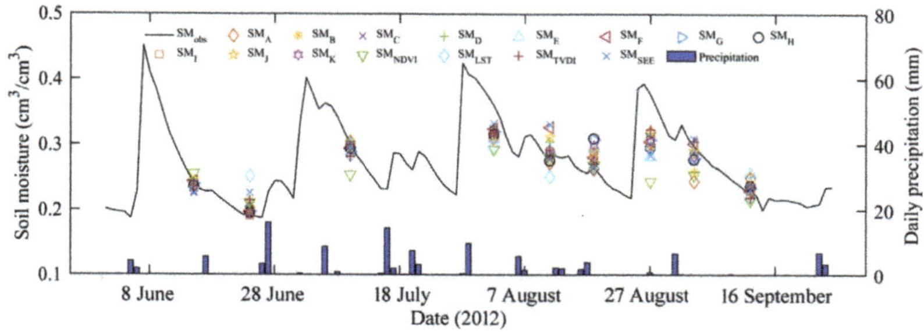

Figure 11. Comparison of the RF mapped with the observed soil moisture at 4 cm of Daman Superstation (SM_{obs}: observed soil moisture, SM_A to SM_K: RF mapped soil moisture with 11 types of multiple auxiliary variables combinations, SM_{NDVI}, SM_{LST}, SM_{TVDI}, and SM_{SEE}: RF-mapped soil moisture with single auxiliary variables including NDVI, LST, TVDI, and SEE separately).

4.4. Discussion

A variety of auxiliary variables have been used to scale soil moisture during the past decades. However, a systematic evaluation of their applicability is still missing. We analyzed four auxiliary variables quantified from ASTER images, including NDVI, LST, TVDI, and SEE from the trend consistency and spatial pattern. The auxiliary variables that reflected soil moisture at the depths of either 4 cm or 10 cm were, in order of performance, SEE > TVDI > LST > NDVI. Using only NDVI or LST did not reflect overall soil moisture because of the strong heterogeneity of the land surface, saturated soil hydraulic conductivity, soil texture, and climate conditions. NDVI saturates at higher vegetation density, not reflecting changes in soil moisture. NDVI often indicates the vegetation amount and chlorophyll content rather than water status and it is a rather conservative indicator of water stress [10]. LST often closely follows vegetation transpiration and can rise rapidly with water stress. TVDI is assumed to reflect soil moisture as it incorporates the advantages of both NDVI and LST. However, the construction of the NDVI/LST feature space requires that the study area is large enough to capture canopy coverages ranging from bare soil to dense vegetation and surface soil moisture varying from dry to saturated. Based on the performance metrics between these auxiliary variables and observed soil moisture and by virtue of the HD maximum and minimum distances between the auxiliary variables and PLMR soil moisture, the SEE, which is controlled by soil surface evaporation and contained more information than the other auxiliary variables, was found to be an optimal auxiliary variable for scaling and mapping of soil moisture.

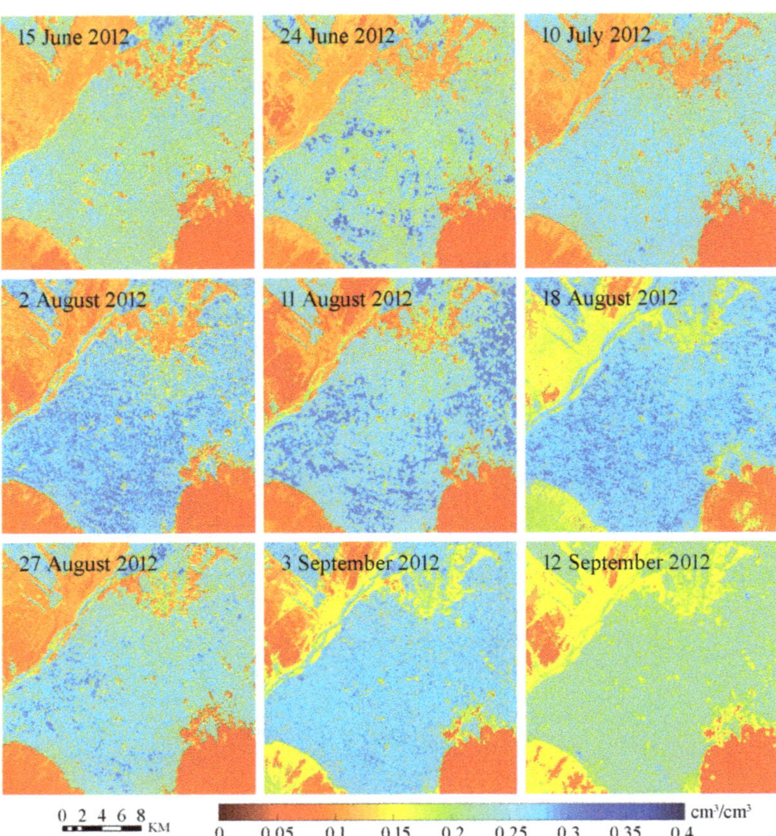

Figure 12. The results of soil moisture mapping by RF based on combination H (LST, NDVI, and SEE).

The RF machine learning method has high prediction accuracy and tolerance for anomalies and noise in the data, with rare over-fitting in practical application. RF has the advantage of learning the relationships between soil moisture and auxiliary variables, especially nonlinear, indirect relationships, based on a large number of samples. Physics-based relationships between the variables and soil moisture and machine learning gradually showed a trend of integration. Physics-based relationships constraining machine learning can improve the training efficiency. Physics-based relationships may guide machine learning methods to obtain high-precision of soil moisture. We introduced the RF both to evaluate the feasibility of the selected optimal auxiliary variable and to map soil moisture. Soil moisture mapped by each single auxiliary variable indicated that an optimal accuracy (R^2 = 0.86, RMSE = 0.035 cm^3/cm^3) existed in the SEE-based mapping soil moisture and in situ observations due to the SEE's close relation to soil evaporation. Soil moisture mapped by multiple auxiliary variables combinations indicated that each combination can reflect soil moisture, with $0.90 < R < 0.93$, 0.033 cm^3/cm^3 < RMSE < 0.038 cm^3/cm^3, and 0.068 cm^3/cm^3 < STD < 0.074 cm^3/cm^3 because multiple auxiliary variables combinations contain more information about soil moisture than a single auxiliary variable. However, combination H (LST, NDVI, SEE) was relatively better (R^2 = 0.86, RMSE = 0.034 cm^3/cm^3) than the others (R^2 < 0.85, RMSE > 0.035 cm^3/cm^3) at reflecting soil moisture with an accuracy of 0.034 cm^3/cm^3, which is less than the SMAP/SMOS required soil moisture accuracy

(0.04 cm^3/cm^3). Soil moisture (15 m) mapped by combination H in the middle reach of the HRB was consistent with the spatiotemporal changes of irrigation or precipitation.

The optimal auxiliary variable (SEE) and the multiple auxiliary variables combination H (LST, NDVI, and SEE) screened in this study are based on the middle reach of HRB, which has complex surface heterogeneity, variable climate, and significant water and heat exchange. The study spans the entire vegetation growing season proved the feasibility of the method by using the optimal auxiliary variable and combination of auxiliary variables selected for soil moisture mapping. Therefore, for other surfaces, the selected optimal auxiliary variable and combination of auxiliary variables are also applicable under the ground data constraints.

5. Conclusions

To support scaling and mapping of soil moisture for producing mid- and high-resolution soil moisture estimates and validating satellite remote sensing products, the applicability of four auxiliary variables (NDVI, LST, TVDI, and SEE) was quantitatively evaluated in the arid region of China.

The auxiliary variables that reflected soil moisture at the depths of either 4 cm or 10 cm were, in order of performance, SEE > TVDI > LST > NDVI. Based on the performance metrics between these auxiliary variables and observed soil moisture and by virtue of the HD maximum and minimum distances between the auxiliary variables and PLMR soil moisture, the SEE was found to be an optimal auxiliary variable for scaling and mapping of soil moisture.

The RF machine learning method was introduced both to evaluate the feasibility of the selected optimal auxiliary variable and to map soil moisture. Soil moisture mapped by SEE and auxiliary variables combination H (LST, NDVI, and SEE) are respectively more optimal than others at reflecting soil moisture with accuracies of 0.035 cm^3/cm^3 and 0.034 cm^3/cm^3, which are less than the SMAP/SMOS required soil moisture accuracy (0.04 cm^3/cm^3). In addition, soil moisture (15 m) mapped by combination H in the middle reach of the HRB was consistent with the spatiotemporal changes of irrigation or precipitation.

SEE is recommended as an optimal auxiliary variable for scaling and mapping of soil moisture. The multiple auxiliary variables combination H (LST, NDVI, and SEE) is recommended for enhancing the scaling and mapping accuracy of soil moisture. Future studies should evaluate the use of SEE for scaling soil moisture. However, two factors may lead to inaccurate SEE calculations. First, from the definition of SEE, directly estimating evaporation using remote sensing observations presents substantial difficulties and uncertainties. Second, the accuracy of surface soil temperature is influenced by vegetation cover. These problems will need to be solved in future research. In addition, multi-source data, such as high-precision remote sensing data as well as evenly distributed and representative in situ observations, should be fused into the RF sample construction to obtain an accurate, high-resolution soil moisture product.

Author Contributions: Conceptualization: Z.Z. and R.J.; methodology: Z.Z. and J.K.; software: Z.Z.; validation: Z.Z., R.J., J.K. and C.M.; formal analysis: J.K. and C.M.; investigation: R.J. and W.W.; resources: R.J. and W.W.; data curation: R.J.; writing—original draft preparation: Z.Z.; writing—review and editing: R.J., J.K., C.M. and W.W.; visualization: Z.Z.; supervision: R.J. and W.W.; project administration: R.J.; funding acquisition: R.J., W.W. and J.K. All authors have read and agreed to the published version of the manuscript.

Funding: This research was funded by the Strategic Priority Research Program of the Chinese Academy of Sciences (grant no. XDA19070104), the National Science and Technology Major Project of China's High Resolution Earth Observation System (grant no. 21-Y20B01-9001-19/22), and the National Natural Science Foundation of China (grant no. 42071347).

Data Availability Statement: The data presented in this study are available upon request from the corresponding author.

Acknowledgments: The authors would like to thank all of the scientists, engineers, and students who participated in the HiWATER experiment, and thank the west data center (http://westdc.westgis.ac.cn/hiwater, accessed on 1 June 2017) for providing the data, and thank anonymous reviewers for supporting this work.

Conflicts of Interest: The authors declare no conflict of interest.

References

1. Wigneron, J.P. Soil moisture retrieval algorithms in the framework of the SMOS mission: Current status and requirements for the EuroSTARRS campaign. *Clin. Psychol. Psychother.* **2003**, *525*, 199–202.
2. Wigneron, J.P.; Calvet, J.C.; Pellarin, T.; Griend, A.A.V.D.; Berger, M.; Ferrazzoli, P. Retrieving near-surface soil moisture from microwave radiometric observations: Current status and future plans. *Remote Sens. Environ.* **2003**, *85*, 489–506. [CrossRef]
3. Ait Hssaine, B.; Merlin, O.; Ezzahar, J.; Ojha, N.; Er-Raki, S.; Khabba, S. An evapotranspiration model self-calibrated from remotely sensed surface soil moisture, land surface temperature and vegetation cover fraction: Application to disaggregated SMOS and MODIS data. *Hydrol. Earth Syst. Sci.* **2020**, *24*, 1781–1803. [CrossRef]
4. Zhang, J.; Zhou, Z.; Yao, F.; Yang, L.; Hao, C. Validating the Modified Perpendicular Drought Index in the North China Region Using In Situ Soil Moisture Measurement. *IEEE Geosci. Remote Sens. Lett.* **2014**, *12*, 542–546. [CrossRef]
5. Alemayehu, T.; Griensven, A.V.; Senay, G.B.; Bauwens, W. Evapotranspiration Mapping in a Heterogeneous Landscape Using Remote Sensing and Global Weather Datasets: Application to the Mara Basin, East Africa. *Remote Sens.* **2017**, *9*, 390. [CrossRef]
6. Jackson, R.D.; Idso, S.B.; Reginato, R.J.; Pinter, P.J. Canopy temperature as a crop water stress indicator. *Water Resour. Res.* **1981**, *17*, 1133–1138. [CrossRef]
7. Lennard, A.T.; Macdonald, N.; Clark, S.; Hooke, J.M. The application of a drought reconstruction in water resource management. *Hydrol. Res.* **2016**, *47*, 646–659. [CrossRef]
8. Smith, K.A.; Barker, L.J.; Tanguy, M.; Parry, S.; Harrigan, S.; Legg, T.P.; Prudhomme, C.; Hannaford, J. A multi-objective ensemble approach to hydrological modelling in the UK: An application to historic drought reconstruction. *Hydrol. Earth Syst. Sci.* **2019**, *23*, 3247–3268. [CrossRef]
9. Krishnan, P.; Black, T.A.; Grant, N.J.; Barr, A.G.; Hogg, E.T.H.; Jassal, R.S.; Morgenstern, K. Impact of changing soil moisture distribution on net ecosystem productivity of a boreal aspen forest during and following drought. *Agric. For. Meteorol.* **2006**, *139*, 208–223. [CrossRef]
10. Rahimzadeh-Bajgiran, P.; Omasa, K.; Shimizu, Y. Comparative evaluation of the Vegetation Dryness Index (VDI), the Temperature Vegetation Dryness Index (TVDI) and the improved TVDI (iTVDI) for water stress detection in semi-arid regions of Iran. *ISPRS J. Photogramm. Remote Sens.* **2012**, *68*, 1–12. [CrossRef]
11. Kang, J.; Jin, R.; Li, X.; Zhang, Y. Mapping High Spatiotemporal-Resolution Soil Moisture by Upscaling Sparse Ground-Based Observations Using a Bayesian Linear Regression Method for Comparison with Microwave Remotely Sensed Soil Moisture Products. *Remote Sens.* **2021**, *13*, 228. [CrossRef]
12. Liu, Y.Y.; Dorigo, W.A.; Parinussa, R.M.; Jeu, R.A.M.D.; Wagner, W.; McCabe, M.F.; Evans, J.P.; Dijk, A.I.J.M.V. Trend-preserving blending of passive and active microwave soil moisture retrievals. *Remote Sens. Environ.* **2012**, *123*, 280–297. [CrossRef]
13. Njoku, E.G.; Jackson, T.J.; Lakshmi, V.; Chan, T.K.; Nghiem, S.V. Soil moisture retrieval from AMSR-E. *IEEE Trans. Geosci. Remote Sens.* **2003**, *41*, 215–229. [CrossRef]
14. Owe, M.; Jeu, R.D.; Holmes, T. Multisensor historical climatology of satellite-derived global land surface moisture. *J. Geophys. Res. Earth Surf.* **2008**, *113*, F01002. [CrossRef]
15. Koike, T.; Nakamura, Y.; Kaihotsu, I.; Davaa, G.; Matsuura, N.; Tamagawa, K.; Fujii, H. Development of an Advanced Microwave Scanning Radiometer (Amsr-E) Algorithm for Soil Moisture and Vegetation Water Content. *Doboku Gakkai Ronbunshuu B* **2004**, *48*, 217–222. [CrossRef]
16. Kerr, Y.H.; Waldteufel, P.; Richaume, P.; Wigneron, J.P.; Ferrazzoli, P.; Mahmoodi, A.; Bitar, A.A.; Cabot, F.; Gruhier, C.; Juglea, S.E. The SMOS Soil Moisture Retrieval Algorithm. *IEEE Trans. Geosci. Remote Sens.* **2012**, *50*, 1384–1403. [CrossRef]
17. Entekhabi, D.; Njoku, E.G.; O'Neill, P.E.; Kellogg, K.H.; Crow, W.T.; Edelstein, W.N.; Entin, J.K.; Goodman, S.D.; Jackson, T.J.; Johnson, J. The Soil Moisture Active Passive (SMAP) Mission. *Proc. IEEE* **2010**, *98*, 704–716. [CrossRef]
18. Chen, Y.; Yang, K.; Qin, J.; Cui, Q.; Lu, H.; Zhu, L.; Han, M.; Tang, W. Evaluation of SMAP, SMOS, and AMSR2 soil moisture retrievals against observations from two networks on the Tibetan Plateau. *J. Geophys. Res. Atmos.* **2017**, *122*, 5780–5792. [CrossRef]
19. Piles, M.; Camps, A.; Vall-Llossera, M.; Corbella, I.; Panciera, R.; Rudiger, C.; Kerr, Y.H.; Walker, J. Downscaling SMOS-Derived Soil Moisture Using MODIS Visible/Infrared Data. *IEEE Trans. Geosci. Remote Sens.* **2011**, *49*, 3156–3166. [CrossRef]
20. Pan, M.; Wood, E.F. Impact of Accuracy, Spatial Availability, and Revisit Time of Satellite-Derived Surface Soil Moisture in a Multiscale Ensemble Data Assimilation System. *IEEE J. Sel. Top. Appl. Earth Obs. Remote Sens.* **2010**, *3*, 49–56. [CrossRef]
21. Waite, W.P.; Sadeghi, A.M.; Scott, H.D. Microwave bistatic reflectivity dependence on the moisture content and matric potential of bare soil. *IEEE Trans. Geosci. Remote Sens.* **1984**, *GE-22*, 394–405. [CrossRef]
22. Merlin, O.; Walker, J.P.; Chehbouni, A.; Kerr, Y. Towards deterministic downscaling of SMOS soil moisture using MODIS derived soil evaporative efficiency. *Remote Sens. Environ.* **2009**, *112*, 3935–3946. [CrossRef]

23. Merlin, O.; Malbéteau, Y.; Notfi, Y.; Bacon, S.; Khabba, S.; Jarlan, L. Performance Metrics for Soil Moisture Downscaling Methods: Application to DISPATCH Data in Central Morocco. *Remote Sens.* **2015**, *7*, 3783–3807. [CrossRef]
24. Alemohammad, S.H.; Kolassa, J.; Prigent, C.; Aires, F.; Gentine, P. Global Downscaling of Remotely-Sensed Soil Moisture using Neural Networks. *Hydrol. Earth Syst. Sci. Discuss.* **2018**, *22*, 5341–5356. [CrossRef]
25. Jin, Y.; Ge, Y.; Wang, J.; Heuvelink, G.B.M. Deriving temporally continuous soil moisture estimations at fine resolution by downscaling remotely sensed product. *Int. J. Appl. Earth Obs. Geoinf.* **2018**, *68*, 8–19. [CrossRef]
26. Chauhan, N.S.; Miller, S.; Ardanuy, P. Spaceborne soil moisture estimation at high resolution: A microwave-optical/IR synergistic approach. *Int. J. Remote Sens.* **2003**, *24*, 4599–4622. [CrossRef]
27. Jin, R.; Li, X.; Yan, B.; Li, X.; Luo, W.; Ma, M.; Guo, J.; Kang, J.; Zhu, Z.; Zhao, S. A Nested Ecohydrological Wireless Sensor Network for Capturing the Surface Heterogeneity in the Midstream Areas of the Heihe River Basin, China. *IEEE Geosci. Remote Sens. Lett.* **2014**, *11*, 2015–2019. [CrossRef]
28. Zreda, M.; Shuttleworth, W.J.; Zeng, X.; Zweck, C.; Franz, T.; Rosolem, R. COsmic-ray Soil Moisture Observing System (COSMOS): Soil moisture and beyond. In Proceedings of the EGU General Assembly Conference, Vienna, Austria, 7–12 April 2013; pp. 4079–4099.
29. Qin, J.; Yang, K.; Lu, N.; Chen, Y.; Zhao, L.; Han, M. Spatial upscaling of in-situ soil moisture measurements based on MODIS-derived apparent thermal inertia. *Remote Sens. Environ.* **2013**, *138*, 1–9. [CrossRef]
30. Kang, J.; Jin, R.; Li, X.; Zhang, Y.; Zhu, Z. Spatial Upscaling of Sparse Soil Moisture Observations Based on Ridge Regression. *Remote Sens.* **2018**, *10*, 192. [CrossRef]
31. Kang, J.; Jin, R.; Li, X.; Ma, C.; Qin, J.; Zhang, Y. High spatio-temporal resolution mapping of soil moisture by integrating wireless sensor network observations and MODIS apparent thermal inertia in the Babao River Basin, China. *Remote Sens. Environ.* **2017**, *191*, 232–245. [CrossRef]
32. Yao, X.; Bojie, F.; Yihe, L.; Feixiang, S.; Shuai, W.; Min, L. Comparison of Four Spatial Interpolation Methods for Estimating Soil Moisture in a Complex Terrain Catchment. *PLoS ONE* **2013**, *8*, e54660. [CrossRef]
33. Rouse, J.W. Monitoring vegetation systems in the great plains with ERTS. *NASA Spec. Publ.* **1974**, *351*, 309.
34. Kogan, F.N. Remote sensing of weather impacts on vegetation in non-homogeneous areas. *Int. J. Remote Sens.* **1990**, *11*, 1405–1419. [CrossRef]
35. Kogan, F.N. Application of vegetation index and brightness temperature for drought detection. *Adv. Space Res.* **1995**, *15*, 91–100. [CrossRef]
36. Chen, W.; Xiao, Q.; Sheng, Y. Application of the anomaly vegetation index to monitoring heavy drought in 1992. *Remote Sens. Environ.* **1994**, *9*, 106–112.
37. Lambin, E.F.; Ehrlich, D. The surface temperature-vegetation index space for land cover and land-cover change analysis. *Int. J. Remote Sens.* **1996**, *17*, 463–487. [CrossRef]
38. Zhao, W.; Li, Z.L. Sensitivity study of soil moisture on the temporal evolution of surface temperature over bare surfaces. *Int. J. Remote Sens.* **2013**, *34*, 3314–3331. [CrossRef]
39. Bartholic, J.F.; Namkem, L.N.; Wiegand, C.L. Aerial Thermal Scanner to Determine Temperatures of Soils and of Crop Canopies Differing in Water Stress. *Agron. J.* **1972**, *64*, 603–608. [CrossRef]
40. Carlson, T.N.; Gillies, R.R.; Perry, E.M. A Method to Make Use of Thermal Infrared Temperature and NDVI Measurements to Infer Surface Soil Water Content and Fractional Vegetation Cover. *Remote Sens. Rev.* **1994**, *9*, 161–173. [CrossRef]
41. Price, J.C. On the analysis of thermal infrared imagery: The limited utility of apparent thermal inertia. *Remote Sens. Environ.* **1985**, *18*, 59–73. [CrossRef]
42. Sandholt, I.; Rasmussen, K.; Andersen, J. A simple interpretation of the surface temperature/vegetation index space for assessment of surface moisture status. *Remote Sens. Environ.* **2002**, *79*, 213–224. [CrossRef]
43. Wang, P.-X.; Li, X.-W.; Gong, J.-Y.; Song, C. Vegetation-Temperature Condition Index and Its Application for Drought Monitoring. In Proceedings of the IEEE International Geoscience and Remote Sensing Symposium (IGARSS), Sydney, Australia, 9–13 July 2001; pp. 141–143.
44. Kang, J.; Jin, R.; Li, X. Regression Kriging-Based Upscaling of Soil Moisture Measurements From a Wireless Sensor Network and Multiresource Remote Sensing Information Over Heterogeneous Cropland. *IEEE Geosci. Remote Sens. Lett.* **2015**, *12*, 92–96. [CrossRef]
45. Peng, J.; Loew, A.; Zhang, S.; Wang, J.; Niesel, J. Spatial Downscaling of Satellite Soil Moisture Data Using a Vegetation Temperature Condition Index. *IEEE Trans. Geosci. Remote Sens.* **2015**, *54*, 558–566. [CrossRef]
46. Nishida, K.; Nemani, R.R.; Glassy, J.M.; Running, S.W. Development of an evapotranspiration index from Aqua/MODIS for monitoring surface moisture status. *IEEE Trans. Geosci. Remote Sens.* **2003**, *41*, 493–501. [CrossRef]
47. Bindlish, R.; Barros, A.P. Subpixel variability of remotely sensed soil moisture: An inter-comparison study of SAR and ESTAR. *IEEE Trans. Geosci. Remote Sens.* **2002**, *40*, 326–337. [CrossRef]
48. Zhan, Z.M.; Qin, Q.M.; Ghulan, A.; Wang, D.D. NIR-red spectral space based new method for soil moisture monitoring. *Sci. China* **2007**, *50*, 283–289. [CrossRef]
49. Choi, M.; Hur, Y. A microwave-optical/infrared disaggregation for improving spatial representation of soil moisture using AMSR-E and MODIS products. *Remote Sens. Environ.* **2012**, *124*, 259–269. [CrossRef]

50. Kim, J.; Hogue, T.S. Improving Spatial Soil Moisture Representation through Integration of AMSR-E and MODIS Products. *IEEE Trans. Geosci. Remote Sens.* **2012**, *50*, 446–460. [CrossRef]
51. Merlin, O.; Albitar, A.; Walker, J.P.; Kerr, Y. An improved algorithm for disaggregating microwave-derived soil moisture based on red, near-infrared and thermal-infrared data. *Remote Sens. Environ.* **2010**, *114*, 2305–2316. [CrossRef]
52. Sánchez-Ruiz, S.; Piles, M.; Sánchez, N.; Martínez-Fernández, J.; Vall-Llossera, M.; Camps, A. Combining SMOS with visible and near/shortwave/thermal infrared satellite data for high resolution soil moisture estimates. *J. Hydrol.* **2014**, *516*, 273–283. [CrossRef]
53. Njoku, E.G.; Wilson, W.J.; Yueh, S.H.; Dinardo, S.J. Observations of soil moisture using a passive and active low-frequency microwave airborne sensor during SGP99. *Geosci. Remote Sens. IEEE Trans.* **2002**, *40*, 2659–2673. [CrossRef]
54. Das, N.N.; Entekhabi, D.; Njoku, E.G. An Algorithm for Merging SMAP Radiometer and Radar Data for High-Resolution Soil-Moisture Retrieval. *IEEE Trans. Geosci. Remote Sens.* **2011**, *49*, 1504–1512. [CrossRef]
55. Knipper, K.R.; Hogue, T.S.; Franz, K.J.; Scott, R.L. Downscaling SMAP and SMOS soil moisture with moderate-resolution imaging spectroradiometer visible and infrared products over southern Arizona. *J. Appl. Remote Sens.* **2017**, *11*, 026021. [CrossRef]
56. Li, X.; Cheng, G.; Liu, S.; Xiao, Q.; Ma, M.; Jin, R.; Che, T.; Liu, Q.; Wang, W.; Qi, Y. Heihe Watershed Allied Telemetry Experimental Research (HiWATER): Scientific Objectives and Experimental Design. *Bull. Am. Meteorol. Soc.* **2013**, *94*, 1145–1160. [CrossRef]
57. Li, H.; Sun, D.; Yu, Y.; Wang, H.; Liu, Y.; Liu, Q.; Du, Y.; Wang, H.; Cao, B. Evaluation of the VIIRS and MODIS LST products in an arid area of Northwest China. *Remote Sens. Environ.* **2014**, *142*, 111–121. [CrossRef]
58. Li, X.; Liu, S.; Xiao, Q.; Ma, M.; Jin, R.; Tao, C.; Wang, W.; Hu, X.; Xu, Z.; Wen, J. A multiscale dataset for understanding complex eco-hydrological processes in a heterogeneous oasis system. *Sci. Data* **2017**, *4*, 170083. [CrossRef]
59. Li, D.; Jin, R.; Zhou, J.; Kang, J. Analysis and Reduction of the Uncertainties in Soil Moisture Estimation with the L-MEB Model Using EFAST and Ensemble Retrieval. *IEEE Geosci. Remote Sens. Lett.* **2017**, *12*, 1337–1341.
60. Liu, S.; Li, X.; Xu, Z.; Che, T.; Xiao, Q.; Ma, M.; Liu, Q.; Jin, R.; Guo, J.; Wang, L. The Heihe Integrated Observatory Network: A basin-scale land surface processes observatory in China. *Vadose Zone J.* **2018**, *17*, 1–21. [CrossRef]
61. Wang, J.F.; Christakos, G.; Hu, M.G. Modeling Spatial Means of Surfaces with Stratified Nonhomogeneity. *IEEE Trans. Geosci. Remote Sens.* **2009**, *47*, 4167–4174. [CrossRef]
62. Kang, J.; Li, X.; Jin, R.; Ge, Y.; Wang, J.; Wang, J. Hybrid optimal design of the eco-hydrological wireless sensor network in the middle reach of the Heihe River Basin, China. *Sensors* **2014**, *14*, 19095. [CrossRef]
63. Goward, S.N.; Cruickshanks, G.D.; Hope, A.S. Observed relation between thermal emission and reflected spectral radiance of a complex vegetated landscape. *Remote Sens. Environ.* **1985**, *18*, 137–146. [CrossRef]
64. Li, Z.L.; Tang, R.L.; Wan, Z.M.; Bi, Y.Y.; Zhou, C.H.; Tang, B.H.; Yan, G.J.; Zhang, X.Y. A review of current methodologies for regional evapotranspiration estimation from remotely sensed data. *Sensors* **2009**, *9*, 3801–3853. [CrossRef] [PubMed]
65. Tang, R.; Li, Z.-L.; Tang, B. An application of the Ts–VI triangle method with enhanced edges determination for evapotranspiration estimation from MODIS data in arid and semi-arid regions: Implementation and validation. *Remote Sens. Environ.* **2010**, *114*, 540–551. [CrossRef]
66. Garcia, M.; Fernández, N.; Villagarcía, L.; Domingo, F.; Puigdefábregas, J.; Sandholt, I. Accuracy of the Temperature–Vegetation Dryness Index using MODIS under water-limited vs. energy-limited evapotranspiration conditions. *Remote Sens. Environ.* **2014**, *149*, 100–117. [CrossRef]
67. Patel, N.R.; Anapashsha, R.; Kumar, S.; Saha, S.K.; Dadhwal, V.K. Assessing potential of MODIS derived temperature/vegetation condition index (TVDI) to infer soil moisture status. *Int. J. Remote Sens.* **2009**, *30*, 23–39. [CrossRef]
68. Song, L.; Liu, S.; Kustas, W.P.; Ji, Z.; Ma, Y. Using the Surface Temperature-Albedo Space to Separate Regional Soil and Vegetation Temperatures from ASTER Data. *Remote Sens.* **2015**, *7*, 5828–5848. [CrossRef]
69. Kustas, W.P.; Norman, J.M. Evaluation of soil and vegetation heat flux predictions using a simple two-source model with radiometric temperatures for partial canopy cover. *Agric. For. Meteorol.* **1999**, *94*, 13–29. [CrossRef]
70. Zhao, W.; Li, A.; Bian, J.; Jin, H.; Zhang, Z. A Synergetic Algorithm for Mid-Morning Land Surface Soil and Vegetation Temperatures Estimation Using MSG-SEVIRI Products and TERRA-MODIS Products. *Remote Sens.* **2014**, *6*, 2213–2238. [CrossRef]
71. Huttenlocher, D.P.; Kedem, K.; Sharir, M. The upper envelope of Voronoi surfaces and its applications. *Discret. Comput. Geom.* **1993**, *9*, 267–291. [CrossRef]
72. Takács, B. Comparing face images using the modified Hausdorff distance. *Pattern Recognit.* **1998**, *31*, 1873–1881. [CrossRef]
73. Gao, Y. Efficiently comparing face images using a modified Hausdorff distance. *IEE Proc. Vis. Image Signal Process.* **2003**, *150*, 346–350. [CrossRef]
74. Breiman, L. Random Forests. *Mach. Learn.* **2001**, *45*, 5–32. [CrossRef]
75. Cutler, A.; Cutler, D.R.; Stevens, J.R. Random Forests. In *Machine Learning*; Zhang, C., Ma, Y., Eds.; Springer: Boston, MA, USA, 2001; pp. 157–175.
76. Breiman, L. Bagging predictors. *Mach. Learn.* **1996**, *24*, 123–140. [CrossRef]
77. Ho, T.K. The random subspace method for constructing decision forests. *IEEE Trans. Pattern Anal. Mach. Intell.* **1998**, *20*, 832–844.
78. Fantazzini, D.; Figini, S. Random Survival Forests Models for SME Credit Risk Measurement. *Methodol. Comput. Appl. Probab.* **2009**, *11*, 29–45. [CrossRef]
79. Umezawa, Y.; Mori, H. Credit Risk Evaluation of Power Market Players with Random Forest. *IEEJ Trans. Power Energy* **2008**, *128*, 165–172. [CrossRef]

80. Zeraatpisheh, M.; Ayoubi, S.; Jafari, A.; Tajik, S.; Finke, P. Digital mapping of soil properties using multiple machine learning in a semi-arid region, central Iran. *Geoderma* **2019**, *338*, 445–452. [CrossRef]
81. Chen, X.W.; Liu, M. Prediction of protein-protein interactions using random decision forest framework. *Bioinformatics* **2005**, *21*, 4394–4400. [CrossRef]
82. Díaz-Uriarte, R.; Andrés, S.A.D. Gene selection and classification of microarray data using random forest. *BMC Bioinform.* **2006**, *7*, 3. [CrossRef]
83. Cutler, R.D.; Edwards, T.C.; Beard, K.H.; Adele, C.; Hess, K.T.; Jacob, G.; Lawler, J.J. Random forests for classification in ecology. *Ecology* **2007**, *88*, 2783–2792. [CrossRef]
84. Gislason, P.O.; Benediktsson, J.A.; Sveinsson, J.R. Random Forests for land cover classification. *Pattern Recognit. Lett.* **2006**, *27*, 294–300. [CrossRef]
85. Wardlow, B.D. The Vegetation Drought Response Index (VegDRI): A New Integrated Approach for Monitoring Drought Stress in Vegetation. *GIsci. Remote Sens.* **2008**, *45*, 16–46.

 *remote sensing* 

Article

Downscaling Satellite Soil Moisture Using a Modular Spatial Inference Framework

Ricardo M. Llamas [1], Leobardo Valera [2], Paula Olaya [2], Michela Taufer [2] and Rodrigo Vargas [1,*]

[1] Department of Plant and Soil Sciences, University of Delaware, Newark, DE 19716, USA; rllamas@udel.edu
[2] Department of Electrical Engineering and Computer Science, University of Tennessee, Knoxville, TN 37996, USA; lvalera@utk.edu (L.V.); polaya@vols.utk.edu (P.O.); taufer@utk.edu (M.T.)
* Correspondence: rvargas@udel.edu; Tel.: +1-302-831-1386

Abstract: Soil moisture is an important parameter that regulates multiple ecosystem processes and provides important information for environmental management and policy decision-making. Spaceborne sensors provide soil moisture information over large areas, but information is commonly available at coarse resolution with spatial and temporal gaps. Here, we present a modular spatial inference framework to downscale satellite-derived soil moisture using terrain parameters and test the performance of two modeling methods (Kernel-Weighted K-Nearest Neighbor <KKNN> and Random Forest <RF>). We generate monthly and weekly gap-free spatial predictions on soil moisture at 1 km using data from the European Space Agency Climate Change Initiative (ESA-CCI; version 6.1) over two regions in the conterminous United States. RF was the method that performed better in cross-validation when comparing with the reference ESA-CCI data, but KKNN showed a slightly higher agreement with ground-truth information as part of independent validation. We postulate that more heterogeneous landscapes (i.e., high topographic variation) may be more challenging for downscaling and predicting soil moisture; therefore, moisture networks should increase monitoring efforts across these complex landscapes. Future opportunities for development of modular cyberinfrastructure tools for downscaling satellite-derived soil moisture are discussed.

Keywords: soil moisture; downscaling; ESA-CCI; SOMOSPIE; spatial inference; KKNN; random forest

Citation: Llamas, R.M.; Valera, L.; Olaya, P.; Taufer, M.; Vargas, R. Downscaling Satellite Soil Moisture Using a Modular Spatial Inference Framework. *Remote Sens.* **2022**, *14*, 3137. https://doi.org/10.3390/rs14133137

Academic Editors: Wei Zhao, Jian Peng, Hongliang Ma, Chunfeng Ma and Jiangyuan Zeng

Received: 20 May 2022
Accepted: 27 June 2022
Published: 29 June 2022

Publisher's Note: MDPI stays neutral with regard to jurisdictional claims in published maps and institutional affiliations.

1. Introduction

The top layer of soil is critical for the root system of plants and the available water that sustains most of the vegetation and controls many soil processes. Due to its importance, soil moisture has been recognized as an Essential Climate Variable [1], and in conjunction with variables, such as land cover, is critical in shaping Earth system dynamics. Soil moisture importance relies not only on its role within the water cycle, but also on its relationship with other ecological processes, such as runoff generation, sediment transport and energy balance [2–4], drought occurrence [5,6], plant and soil respiration [7–9], regulation of greenhouse gas fluxes from soils to the atmosphere [10–12], and plant growth, which influences the terrestrial carbon budget [4,7,13]. Water content in the top centimeters of the soil also serves as a retardant for wildfires, regulates runoff during extreme rain events, and provides information for flash floods and drought early warning systems [14–17]. Additionally, soil moisture information is a key input for agricultural planning [6,18], regional stewardship [19], and multiple models used in weather forecasting or climate variability and change [20–22].

Traditionally, soil moisture information was acquired from point measurements using instruments, such as Time–Domain Reflectometers (TDR), which offer instantaneous values of soil water content based on information of electric and dielectric properties within a small volume of soil [23]. However, the availability of soil moisture data from these ground sensors across large areas is often limited [24,25]. At the global scale, the International Soil Moisture Network [26,27] provides ground-truth information, and within the United States, the Soil Climate Analysis Network (SCAN) [28] and the North American Soil Moisture Database (NASMD) [29] provide soil moisture information derived from ground sensors. However, due to large spatial and temporal variability in soil moisture, this information, although invaluable, is not enough to address multiple applications where detailed spatial and temporal variability in soil moisture is required.

To address the limited spatial coverage of ground-based soil moisture networks, alternative approaches can be applied to estimate soil moisture. Satellite-based sensors offer a feasible way to estimate soil moisture over large areas on a regular basis, ranging from 3 to ~36 km [30–33]. Satellite sensors estimate soil moisture using radar instruments or radiometers, which are based on the dielectric constant and temperature emissivity of the soil, respectively [33,34]. Various satellite sensors are used to estimate soil moisture, some specifically conceived for this purpose, such as SMAP (Soil Moisture Active Passive) [30] or SMOS (Soil Moisture and Ocean Salinity mission) [35], while others, such as the European Space Agency Climate Change Initiative (ESA-CCI) soil moisture [15], Sentinel [36] and GPS-aided values [37], can be used to indirectly derive soil moisture information. These satellite-based efforts aim to provide global soil moisture values at high temporal resolution (1~3 days). The ESA-CCI offers the longest available global records at the daily scale, beginning in November 1978, with improved accuracy since 1991 due to a combination of information from active and passive sensors [38]. These efforts have provided unprecedented information, but they have two important limitations: they have coarse spatial resolution, and they have spatial and temporal gaps.

Various approaches have been used to downscale satellite-derived soil moisture values. These approaches can be categorized as (1) satellite-based, (2) geoinformation-based, and (3) model-based [39]. Satellite-based approaches include various techniques, such as *Active and Passive Microwave Data Fusion* and *Optical/Thermal and Microwave Fusion* [39]. Geoinformation-based methods have explored the known correlation of soil moisture with *topography*, *soil attributes*, and *vegetation characteristics* [39]. Model-based methods include other approaches, such as *statistical models*, *integration of a Land Surface Model*, *statistical downscaling*, and *data assimilation* [39].

Here, we present a geoinformation-based approach, considering the relationship between soil moisture and topography to downscale and gap-fill satellite-based soil moisture information at the regional scale [39,40]. Topography has been explored previously as a meaningful environmental variable for downscaling soil moisture at the catchment scale [41–43] and across the United States [44]. We used a modular spatial inference framework, which is the foundation of a cyberinfrastructure tool named SOil Moisture SPatial Inference Engine (SOMOSPIE) [45–47]. We tested the performance of two modeling methods coupled with geoinformation from terrain parameters to downscale satellite-derived soil moisture. Specifically, SOMOSPIE framework combines publicly available satellite-derived soil moisture information to generate fine-grained and gap-free predictions (from 0.25 degrees (which is about 27 km) to 1 km) using different modeling methods: a kernel-based approach (Kernel-Weighted k-Nearest Neighbors (KKNN), and a tree-based approach (Random Forests or RF).

We tested our framework across two contrasting regions of interest (ROIs) within the conterminous United States at monthly and weekly time scales in 2010 and 1 km spatial resolution. We found that RF was consistently the method that performed better at the monthly and weekly scales when compared with the reference ESA-CCI data. In contrast, KKNN showed a slightly higher agreement with ground-truth information as part of independent validation. We postulate that differences in model performance are influenced by the multivariate space of topographic features, where more heterogeneous landscapes (i.e., high topographic variation) may be more challenging to downscale and predict soil moisture. Finally, we demonstrate that our framework is a flexible, transparent, and replicable approach to downscale satellite-derived soil moisture at different temporal scales.

2. Materials and Methods

2.1. Regions of Interest

Our study was conducted over two regions of interest (ROI) within the conterminous United States (CONUS; Figure 1a). Each region encompasses a polygon of 7.5° × 3.75° (450 pixels with 30 columns and 15 rows in the native resolution of the ESA-CCI soil moisture product), and each ROI was aligned to the original edges of the ESA-CCI grid. Both areas were selected as they offer a contrast in climatic and topographic conditions, and anthropogenic activities such as different agricultural and forestry practices.

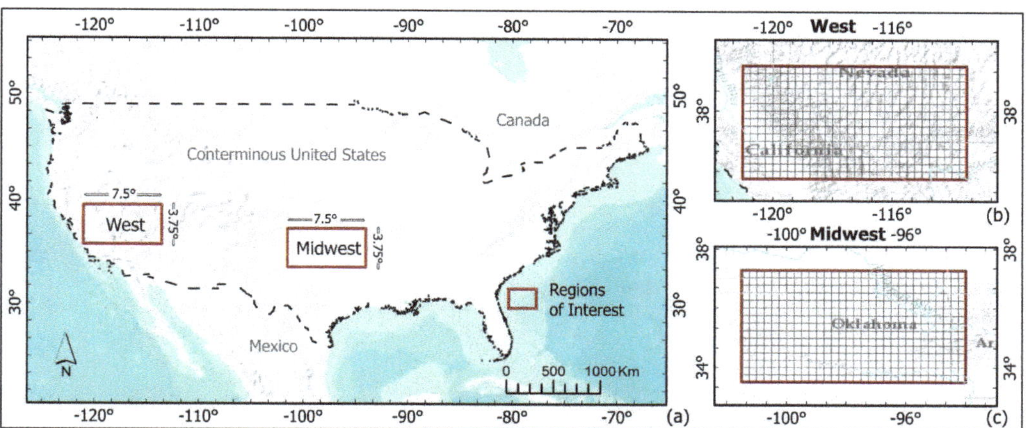

Figure 1. (**a**) Regions of interest (ROIs) for soil moisture downscaling; (**b**) West ROI; (**c**) Midwest ROI.

The West region (Figure 1b) comprises an area of 275,516 km^2 with heterogeneous topographic features and a wide diversity of climate conditions ranging from the central valley of California in the West, passing through the densely forested areas in the Rocky Mountains, and water-limited ecosystems across California, Nevada, Utah, and Arizona.

The Midwest region (Figure 1c) comprises an area of 283,499 km^2. This region lacks extensive mountainous areas (except for the Ouachita Mountains) and has a large influence of agricultural activity that strongly influences the dynamics of soil moisture. This region was also selected because of the extensive availability of ground-truth data [48] from the monitoring network MESONET [49], mainly over Oklahoma.

2.2. Input Data

2.2.1. Satellite-Derived Soil Moisture Data

We use information from the ESA-CCI soil moisture product Version 6.1 (revised in September 2021) which is the latest release by ESA-CCI [50]. ESA-CCI product merges daily data derived from C-band scatterometers (e.g., ERS-$\frac{1}{2}$, METOP) and data from multi-frequency radiometers (e.g., SMMR, SSM/I, TMI, AMSR-E, Windsat, AMSR-2, SMOS, SMAP, GPM, and FengYun-3B) at 0.25 degrees spatial resolution [51]. Based on daily soil moisture values, we calculated mean values for each pixel at the monthly and weekly scales for each ROI. Thus, obtaining 12 monthly layers and 52 weekly layers of mean soil moisture for the year 2010.

2.2.2. Terrain Parameters

Topographic information was derived from a digital elevation model (DEM) [52] and we extracted hydrologically meaningful terrain parameters for each ROI following a standardized approach [53]. Briefly, an initial set of 15 terrain parameters was calculated using the terrain analysis module in RSAGA [54], which implements SAGA GIS [55] in R statistical platform [56]. The original terrain parameters were: *Aspect, Analytical Hillshading,*

Channel Network Base Level, Convergence Index, Cross Sectional Curvature, Catchment Area, Elevation, Flow Accumulation, Longitudinal Curvature, Length-Slope Factor, Relative Slope Position, Slope, Topographic Wetness Index, Valley Depth, and *Vertical Distance to Channel Network.* To reduce model complexity, identify the best prediction parameters, and avoid redundancy of information, we predicted soil moisture at 1 km over CONUS using different combinations of terrain parameters and geographic coordinates (i.e., latitude and longitude). This test was performed using a KKNN algorithm, combinations of the aforementioned predictors, and the ESA-CCI soil moisture annual mean of 2010 as the training dataset. Based on correlation and error values from cross-validation automatically performed during model training and evaluation, we identified the combination of predictors that best represented soil moisture reference values. Our results identified geographic coordinates (*latitude* and *longitude*) and 4 terrain parameters (*elevation, aspect, slope,* and *topographic wetness index*) as the best predictors for our study. Results of cross-validation from all the predictor combinations tested are included in Supplementary Material S1.

2.2.3. Data Used for Independent Validation

We validated downscaled soil moisture predictions using independent data from ground-truth soil moisture records from the North American Soil Moisture Database (NASMD). The NASMD integrates data from 33 observation networks, as well as 2 short-term monitoring campaigns that put together over 1800 observation sites across the United States, Canada, and Mexico [29]. We reiterate that data from the NASMD was not used for downscaling satellite-derived soil moisture, and only used for independent validation purposes.

We selected all the available stations for the year 2010 with daily records of soil moisture in the top 5 cm of the soil layer for the two ROIs. The maximum number of available stations within CONUS was 743 (Figure 2a), while a maximum of 39 stations were available for the West region (Figure 2b) and a maximum of 116 were available for the Midwest region (Figure 2c). The number of stations available at the monthly and weekly scales ranged from ~26 to 39 in the West region, and from ~110 to 116 in the Midwest region (Supplementary Material S2). Monthly and weekly means of top 5 cm soil moisture records were calculated for each field station, to generate the reference data to validate monthly and weekly downscaled soil moisture predictions.

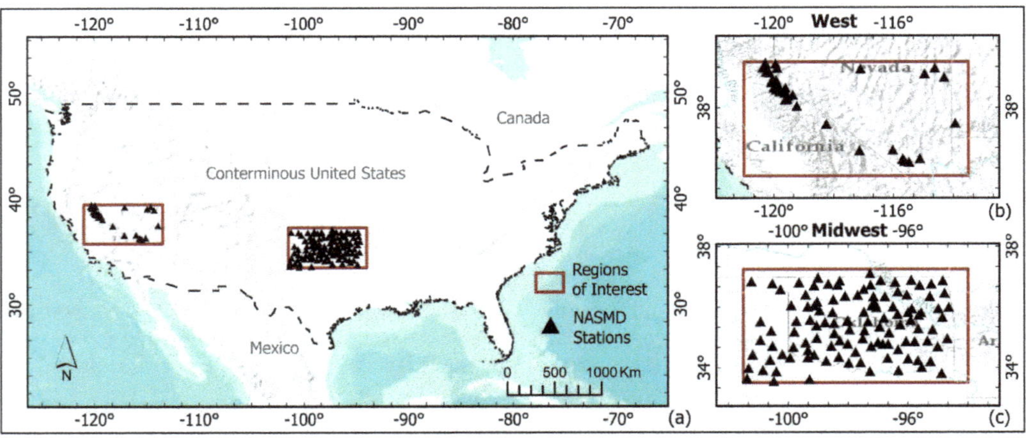

Figure 2. (**a**) North American Soil Moisture Database (NASMD) stations over the two ROIs available in 2010; (**b**) West ROI; and (**c**) Midwest ROI.

2.3. Data Preparation

2.3.1. Training Matrices

We generated a set of training matrices to obtain model parameters required by KKNN and RF. We selected the coordinates of the centroid of each original pixel (0.25 degrees) from the ESA-CCI product and assigned the soil moisture values to those coordinates. Then, we extracted the values of the 4 predefined terrain parameters at the finer resolution (1 km) that overlapped the ESA-CCI pixels centroids, and we added them to the training matrix. In each matrix, 70% of the available sampling points were randomly selected to conform the training dataset to build the models, and the 30% of remaining sampling points were set aside for further validation of models' outputs.

Our final training matrices represent 12 monthly and 52 weekly files for each ROI, containing up to 315 records (70% of the maximum number of pixels available for each ROI that included soil moisture values and 6 predictors (4 terrain parameters, and latitude and longitude values)).

2.3.2. Prediction Matrices

We generated one matrix for each ROI to predict soil moisture at 1 km spatial resolution. We extracted all available records of the 4 predefined terrain parameters (predictors) at 1 km and added their corresponding coordinates to the prediction matrices. We integrated a total of 273,840 point locations into each of the two final prediction matrices; this number corresponds to the extension of the two ROIs in square kilometers, encompassing areas of 652 km (X-axis) by 420 km (Y-axis; Figure 1).

2.4. Downscaling Soil Moisture

We used the modular framework of SOMOSPIE to predict soil moisture on a user-defined temporal (e.g., daily, monthly, annual) and spatial resolution (i.e., spatial granularity) to provide gap-free information within an ROI. The SOMOSPIE framework is composed of three main modules that include (1) preprocessing data from: satellite-derived soil moisture, predictive terrain parameters in the target resolution for downscaling (e.g., 1 km spatial resolution), and ground-truth reference data for independent validation purposes; (2) model construction: definition of optimal parameters for each modeling method (i.e., KKNN, RF); and (3) soil moisture prediction: application of model parameters defined in the previous module to predict soil moisture at the target resolution, as well as cross-validation and independent ground-truth validation (Figure 3).

We implemented our framework with two modeling methods (i.e., Kernel-Weighted K-Nearest Neighbors (KKNN) and Random Forest (RF)) to downscale soil moisture at 1 km over the two ROIs at monthly and weekly scales. We used the cloud-based cluster "Caviness" at the University of Delaware High Performance Computing (HPC) [57]. Caviness is a distributed-memory Linux cluster with 126 compute nodes representing 4536 cores with 24.6 TiB of RAM and 200 TB of storage.

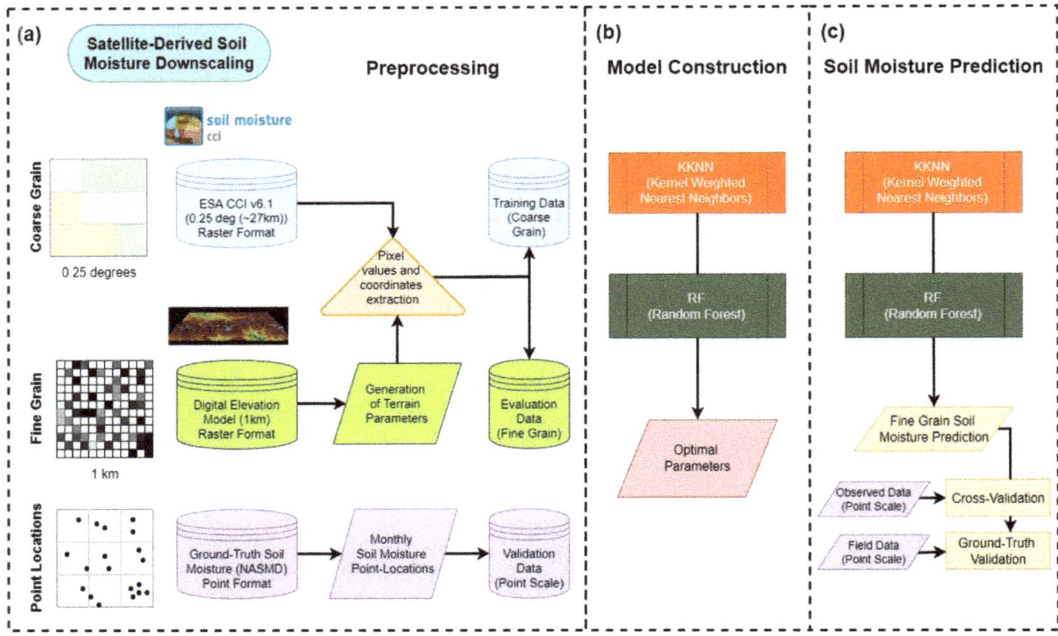

Figure 3. Framework for soil moisture prediction at 1 km spatial resolution derived from coarse resolution ESA-CCI values; (**a**) data preprocessing; (**b**) model construction; (**c**) soil moisture prediction and validation.

2.4.1. Kernel-Weighted K-Nearest Neighbors (KKNN)

K-nearest neighbors (KKNN) in its traditional form is a regression technique that builds many simple models from local data [58], and is based upon decision rules that classify an unsampled point, based on the values of the nearest set of previously classified points or reference values in the sampling space [59]. This method assumes a different level of influence in the prediction space, where the nearest k-points to the target location are the ones with the most relevant influence, while the influence in the construction of the prediction model decreases with distance [45]. To assign distance-related relevance to predict soil moisture, a weighted mean of the k-nearest soil moisture ratios is calculated. This variant is based on the definition of kernel functions (i.e., Triangular, Epanechnikov, Gaussian, Optimal) that serve to find the number of neighbors (k) to be used in the prediction. The number of neighbors and the optimal kernel function are automatically selected through 10-fold cross validation [44,45].

The KKNN code used in the SOMOSPIE framework has been described previously [45] and has been successfully used to downscale satellite-derived soil moisture at different spatial scales [44]. The code is based on the 'kknn' package [60] developed for the R-statistical platform [56]. The definitions of optimal parameters found for each monthly and weekly layer in 2010, over the two ROIs, are shown in Supplementary Material S2.

2.4.2. Random Forest (RF)

Random Forest (RF) in the SOMOSPIE framework has been described previously [45] and is based on the 'quantregForest' package [61] developed for the R-statistical platform [56]. It is based on an ensemble of decision trees through a "bootstrap aggregation" process (bagging), which is a method to generate multiple versions of a predictor and then uses these versions to generate an aggregated predictor that depends on the values of a random vector independently sampled and weighed [62,63]. To predict values at an unsampled location, all decision trees in the ensemble are queried and their prediction outputs are combined through a weighted arithmetic mean. Techniques such as RF do not assume any particular geometric or functional form of the model and are suitable for sampling spaces with sparse data [45].

The definition of optimal parameters for soil moisture prediction with RF in SOMOSPIE considers two main values: (1) the number of trees to grow in the ensemble of regression trees and (2) the number of covariates randomly selected at each level of tree growth. The maximum number of trees allowed was 500, while the number of covariates changes in relation to the number of predictors defined as input (6 predictors for this study: *latitude, longitude, elevation, aspect, slope,* and *topographic wetness index*). The automatic variable selection is performed by 'quantregForest' through a cross-validation process. The optimal parameters selected for each monthly and weekly layer of 2010 over the two ROIs are reported in Supplementary Material S2.

2.5. Validation

To test the two modeling methods (i.e., KKNN and RF), we first used cross-validation with reference satellite-derived soil moisture data not used in the construction of the models, and then we used independent ground-truth soil moisture from the NASMD. We reiterate that the NASMD data was not used to parameterize any model and was only used for independent validation. Predicted soil moisture values were extracted from the 12 monthly and 52 weekly layers over the two ROIs, taking overlapping locations with the centroids of the ESA-CCI soil moisture reference data, and the point-locations of the NASMD available stations for each month and week, respectively.

2.5.1. Cross-Validation with Reference Satellite-Derived Soil Moisture Data

We calculated the correlation and root mean square error (RMSE) values based on matrices containing the predicted and reference values (from ESA-CCI data). The input data for this validation approach corresponds with the 30% of the sampling points set aside during the generation of the training matrices and were not used in the definition of the models' parameters. The cross-validation data matrices contained up to 135 records, depending on the number of available reference points from the ESA-CCI mean values for each month and week.

The values of each predicted soil moisture pixel at a finer spatial resolution (i.e., 1 km) were compared with the reference values of satellite-derived soil moisture values at their original spatial resolution. The results from these analyses for each month and week over the two ROIs are reported in Supplementary Material S3.

2.5.2. Independent Validation with Ground-Truth Data

For these independent analyses, we calculated the overall correlation and RMSE between the predicted downscaled values from each method with the point-based ground-truth data from the NASMD. The results of correlation and RMSE between fine spatial resolution predicted soil moisture values and the point-based ground-truth data for each month and week over the two ROIs are reported in Supplementary Material S3.

2.5.3. Spatial Distribution of Prediction Outputs and Errors

To evaluate the performance of the two methods, we compared the mean values of all monthly and weekly predictions (12 monthly and 52 weekly outputs) in the two ROIs. We generated maps showing the mean values of ESA-CCI values at 0.25 degrees of spatial resolution and the mean values of our 1 km predictions over the set of 30% sampling points set aside for testing in each monthly and weekly scale. Thus, none of the points used in this approach to describe the spatial distribution of error were used to define the models' parameters. We calculated the absolute difference between the mean of predicted soil moisture and the mean of ESA-CCI values at all our monthly and weekly scales over all the centroid coordinates of the ESA-CCI pixels. In a similar approach for all monthly and weekly scales, we calculated the absolute difference between the mean predicted soil moisture at 1 km and the mean values of the point-scale ground-truth records at the coordinates of all available NASMD stations during our time frame. Thus, we aim to observe the similarities in the spatial distribution between ESA-CCI data and the outputs of the two methods tested, as well as the distribution of the prediction errors.

3. Results

In this section, we present our 1 km soil moisture prediction results and evaluate the performance of the two methods used. We compared the predicted soil moisture values with the reference ESA-CCI values, and with independent values from the NASMD. The final soil moisture predictions at monthly and weekly scales over the two ROIs are available at the Consortium of Universities for the Advancement of Hydrologic Science data repository (HydroShare; doi:10.4211/hs.96eeb0d796a64b578f24e8154c166988) [64].

3.1. *Optimal Model Parameters for Each Method*

In the case of KKNN, we found that the automatic generation of model parameters defined a number of K-neighbors between 6 and 29 in the Midwest ROI for all models at monthly and weekly scales. Correlation ranged from 0.489 to 0.894, and RMSE from 0.03 to 0.046. In the West ROI, the number of K-neighbors ranged from 3 to 49, with correlation from 0.244 to 0.785, and RMSE from 0.025 to 0.055.

In the generation of RF models, we found that the number of covariates used as predictors in every model in the Midwest ROI ranged from two to six (out of six possible predefined predictors for this study). Correlation ranged from 0.537 to 0919, and RMSE from 0.028 to 0.043. In the West ROI, the number of covariates ranged from two to six. Correlation ranged from 0.413 to 0.833, and RMSE from 0.023 to 0.047.

All individual KKNN and RF models' parameters are included in Supplementary Material S2.

3.2. *Evaluation of Models' Outputs*

To evaluate the performance of each method tested, we present a series of Taylor Diagrams [65] that show the similarity of our predictions with both data from the ESA-CCI soil moisture values and independent ground-truth records from the NASMD. Taylor diagrams quantify the correspondence between reference observed data and predicted values by means of Pearson correlation coefficient, RMSE and the standard deviation.

3.2.1. Evaluation with Reference Satellite-Derived Soil Moisture Values

We found that RF was consistently the best method in predicting monthly soil moisture when compared against the reference values from the ESA-CCI values (Figure 4). RF correlation and RMSE values ranged from 0.566 to 0.856, and from 0.027 to 0.037, respectively, in the Midwest ROI. In the West ROI, RF correlation and RMSE values ranged from 0.443 to 0.78, and from 0.023 to 0.056, respectively. Regardless of the ROI, values predicted with RF showed the highest correlation and the lowest RMSE in every month, except in January in the West ROI.

Predictions with KKNN showed a consistent lower prediction performance than RF, with monthly correlation and RMSE values ranging from 0.508 to 0.844 and, 0.028 to 0.037, respectively, in the Midwest ROI. KKNN correlation and RMSE values in the West ROI ranged from 0.405 to 0.712 and from 0.023 to 0.054, respectively.

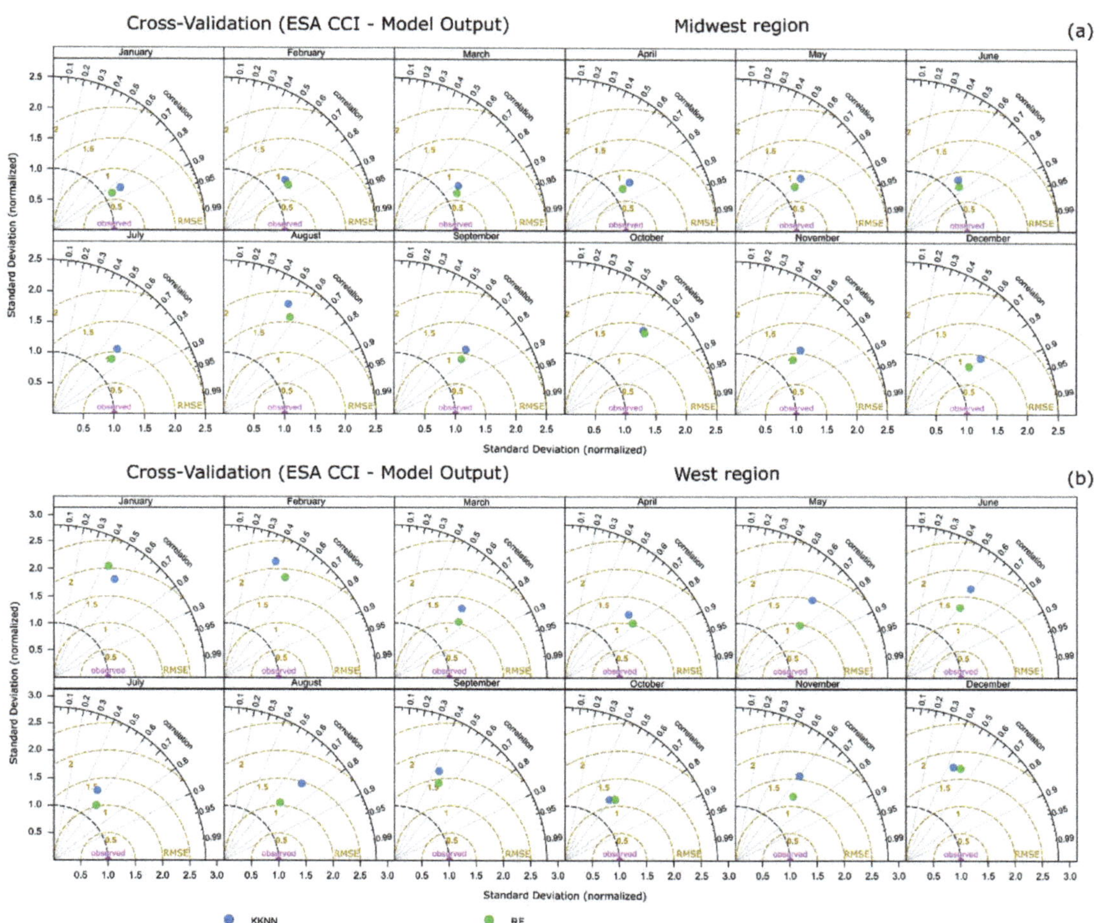

Figure 4. Taylor diagrams showing cross-validation between monthly 1 km predicted soil moisture and ESA-CCI reference data; (**a**) monthly cross-validation of the Midwest ROI; (**b**) monthly cross-validation of the West ROI.

Similar to monthly predictions, we report the weekly performance of the two methods tested, grouping 52 weeks into four 3-month periods (Figure 5). Like monthly predictions, RF consistently showed better performance in all 3-month periods and in both ROIs. Correlation and RMSE values with RF ranged from 0.764 to 0.846, and 0.031 to 0.033, respectively, in the Midwest ROI, and from 0.634 to 0.785, and 0.026 to 0.041 in the West ROI. In contrast, correlation and RMSE values with KKNN in the Midwest region ranged from 0.726 to 0.823, and 0.033 to 0.036, while in the West ROI, these values ranged from 0.555 to 0.746, and 0.028 to 0.043, respectively.

All correlation and RMSE values shown in Figures 4 and 5 are included in Supplementary Material S3.

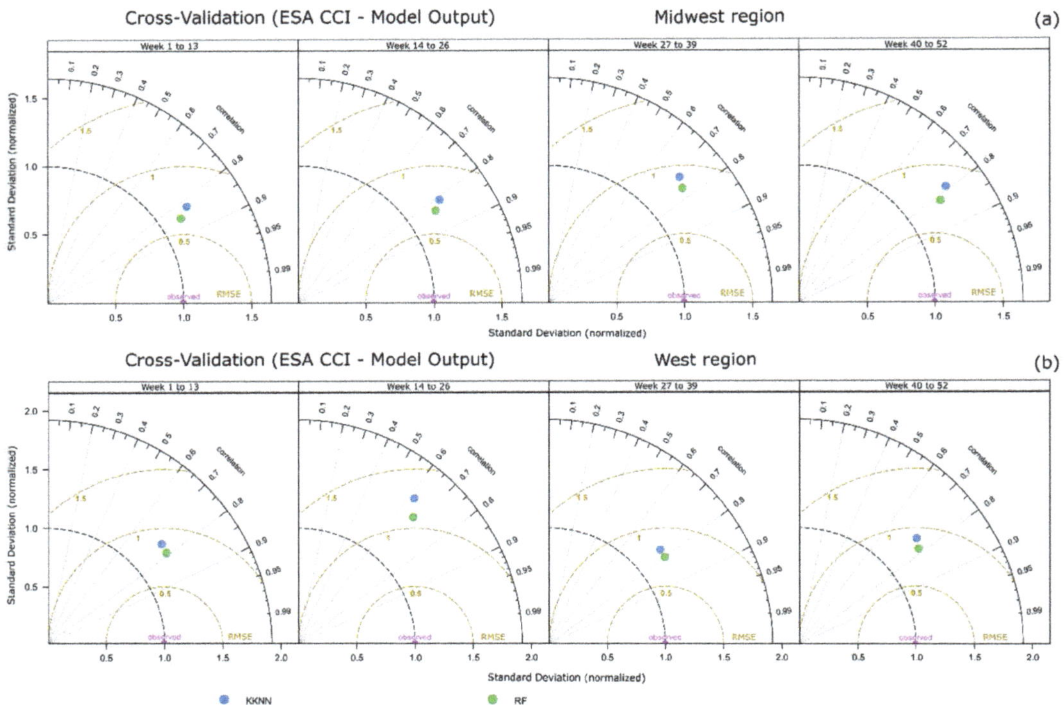

Figure 5. Taylor diagrams showing cross-validation between weekly 1 km predicted soil moisture and ESA-CCI reference data, the 52 weekly predictions are grouped in four 3-month periods; (**a**) weekly cross-validation of the Midwest ROI; (**b**) weekly cross-validation of the West ROI.

3.2.2. Evaluation with Independent Ground-Truth Information

In Figure 6, we show the results of independent validation of monthly soil moisture predictions with ground-truth information from the NASMD. In the Midwest ROI, a similar correspondence between our predicted values and the reference data in all months was clear, except in August, where the ESA-CCI reference better corresponded with ground-truth records. Although the correlation and RMSE values for our two methods are consistently clustered in Figure 6a, RF showed a better correspondence with ground-truth data, and it was closer to the correlation and RMSE values of the reference satellite-derived values. A similar prediction performance was obtained for the West ROI (Figure 6b), where RF had consistently better agreement with the ground-truth reference data. However, the general agreement between ground-truth data, the reference satellite derived data and the models' outputs was evidently lower in the West ROI.

The reference satellite-derived data monthly correlation and RMSE values with the ground-truth data ranged from 0.331 to 0.637 and 0.054 to 0.07 in the Midwest ROI, and from −0.953 to 0.272, and 0.078 to 0.167 in the West ROI, respectively. Monthly RF correlation and RMSE values in the Midwest ROI ranged from 0.216 to 0.55, and 0.052 to 0.073, while in the West ROI, these values ranged from −0.194 to 0.279, and 0.079 to 0.137, respectively. KKNN consistently showed the lowest correspondence with ground-truth data, except in October in the West ROI. KKNN correlation and RMSE values ranged from 0.3 to 0.603, and 0.051 to 0.069 in the Midwest ROI, and from −0.173 to 0.259, and 0.077 to 0.147 in the West ROI.

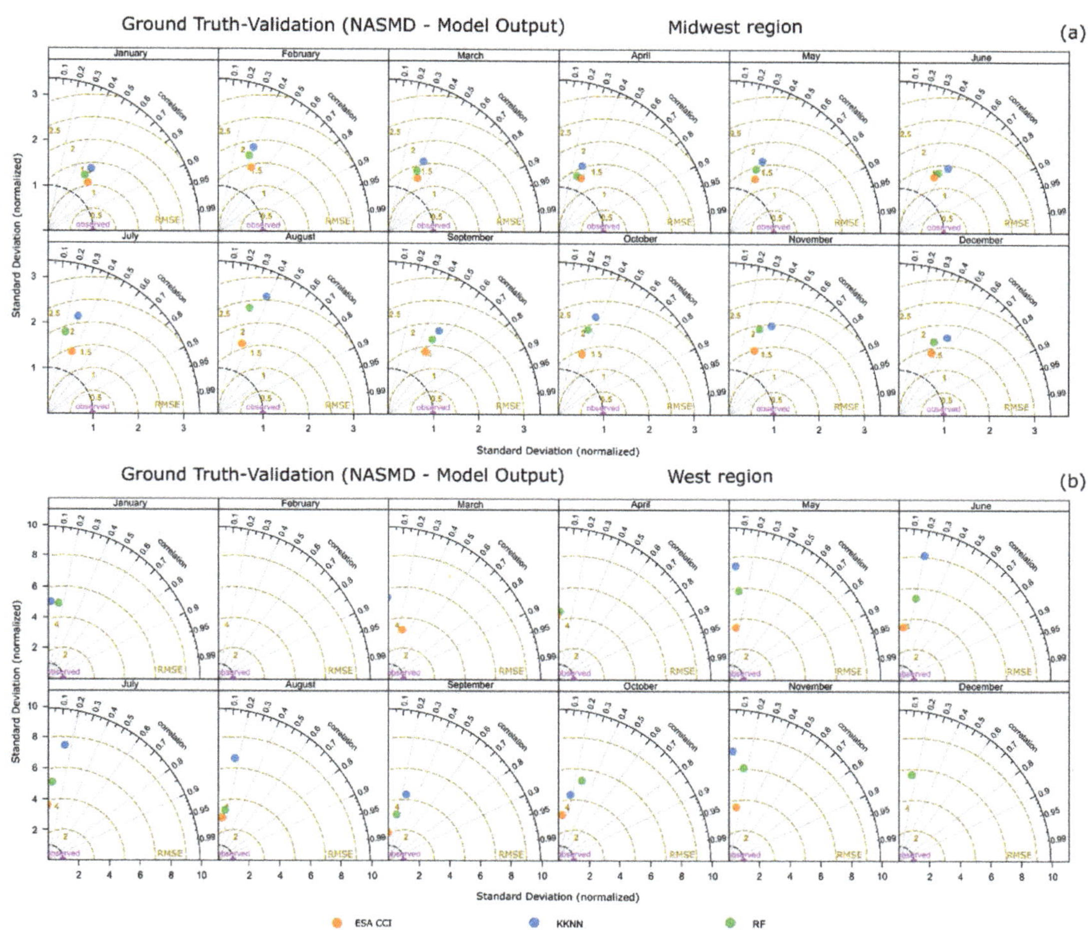

Figure 6. Taylor diagrams showing validation between monthly 1 km predicted soil moisture and ESA-CCI values, and ground-truth data from the NASMD; (**a**) monthly ground-truth validation of the Midwest ROI; (**b**) monthly ground-truth validation of the West ROI.

In the ground-truth validation of the weekly predictions (Figure 7), we found that the two methods showed similar correlation and RMSE values with ground truth data as the reference ESA-CCI in the Midwest ROI. Although there was not a clear pattern of better performance for either of the two methods tested, RF showed slightly better performance for the four 3-month periods in the Midwest ROI. In the West ROI, there was a consistent decrease in the correspondence between ground-truth data, our predictions, and the ESA-CCI values, although RF still showed a better performance in three of the four 3-month periods.

For weekly validation, ESA-CCI reference values exhibited the best correspondence with ground-truth data, with correlation and RMSE values ranging from 0.46 to 0.53, and 0.064 to 0.07 in the Midwest ROI, and from −0.195 to 0.166, and 0.097 to 0.132 in the West ROI. RF correlation and RMSE values ranged from 0.445, to 0.46, and 0.062 to 0.071 in the Midwest ROI, and from −0.041 to 0.158, and 0.091 to 0.126 in the West ROI. KKNN correlation and RMSE values, ranged from 0.464 to 0.494, and 0.06 to 0.069 in the Midwest ROI, and −0.077 to 0.154, and 0.09 to 0.126 in the West ROI.

All correlation and RMSE values shown in Figures 6 and 7 are included in Supplementary Material S3.

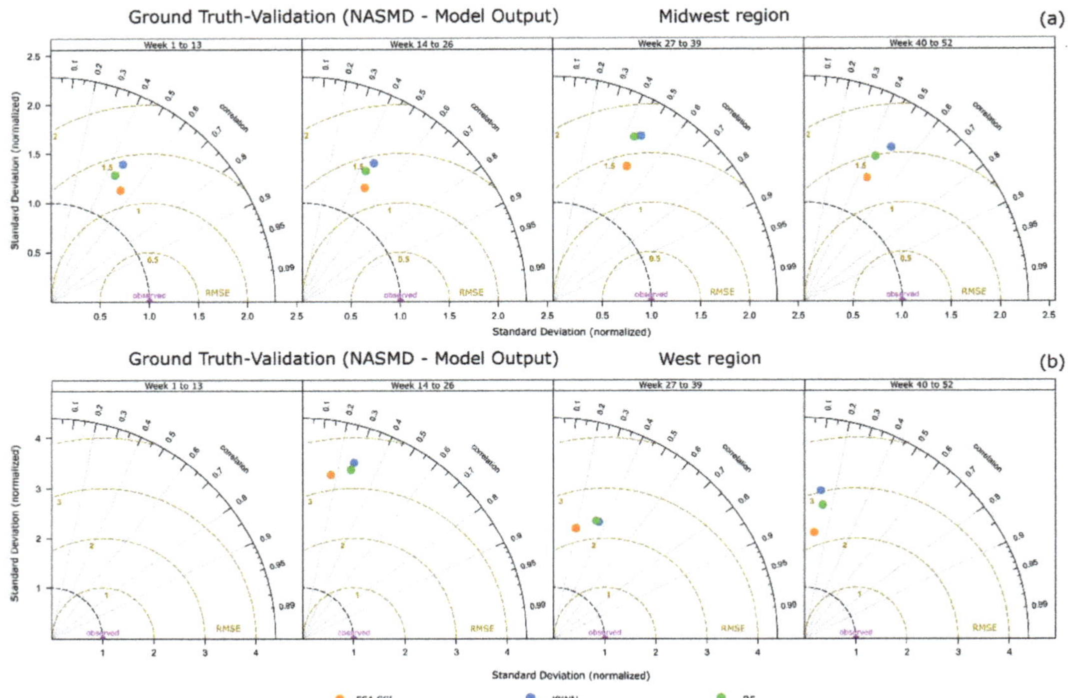

Figure 7. Taylor diagrams showing validation between weekly 1 km predicted soil moisture and ESA-CCI values, and ground-truth data from the NASMD, the 52 weekly layers are grouped in four 3-month periods; (**a**) weekly ground-truth validation of the Midwest ROI; (**b**) weekly ground-truth validation of the West ROI (correlation and RMSE values in the week 1 to 13 period were consistently negative and values are described in Section 3.2.2).

3.3. Spatial Distribution of Prediction Errors

As we display in Figure 8c,d for the Midwest ROI, the spatial patterns of soil moisture values exhibited a similar behavior as the reference ESA-CCI values (Figure 8b). Similar to the ESA-CCI, the lowest soil moisture values were distributed over the west part of the ROI, and highest values over the east section. Low values were also consistent in the south-central portion, and high values in the central-north. The absolute differences between the 30% of sampling points set aside for testing in all layers derived from ESA-CCI values at 0.25 degrees and their spatially correspondent predicted soil moisture values in all layers at 1 km using the two methods tested are shown in Figure 8e,f. Difference values were distributed between 0 and 0.03 for both methods, with highest values in the western portion of the ROI. KKNN was the method with the lowest difference values over most of the ROI. In Figure 8g,h, we present the absolute differences between predicted soil moisture and ground-truth data. Difference values were constantly higher for the two methods in the Midwest ROI. Unlike the comparison between predicted soil moisture and reference ESA-CCI data, the performance of the two methods was similar when compared to ground-truth information. The lowest differences ranged between 0 and 0.04 $m^3 m^{-3}$, and the highest values were up to 0.14 $m^3 m^{-3}$. Although there was not a clear spatial distribution of the absolute differences, the distribution of low and high values was similar across the two methods.

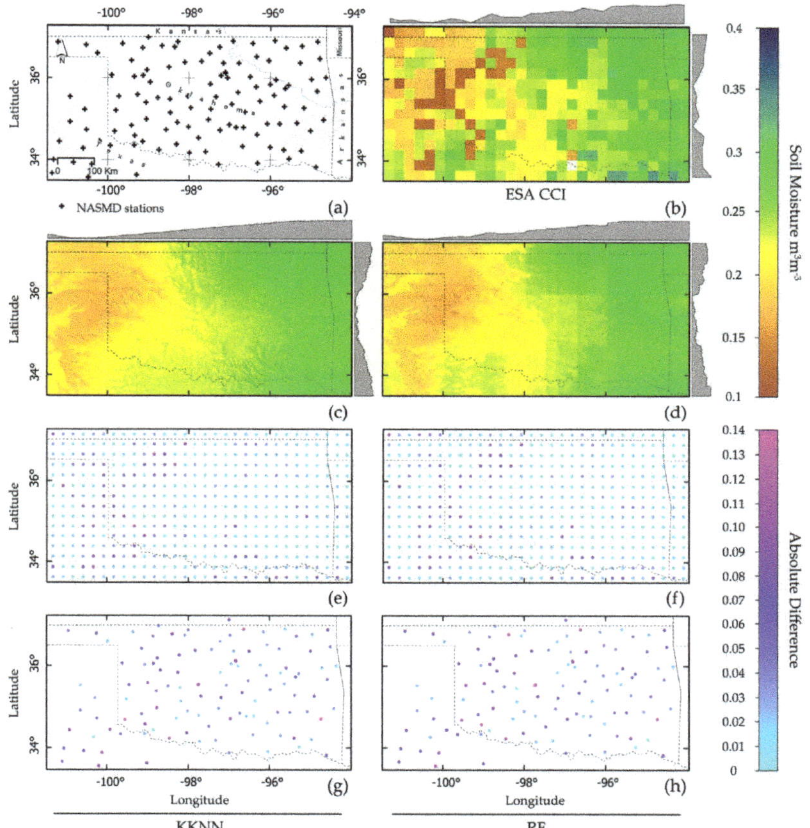

Figure 8. (**a**) Midwest ROI and distribution of NASMD stations throughout 2010; (**b**) mean soil moisture values of 12 monthly and 52 weekly layers based on the reference ESA-CCI values at 0.25 degrees of spatial resolution; (**c**,**d**) mean values of 1 km soil moisture predictions with KKNN and RF; (**e**,**f**) spatial distribution of mean absolute differences between ESA-CCI sampling points at 0.25 degrees and their spatially correspondent predicted soil moisture values in all layers at 1 km with KKNN and RF; (**g**,**h**) spatial distribution of mean absolute differences between all monthly and weekly soil moisture values from NASMD and predicted values at 1 km using the two methods tested.

Figure 9 shows the spatial distribution of soil moisture predicted values and absolute differences with ESA-CCI values, and ground-truth data in the West ROI. Similar to ESA-CCI soil moisture, the lowest predicted values were distributed from the south-center to the north-west of the ROI (Figure 9c,d). However, low soil moisture values described a pattern not as dry as in the ESA-CCI data (between 0.05 and 0.1 m^3 m^{-3}). The highest predicted values with both methods were consistently located in two south-east to north-west lines, along the highest elevations of the Rocky Mountains and the central valley of California, ranging from 0.18 to 0.28 m^3 m^{-3}. Absolute differences between the 30% of test sampling points from ESA-CCI values at 0.25 degrees and their spatially correspondent prediction output values in all layers at 1 km in the West ROI can be observed in Figure 9e,f. Overall, the differences were consistently higher in the West ROI than in the Midwest ROI. The lowest difference values in the West ROI ranged between 0 and 0.045 m^3 m^{-3}, and highest values reached an absolute difference of 0.13 $m^3\ m^{-3}$. Unlike the absolute differences shown in the Midwest ROI, in the West ROI, there was not a clear pattern in the spatial distribution of errors between ESCA-CCI and predicted values with our two methods. Absolute differences between predicted soil moisture and ground-truth data were consistently higher, regardless of the method used (Figure 9g,h). The distribution of the absolute differences across the locations with ground-truth data was similar for the two methods, although RF generally showed lower differences than KKNN. In contrast to the Midwest ROI, the absolute differences between predicted soil moisture and ground-truth information were significantly higher, ranging from 0.015 up to 0.21 $m^3\ m^{-3}$.

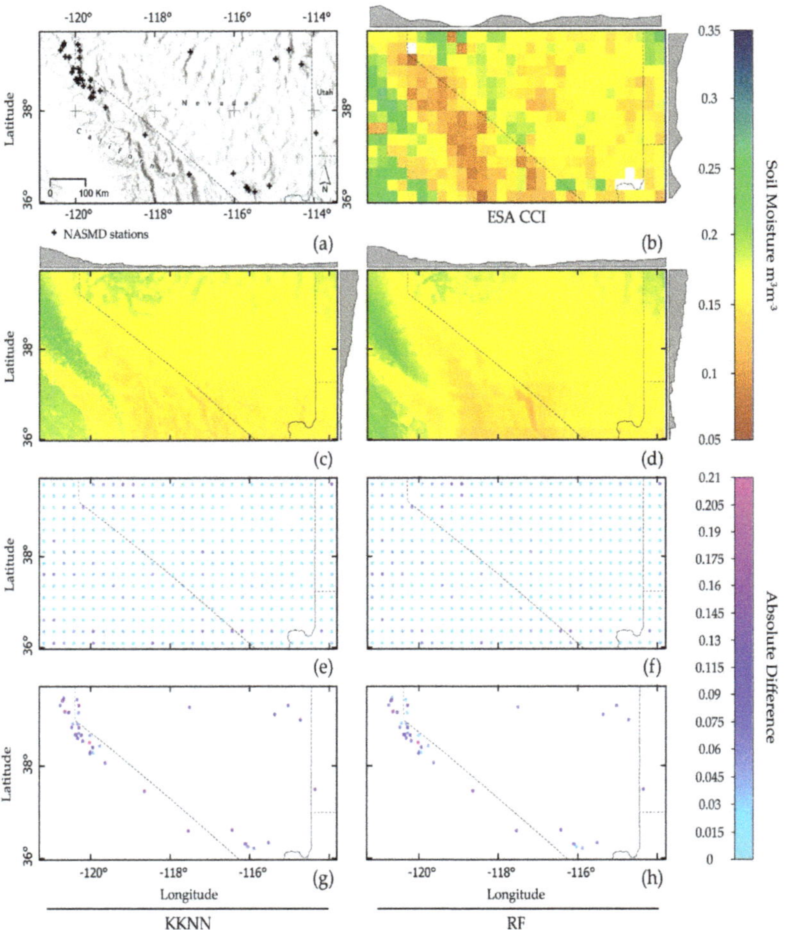

Figure 9. (**a**) West ROI and distribution of NASMD stations throughout 2010; (**b**) mean soil moisture values of 12 monthly and 52 weekly layers based on the reference ESA-CCI values at 0.25 degrees of spatial resolution; (**c**,**d**) mean values of 1 km soil moisture predictions with KKNN and RF; (**e**,**f**) spatial distribution of mean absolute differences between ESA-CCI sampling points at 0.25 degrees and their spatially correspondent predicted soil moisture values in all layers at 1 km with KKNN and RF; (**g**,**h**) spatial distribution of mean absolute differences between all monthly and weekly soil moisture values from NASMD and predicted values at 1 km using the two methods tested.

4. Discussion

Our work shows the performance of two methods within the SOMOSPIE framework for downscaling satellite-derived soil moisture values. We used two ROIs with different topographic and climatic characteristics to compare the performance of the framework. Given the limitations in obtaining field-based measurements of soil moisture over large areas, flexible and adaptable frameworks are alternatives to obtain spatially and temporally detailed information. The SOMOSPIE framework offers an alternative approach to downscale satellite-derived soil moisture and to traditional predictions based on simple extrapolation and interpolation using information from monitoring networks [14,66,67].

Our framework demonstrates that it is possible to obtain soil moisture across different spatial and temporal scales, in relation to the resolution of the predictors and the temporal availability of the input satellite data. In our work, we used 1 km terrain parameters as predictors, but this framework could be extended to use topographic information at different spatial resolutions as input for further predictions. It is known that topography has different levels of influence on the spatial distribution of soil moisture [39], as previous studies have explored the impact of terrain characteristics at watershed and regional scales [40,42,44,45,68], and here, we showed that terrain parameters are suitable predictors at the regional scale. Although other environmental covariates, such as soil texture, surface temperature, and vegetation characteristics, are known to be correlated with the spatial and temporal distribution of soil moisture [3,39,40,69–72], these covariates did not offer significant advantages in our approach. First, soil texture is highly dependent on site-specific conditions [69] rather than our regional approach, while surface temperature and vegetation features might introduce bias that would hinder the effect of using solely terrain parameters as downscaling predictors [44].

We identified that latitude and longitude values, along with Aspect, Elevation, and Topographic Wetness Index, were the most suitable parameters to predict soil moisture at 1 km when using the two proposed methods. This aligns with previous studies that identified similar terrain parameters as relevant factors to derive soil moisture based on their relation with lateral distribution of water in the surface soil layer [40,43,73–76]. In general, we obtained better results with both algorithms in the Midwest ROI, where topographic characteristics are more homogenous than in the West ROI, with more complex terrain. Additionally, we saw similar patterns of soil moisture spatial distribution across coarse and fine scales, supporting previous work in downscaling satellite-derived soil moisture that found that spatial variability agrees with landscape heterogeneity [77]. We highlight that there is increasing evidence on how terrain parameters are useful for modeling soil moisture [39,74], but other environmental factors, such as precipitation, temperature, land cover, and soil properties [69,70,78], should be considered across different scenarios.

The SOMOSPIE framework takes advantage of daily values from the ESA-CCI soil moisture product, being able to predict soil moisture at different temporal scales (e.g., monthly, weekly). The comparison of predicted soil moisture across different periods helps to identify any temporal biases or patterns related to different environmental conditions throughout the year and identify emerging relationships with environmental factors at different points during wet-up and dry-down cycles [79,80]. In autumn and spring, topography becomes a more relevant indicator, whereas its importance decreases during summer and winter due to the influence of evapotranspiration, as well as extensive saturation and porosity control, respectively [74]. This might support the lower prediction performance observed during January and February in the West ROI, where topography plays a more important role in the spatial variability. Additionally, several studies have shown that more homogenous patterns of satellite-derived soil moisture occur under dry conditions, leading to an improved accuracy in satellite retrievals [81,82]. In this regard, the higher prediction accuracy we observed in the Midwest ROI might be linked to a lower retrieval error from ESA-CCI. This contrasts with the prediction accuracy in the West ROI, which might be impacted by a higher retrieval error of ESA-CCI, linked to more heterogeneous environmental conditions.

In general, we found that RF performed better at the monthly and weekly scales across both ROIs. This could be explained because this technique does not assume any particular geometric or functional form of the model. Furthermore, it is suitable in sampling spaces with sparse data [45], such as satellite-derived soil moisture in a coarse resolution, where the distance between pixels' centroids yields substantial separation between data points. In contrast, although KKNN showed a lower prediction performance than RF, this technique still offers advantages for soil moisture downscaling in other regions with high density of sample points based on its ability to build many simple models when more data are available [59].

We observed that the two methods tested showed a similar correspondence to ground-truth information as the original ESA-CCI values in most of the monthly and weekly periods in our experiments. However, KKNN predictions showed a slightly better correspondence with ground-truth information in comparison with RF (values reporting the absolute correlation and RMSE differences between ground-truth information and ESA-CCI, as well as ground-truth and KKNN and RF outputs, are presented in Supplementary Materials S3). Differences in correlation and RMSE values between the two ROIs might be related to the sparse and uneven spatial distribution of available ground-truth stations in the West region (Figure 2). Previous studies found that the optimal number of ground-truth points for validating satellite-derived soil moisture products ranges from 10 to 20 per pixel [75], which is far from the desirable distribution of field stations available in the West ROI.

Although our work aimed at identifying the effect of terrain parameters in downscaling satellite-derived soil moisture information, other parameters, such as surface temperature, vegetation indexes, surface albedo, land cover, and rainfall, have been widely considered in previous research [3,39,40,71,72,75,83,84] and represent an opportunity to evaluate the flexibility of the SOMOSPIE framework.

5. Conclusions

Based on our analysis, we conclude that there is no "best" method that can be defined for every place in the world, as different methods perform differently in each ROI. As has been acknowledged in previous research, different downscaling methods have their own applicability under certain purposes, closely linked to differences in surface and climate conditions, and every method must be calibrated before its implementation elsewhere [39]. Thus, we believe that SOMOSPIE is a flexible framework that should include the methods tested in our work but is able to expand to incorporate additional methods to be tested in other regions around the world.

Despite the advantages of modeling techniques, such as KKNN and RF, in predicting soil moisture at a fine spatial resolution, it is also important to consider the computational resources needed when selecting these methods. When the ROI does not represent a large number of locations where soil moisture will be predicted, the two methods can be applied with no major challenges, but when the sampling space surpasses hundreds of thousands of locations, the selection of the modeling method and the use of computational resources become more important. The understanding of suitable cyberinfrastructure to work with more extensive regions and soil moisture predictions at finer spatial scales (e.g., 100 m, 30 m), along with the implementation of additional modeling methods in SOMOSPIE, is still being addressed through current efforts.

Our research contributes an alternative approach for downscaling satellite-derived soil moisture using a modular spatial inference framework. Here, we tested two methods, but the framework is flexible so multiple algorithms can be included [58,85]. Additional efforts to improve the SOMOSPIE framework include developing a containerized environment that will facilitate the deployment and management of the entire workflow in High-Performance Computing (HPC) or cloud environments [86].

Supplementary Materials: The following are available online at https://www.mdpi.com/article/10.3390/rs14133137/s1, Supplementary Materials S1: Selection of most relevant terrain parameters used as predictors to estimate soil moisture at 1 km spatial resolution over the conterminous United States. Refs. [44,45,52,54–60,87] are cited in the Supplementary Materials S1. Supplementary Materials S2: Number of North American Soil Moisture Database available stations in 2010 over the two regions of interest. Supplementary Materials S3: Cross-validation and ground-truth validation tables of monthly and weekly soil moisture predictions.

Author Contributions: R.M.L., L.V., M.T. and R.V. conceived and designed the research. R.M.L. and L.V. performed the experiments and analysis. R.M.L. wrote the first draft of the manuscript with input from L.V., P.O., M.T. and R.V.; R.M.L. wrote the code for data and analysis visualization and made the cartographic edition of map figures. P.O. contributed to the optimization of analyses performance in cloud-based computing environments. All authors contributed to interpretation of the results, reviewed, and approved the manuscript. R.V. and M.T. supervised and coordinated the research team and managed funding acquisition. All authors have read and agreed to the published version of the manuscript.

Funding: This study was funded by a University of Delaware Strategic Initiative research grant and the NSF (OAC grants #2103854 and #2103836 "Software Ecosystem for kNowledge diScOveRY-a data-driven framework for soil moisture applications").

Data Availability Statement: Monthly and weekly soil moisture predictions at 1 km spatial resolution over the two regions of interested defined in this work can be accessed through HydroShare, https://doi.org/10.4211/hs.96eeb0d796a64b578f24e8154c166988 (accessed on 10 May 2022).

Acknowledgments: The authors want to acknowledge the support from the College of Agriculture and Natural Resources at the University of Delaware for supporting and facilitating research efforts among young scientists. We thank the UDIT Research Cyberinfrastructure unit for facilitating the analyses performed in this work through the Community Cluster Program. We are thankful for the valuable comments and contributions from current and former members of the Global computing Lab at the University of Tennessee, Knoxville: Danny Rorabaugh, Ria Patel, and Travis Johnston. RML and RV are thankful for the valuable comments and contributions from Mario Guevara.

Conflicts of Interest: The authors declare no conflict of interest.

References

1. Ward, S. *The Earth Observation Handbook—Climate Change Special Edition 2008*; Bond, P., Ed.; Committee on Earth Observation Satellites, European Space Agency: Noordwijk, The Netherlands, 2008; ISBN 978-92-9221-408-1.
2. Crow, W.T.; Wood, E.F. The Value of Coarse-Scale Soil Moisture Observations for Regional Surface Energy Balance Modeling. *J. Hydrometeorol.* **2002**, *3*, 467–482. [CrossRef]
3. Legates, D.R.; Mahmood, R.; Levia, D.F.; DeLiberty, T.L.; Quiring, S.M.; Houser, C.; Nelson, F.E. Soil moisture: A central and unifying theme in physical geography. *Prog. Phys. Geogr.* **2010**, *35*, 65–86. [CrossRef]
4. Williams, C.A.; Albertson, J.D. Soil moisture controls on canopy-scale water and carbon fluxes in an African savanna. *Water Resour. Res.* **2004**, *40*, 1–14. [CrossRef]
5. Hamlet, A.F.; Mote, P.W.; Clark, M.P.; Lettenmaier, D.P. Twentieth-Century Trends in Runoff, Evapotranspiration, and Soil Moisture in the Western United States. *J. Clim.* **2007**, *20*, 1468–1486. [CrossRef]
6. Narasimhan, B.; Srinivasan, R. Development and evaluation of Soil Moisture Deficit Index (SMDI) and Evapotranspiration Deficit Index (ETDI) for agricultural drought monitoring. *Agric. For. Meteorol.* **2005**, *133*, 69–88. [CrossRef]
7. Davidson, E.A.; Belk, E.; Boone, R.D. Soil water content and temperature as independent or confounded factors controlling soil respiration in a temperate mixed hardwood forest. *Glob. Chang. Biol.* **1998**, *4*, 217–227. [CrossRef]
8. Falloon, P.; Jones, C.D.; Ades, M.; Paul, K. Direct soil moisture controls of future global soil carbon changes: An important source of uncertainty. *Glob. Biogeochem. Cycles* **2011**, *25*, 1–14. [CrossRef]
9. Vargas, R.; Allen, M.F. Environmental controls and the influence of vegetation type, fine roots and rhizomorphs on diel and seasonal variation in soil respiration. *New Phytol.* **2008**, *179*, 460–471. [CrossRef]
10. Schaufler, G.; Kitzler, B.; Schindlbacher, A.; Skiba, U.; Sutton, M.A.; Zechmeister-Boltenstern, S. Greenhouse gas emissions from European soils under different land use: Effects of soil moisture and temperature. *Eur. J. Soil Sci.* **2010**, *61*, 683–696. [CrossRef]
11. Vargas, R.; Baldocchi, D.D.; Allen, M.F.; Bahn, M.; Black, T.A.; Collins, S.L.; Yuste, J.C.; Hirano, T.; Jassal, R.S.; Pumpanen, J.; et al. Looking deeper into the soil: Biophysical controls and seasonal lags of soil CO_2 production and efflux. *Ecol. Appl.* **2010**, *20*, 1569–1582. [CrossRef]
12. Vargas, R.; Detto, M.; Baldocchi, D.D.; Allen, M.F. Multiscale analysis of temporal variability of soil CO_2 production as influenced by weather and vegetation. *Glob. Chang. Biol.* **2010**, *16*, 1589–1605. [CrossRef]

13. Baldocchi, D. "Breathing" of the terrestrial biosphere: Lessons learned from a global network of carbon dioxide flux measurement systems. *Aust. J. Bot.* **2008**, *56*, 1–26. [CrossRef]
14. Chen, H.; Fan, L.; Wu, W.; Liu, H.-B. Comparison of spatial interpolation methods for soil moisture and its application for monitoring drought. *Environ. Monit. Assess.* **2017**, *189*, 525. [CrossRef] [PubMed]
15. Dorigo, W.; Wagner, W.; Albergel, C.; Albrecht, F.; Balsamo, G.; Brocca, L.; Chung, D.; Ertl, M.; Forkel, M.; Gruber, A.; et al. ESA CCI Soil Moisture for improved Earth system understanding: State-of-the art and future directions. *Remote Sens. Environ.* **2017**, *203*, 185–215. [CrossRef]
16. Martínez-Fernández, J.; González-Zamora, A.; Sánchez, N.; Gumuzzio, A.; Herrero-Jiménez, C.M. Satellite soil moisture for agricultural drought monitoring: Assessment of the SMOS derived Soil Water Deficit Index. *Remote Sens. Environ.* **2016**, *177*, 277–286. [CrossRef]
17. Crow, W.T. Utility of soil moisture data products for natural disaster applications. In *Extreme Hydroclimatic Events and Multivariate Hazards in a Changing Environment*; Maggioni, V., Nassari, C., Eds.; Elsevier: San Diego, CA, USA, 2019; pp. 65–85.
18. Engman, E.T. Applications of microwave remote sensing of soil moisture for water resources and agriculture. *Remote Sens. Environ.* **1991**, *35*, 213–226. [CrossRef]
19. Kimmins, J.P. From science to stewardship: Harnessing forest ecology in the service of society. *For. Ecol. Manag.* **2008**, *256*, 1625–1635. [CrossRef]
20. Koster, R.D.; Suarez, M.J. Soil Moisture Memory in Climate Models. *J. Hydrometeorol.* **2001**, *2*, 558–570. [CrossRef]
21. Meehl, G.A.; Washington, W.M. A Comparison of Soil-Moisture Sensitivity in Two Global Climate Models. *J. Atmos. Sci.* **1988**, *45*, 1476–1492. [CrossRef]
22. Seneviratne, S.I.; Corti, T.; Davin, E.L.; Hirschi, M.; Jaeger, E.B.; Lehner, I.; Orlowsky, B.; Teuling, A.J. Investigating soil moisture-climate interactions in a changing climate: A review. *Earth-Sci. Rev.* **2010**, *99*, 125–161. [CrossRef]
23. Walker, J.P.; Willgoose, G.R.; Kalma, J.D. In situ measurement of soil moisture: A comparison of techniques. *J. Hydrol.* **2004**, *293*, 85–99. [CrossRef]
24. Martínez-Fernández, J.; Ceballos, A. Temporal Stability of Soil Moisture in a Large-Field Experiment in Spain. *Soil Sci. Soc. Am. J.* **2003**, *67*, 1647–1656. [CrossRef]
25. Robock, A.; Mu, M.; Vinnikov, K.; Trofimova, I.V.; Adamenko, T.I. Forty-five years of observed soil moisture in the Ukraine: No summer desiccation (yet). *Geophys. Res. Lett.* **2005**, *32*, L03401. [CrossRef]
26. Dorigo, W.A.; Xaver, A.; Vreugdenhil, M.; Gruber, A.; Hegyiová, A.; Sanchis-Dufau, A.D.; Zamojski, D.; Cordes, C.; Wagner, W.; Drusch, M. Global Automated Quality Control of In Situ Soil Moisture Data from the International Soil Moisture Network. *Vadose Zone J.* **2013**, *12*, 1–21. [CrossRef]
27. Dorigo, W.A.; Wagner, W.; Hohensinn, R.; Hahn, S.; Paulik, C.; Xaver, A.; Gruber, A.; Drusch, M.; Mecklenburg, S.; van Oevelen, P.; et al. The International Soil Moisture Network: A data hosting facility for global in situ soil moisture measurements. *Hydrol. Earth Syst. Sci.* **2011**, *15*, 1675–1698. [CrossRef]
28. Schaefer, G.L.; Cosh, M.H.; Jackson, T.J. The USDA Natural Resources Conservation Service Soil Climate Analysis Network (SCAN). *J. Atmos. Ocean. Technol.* **2007**, *24*, 2073–2077. [CrossRef]
29. Quiring, S.M.; Ford, T.W.; Wang, J.K.; Khong, A.; Harris, E.; Lindgren, T.; Goldberg, D.W.; Li, Z. The North American Soil Moisture Database: Development and Applications. *Bull. Am. Meteorol. Soc.* **2016**, *97*, 1441–1459. [CrossRef]
30. Entekhabi, D.; Njoku, E.G.; O'Neill, P.E.; Kellogg, K.H.; Crow, W.T.; Edelstein, W.N.; Entin, J.K.; Goodman, S.D.; Jackson, T.J.; Johnson, J.; et al. The Soil Moisture Active Passive (SMAP) Mission. *Proc. IEEE* **2010**, *98*, 704–716. [CrossRef]
31. Das, N.N.; Entekhabi, D.; Dunbar, R.S.; Colliander, A.; Chen, F.; Crow, W.; Jackson, T.J.; Berg, A.; Bosch, D.D.; Caldwell, T.; et al. The SMAP mission combined active-passive soil moisture product at 9 km and 3 km spatial resolutions. *Remote Sens. Environ.* **2018**, *211*, 204–217. [CrossRef]
32. Liu, Y.Y.; Parinussa, R.M.; Dorigo, W.A.; De Jeu, R.A.M.; Wagner, W.; van Dijk, A.I.J.M.; McCabe, M.F.; Evans, J.P. Developing an improved soil moisture dataset by blending passive and active microwave satellite-based retrievals. *Hydrol. Earth Syst. Sci.* **2011**, *15*, 425–436. [CrossRef]
33. Peng, J.; Loew, A. Recent Advances in Soil Moisture Estimation from Remote Sensing. *Water* **2017**, *9*, 530. [CrossRef]
34. Mohanty, B.P.; Cosh, M.H.; Lakshmi, V.; Montzka, C. Soil Moisture Remote Sensing: State-of-the-Science. *Vadose Zone J.* **2017**, *16*, 1–9. [CrossRef]
35. Barre, H.M.J.P.; Duesmann, B.; Kerr, Y.H. SMOS: The Mission and the System. *IEEE Trans. Geosci. Remote Sens.* **2008**, *46*, 587–593. [CrossRef]
36. Paloscia, S.; Pettinato, S.; Santi, E.; Notarnicola, C.; Pasolli, L.; Reppucci, A. Soil moisture mapping using Sentinel-1 images: Algorithm and preliminary validation. *Remote Sens. Environ.* **2013**, *134*, 234–248. [CrossRef]
37. Srivastava, P.K.; Pandey, P.C.; Petropoulos, G.P.; Kourgialas, N.N.; Pandey, V.; Singh, U. GIS and Remote Sensing Aided Information for Soil Moisture Estimation: A Comparative Study of Interpolation Techniques. *Resources* **2019**, *8*, 70. [CrossRef]
38. Dorigo, W.A.; Gruber, A.; De Jeu, R.A.M.; Wagner, W.; Stacke, T.; Loew, A.; Albergel, C.; Brocca, L.; Chung, D.; Parinussa, R.M.; et al. Evaluation of the ESA CCI soil moisture product using ground-based observations. *Remote Sens. Environ.* **2015**, *162*, 380–395. [CrossRef]
39. Peng, J.; Loew, A.; Merlin, O.; Verhoest, N.E.C. A review of spatial downscaling of satellite remotely sensed soil moisture. *Rev. Geophys.* **2017**, *55*, 341–366. [CrossRef]

40. Busch, F.A.; Niemann, J.D.; Coleman, M. Evaluation of an empirical orthogonal function-based method to downscale soil moisture patterns based on topographical attributes. *Hydrol. Process.* **2012**, *26*, 2696–2709. [CrossRef]
41. Ranney, K.J.; Niemann, J.D.; Lehman, B.M.; Green, T.R.; Jones, A.S. A method to downscale soil moisture to fine resolutions using topographic, vegetation, and soil data. *Adv. Water Resour.* **2015**, *76*, 81–96. [CrossRef]
42. Coleman, M.L.; Niemann, J.D. Controls on topographic dependence and temporal instability in catchment-scale soil moisture patterns. *Water Resour. Res.* **2013**, *49*, 1625–1642. [CrossRef]
43. Droesen, J.M. Downscaling Soil Moisture Using Topography—The Evaluation and Optimisation of a Downscaling Approach. Master's Thesis, Wageningen University, Wageningen, The Netherlands, 2016.
44. Guevara, M.; Vargas, R. Downscaling satellite soil moisture using geomorphometry and machine learning. *PLoS ONE* **2019**, *14*, e0219639. [CrossRef] [PubMed]
45. Rorabaugh, D.; Guevara, M.; Llamas, R.; Kitson, J.; Vargas, R.; Taufer, M. SOMOSPIE: A Modular SOil MOisture SPatial Inference Engine Based on Data-Driven Decisions. In Proceedings of the 2019 15th International Conference on eScience (eScience), IEEE, San Diego, CA, USA, 24–27 September 2019; pp. 1–10. [CrossRef]
46. Kitson, T.; Olaya, P.; Racca, E.; Wyatt, M.R.; Guevara, M.; Vargas, R.; Taufer, M. Data analytics for modeling soil moisture patterns across united states ecoclimatic domains. In Proceedings of the 2017 IEEE International Conference on Big Data (Big Data), Boston, MA, USA, 11–14 December 2017; pp. 4768–4770.
47. McKinney, R.; Pallipuram, V.K.; Vargas, R.; Taufer, M. From HPC Performance to Climate Modeling: Transforming Methods for HPC Predictions into Models of Extreme Climate Conditions. In Proceedings of the 2015 IEEE 11th International Conference on e-Science, Munich, Germany, 31 August–4 September 2015; pp. 108–117.
48. Llamas, R.M.; Guevara, M.; Rorabaugh, D.; Taufer, M.; Vargas, R. Spatial Gap-Filling of ESA CCI Satellite-Derived Soil Moisture Based on Geostatistical Techniques and Multiple Regression. *Remote Sens.* **2020**, *12*, 665. [CrossRef]
49. Brock, F.V.; Crawford, K.C.; Elliott, R.L.; Cuperus, G.W.; Stadler, S.J.; Johnson, H.L.; Eilts, M.D. The Oklahoma Mesonet: A Technical Overview. *J. Atmos. Ocean. Technol.* **1995**, *12*, 5–19. [CrossRef]
50. Hirschi, M.; Nicolai-Shaw, N.; Preimesberger, W.; Scanlon, T.; Dorigo, W.; Kidd, R. *Product Validation and Intercomparison Report (PVIR), Supporting Product, version v06.1*; European Space Agency: Vienna, Austria, 2021.
51. van der Schalie, R.; Preimesberger, W.; Pasik, A.; Scanlon, T.; Kidd, R. *ESA Climate Change Initiative Plus Soil Moisture, Product User Guide, Supporting Product, version v06.1*; European Space Agency: Vienna, Austria, 2021.
52. Becker, J.J.; Sandwell, D.T.; Smith, W.H.F.; Braud, J.; Binder, B.; Depner, J.; Fabre, D.; Factor, J.; Ingalls, S.; Kim, S.-H.; et al. Global Bathymetry and Elevation Data at 30 Arc Seconds Resolution: SRTM30_PLUS. *Mar. Geod.* **2009**, *32*, 355–371. [CrossRef]
53. Guevara, M.; Vargas, R. Annual Soil Moisture Predictions across Conterminous United States Using Remote Sensing and Terrain Analysis across 1 km Grids (1991–2016). 2019. Available online: https://doi.org/10.4211/hs.b8f6eae9d89241cf8b5904033460af61 (accessed on 17 February 2022).
54. Brenning, A.; Bangs, D.; Becker, M. RSAGA: SAGA Geoprocessing and Terrain Analysis in R (1.3.0). 2008. Available online: https://github.com/r-spatial/RSAGA (accessed on 23 July 2021).
55. Conrad, O.; Bechtel, B.; Bock, M.; Dietrich, H.; Fischer, E.; Gerlitz, L.; Wehberg, J.; Wichmann, V.; Böhner, J. System for Automated Geoscientific Analyses (SAGA) v. 2.1.4. *Geosci. Model Dev.* **2015**, *8*, 1991–2007. [CrossRef]
56. R Core Team. *R: A Language and Environment for Statistical Computing (4.0.3)*; R Foundation for Statistical Computing: Vienna, Austria, 2020; Available online: https://www.r-project.org/ (accessed on 27 August 2021).
57. UDIT Research CyberInfrastructure CAVINESS, Supporting Researchers at University of Delaware. Available online: https://sites.udel.edu/it-rci/compute/community-cluster-program/caviness/ (accessed on 23 August 2021).
58. Johnston, T.; Zanin, C.; Taufer, M. HYPPO: A Hybrid, Piecewise Polynomial Modeling Technique for Non-Smooth Surfaces. In Proceedings of the 2016 28th International Symposium on Computer Architecture and High Performance Computing (SBAC-PAD), Los Angeles, CA, USA, 26–28 October 2016; pp. 26–33. [CrossRef]
59. Cover, T.; Hart, P. Nearest neighbor pattern classification. *IEEE Trans. Inf. Theory* **1967**, *13*, 21–27. [CrossRef]
60. Hechenbichler, K.; Schliep, K. *Weighted k-Nearest-Neighbor Techniques and Ordinal Classification*; Collaborative Research Center 386, Discussion Paper 399; Ludwig-Maximilians-Universität München: Munich, Germany, 2004. [CrossRef]
61. Meinshausen, N. Quantile Regression Forests. *J. Mach. Learn. Res.* **2006**, *7*, 983–999.
62. Breiman, L. Bagging Predictors. *Mach. Learn.* **1996**, *24*, 123–140. [CrossRef]
63. Breiman, L. Random forests. *Mach. Learn.* **2001**, *45*, 5–32. [CrossRef]
64. Llamas, R.M.; Valera, L.; Olaya, P.; Taufer, M.; Vargas, R. 1-km Soil Moisture Predictions in the United States with SOMOSPIE Framework. 2022. Available online: https://doi.org/10.4211/hs.96eeb0d796a64b578f24e8154c166988 (accessed on 10 May 2022).
65. Taylor, K.E. Summarizing multiple aspects of model performance in a single diagram. *J. Geophys. Res.* **2001**, *106*, 7183–7192. [CrossRef]
66. Bárdossy, A.; Lehmann, W. Spatial distribution of soil moisture in a small catchment. Part 1: Geostatistical analysis. *J. Hydrol.* **1998**, *206*, 1–15. [CrossRef]
67. Ding, Y.; Wang, Y.; Miao, Q. Research on the spatial interpolation methods of soil moisture based on GIS. In Proceedings of the International Conference on Information Science and Technology, ICIST 2011, Nanjing, China, 26–28 March 2011; pp. 709–711. [CrossRef]

68. Escorihuela, M.J.; Quintana-Seguí, P. Comparison of remote sensing and simulated soil moisture datasets in Mediterranean landscapes. *Remote Sens. Environ.* **2016**, *180*, 99–114. [CrossRef]
69. Loew, A.; Mauser, W. On the Disaggregation of Passive Microwave Soil Moisture Data using a Priori Knowledge of Temporally Persistent Soil Moisture Fields. In Proceedings of the IGARSS 2008—2008 IEEE International Geoscience and Remote Sensing Symposium, Boston, MA, USA, 7–11 July 2008; Volume 3, pp. III-226–III-229.
70. Mattikalli, N.M.; Engman, E.T.; Jackson, T.J.; Ahuja, L.R. Microwave remote sensing of temporal variations of brightness temperature and near-surface soil water content during a watershed-scale field experiment, and its application to the estimation of soil physical properties. *Water Resour. Res.* **1998**, *34*, 2289–2299. [CrossRef]
71. Merlin, O.; Chehbouni, A.; Kerr, Y.H.; Goodrich, D.C. A downscaling method for distributing surface soil moisture within a microwave pixel: Application to the Monsoon '90 data. *Remote Sens. Environ.* **2006**, *101*, 379–389. [CrossRef]
72. Kovačević, J.; Cvijetinović, Ž.; Stančić, N.; Brodić, N.; Mihajlović, D. New downscaling approach using ESA CCI SM products for obtaining high resolution surface soil moisture. *Remote Sens.* **2020**, *12*, 1119. [CrossRef]
73. Western, A.W.; Blöschl, G. On the spatial scaling of soil moisture. *J. Hydrol.* **1999**, *217*, 203–224. [CrossRef]
74. Western, A.W.; Grayson, R.B.; Blöschl, G.; Willgoose, G.R.; McMahon, T.A. Observed spatial organization of soil moisture and its relation to terrain indices. *Water Resour. Res.* **1999**, *35*, 797–810. [CrossRef]
75. Crow, W.T.; Berg, A.A.; Cosh, M.H.; Loew, A.; Mohanty, B.P.; Panciera, R.; de Rosnay, P.; Ryu, D.; Walker, J.P. Upscaling sparse ground-based soil moisture observations for the validation of coarse-resolution satellite soil moisture products. *Rev. Geophys.* **2012**, *50*, 1–20. [CrossRef]
76. Julien, P.Y.; Moglen, G.E. Similarity and length scale for spatially varied overland flow. *Water Resour. Res.* **1990**, *26*, 1819–1832. [CrossRef]
77. van der Velde, R.; Salama, M.S.; Eweys, O.A.; Wen, J.; Wang, Q. Soil Moisture Mapping Using Combined Active/Passive Microwave Observations Over the East of the Netherlands. *IEEE J. Sel. Top. Appl. Earth Obs. Remote Sens.* **2015**, *8*, 4355–4372. [CrossRef]
78. Kim, G.; Chung, J.; Kim, J. Spatial characterization of soil moisture estimates from the Southern Great Plain (SGP 97) hydrology experiment. *KSCE J. Civ. Eng.* **2002**, *6*, 177–184. [CrossRef]
79. Panciera, R. Effect of Land Surface Heterogeneity on Satellite Near-Surface Soil Moisture Observations. Ph.D. Thesis, University of Melbourne, Melbourne, Australia, 2009.
80. Vachaud, G.; Passerat De Silans, A.; Balabanis, P.; Vauclin, M. Temporal Stability of Spatially Measured Soil Water Probability Density Function. *Soil Sci. Soc. Am. J.* **1985**, *49*, 822–828. [CrossRef]
81. Wigneron, J.-P.; Waldteufel, P.; Chanzy, A.; Calvet, J.-C.; Kerr, Y. Two-Dimensional Microwave Interferometer Retrieval Capabilities over Land Surfaces (SMOS Mission). *Remote Sens. Environ.* **2000**, *73*, 270–282. [CrossRef]
82. Friesen, J.; Rodgers, C.; Ogunrunde, P.G.; Hendrickx, J.M.H.; Van De Giesen, N. Hydrotope-based protocol to determine average soil moisture over large areas for satellite calibration and validation with results from an observation campaign in the Volta Basin, West Africa. *IEEE Trans. Geosci. Remote Sens.* **2008**, *46*, 1995–2004. [CrossRef]
83. Kim, G.; Barros, A. Downscaling of remotely sensed soil moisture with a modified fractal interpolation method using contraction mapping and ancillary data. *Remote Sens. Environ.* **2002**, *83*, 400–413. [CrossRef]
84. Temimi, M.; Leconte, R.; Chaouch, N.; Sukumal, P.; Khanbilvardi, R.; Brissette, F. A combination of remote sensing data and topographic attributes for the spatial and temporal monitoring of soil wetness. *J. Hydrol.* **2010**, *388*, 28–40. [CrossRef]
85. Johnston, T.; Alsulmi, M.; Cicotti, P.; Taufer, M. Performance tuning of MapReduce jobs using surrogate-based modeling. *Procedia Comput. Sci.* **2015**, *51*, 49–59. [CrossRef]
86. Olaya, P.; Kennedy, D.; Llamas, R.; Valera, L.; Vargas, R.; Lofstead, J.; Taufer, M. Building Trust in Earth Science Findings through Data Traceability and Results Explainability. *Trans. Parallel Distrib. Syst.* **2022**. submitted.
87. Hallema, D.W.; Moussa, R.; Sun, G.; Mcnulty, S.G. Surface storm flow prediction on hillslopes based on topography and hydrologic connectivity. *Ecol. Process.* **2016**, *5*, 13. [CrossRef]

Article

A Novel Method for Long Time Series Passive Microwave Soil Moisture Downscaling over Central Tibet Plateau

Hongtao Jiang [1,2], Sanxiong Chen [3,*], Xinghua Li [4], Jingan Wu [5], Jing Zhang [1] and Longfeng Wu [3]

1 College of Resources and Environment, Zhongkai University of Agriculture and Engineering, Guangzhou 510230, China; jianghongtao@zhku.edu.cn (H.J.); zhangjing201306@sina.com (J.Z.)
2 Academy of Intelligent Agricultural Engineering Innovations, Zhongkai University of Agriculture and Engineering, Guangzhou 510230, China
3 College of Horticulture and Landscape Architecture, Zhongkai University of Agriculture and Engineering, Guangzhou 510230, China; wulongfeng@zhku.edu.cn
4 School of Resources and Environmental Sciences, Wuhan University, Wuhan 430070, China; lixinghua5540@whu.edu.cn
5 School of Geospatial Engineering and Science, Sun Yat-sen University, Zhuhai 519082, China; wujg5@mail.sysu.edu.cn
* Correspondence: chensanxiong@zhku.edu.cn

Citation: Jiang, H.; Chen, S.; Li, X.; Wu, J.; Zhang, J.; Wu, L. A Novel Method for Long Time Series Passive Microwave Soil Moisture Downscaling over Central Tibet Plateau. *Remote Sens.* **2022**, *14*, 2902. https://doi.org/10.3390/rs14122902

Academic Editors: Wei Zhao, Jian Peng, Hongliang Ma, Chunfeng Ma and Jiangyuan Zeng

Received: 8 March 2022
Accepted: 13 June 2022
Published: 17 June 2022

Abstract: The coarse scale of passive microwave surface soil moisture (SSM) is not suitable for regional agricultural and hydrological applications such as drought monitoring and irrigation management. The optical/thermal infrared (OTI) data-based passive microwave SSM downscaling method can effectively improve its spatial resolution to fine scale for regional applications. However, the estimation capability of SSM with long time series is limited by OTI data, which are heavily polluted by clouds. To reduce the dependence of the method on OTI data, an SSM retrieval and spatio-temporal fusion model (SMRFM) is proposed in the study. Specifically, a model coupling in situ data, MODerate-resolution Imaging Spectro-radiometer (MODIS) OTI data, and topographic information is developed to retrieve MODIS SSM (1 km) using the least squares method. Then the retrieved MODIS SSM and the spatio-temporal fusion model are employed to downscale the passive microwave SSM from coarse scale to 1 km. The proposed SMRFM is implemented in a grassland dominated area over Naqu, central Tibet Plateau, for Advanced Microwave Scanning Radiometer—Earth Observing System sensor (AMSR-E) SSM downscaling in unfrozen period. The in situ SSM and Noah land surface model 0.01° SSM are used to validate the estimated MODIS SSM with long time series. The evaluations show that the estimated MODIS SSM has the same temporal resolution with AMSR-E and obtains significantly improved detailed spatial information. Moreover, the temporal accuracy of estimated MODIS SSM against in situ data ($r = 0.673$, $\mu bRMSE = 0.070$ m^3/m^3) is better than the AMSR-E ($r = 0.661$, $\mu bRMSE = 0.111$ m^3/m^3). In addition, the temporal r of estimated MODIS SSM is obviously higher than that of Noah data. Therefore, this suggests that the SMRFM can be used to estimate MODIS SSM with long time series by AMSR-E SSM downscaling in the study. Overall, the study can provide help for the development and application of microwave SSM-related scientific research at the regional scale.

Keywords: long time series; microwave surface soil moisture downscaling; MODIS scale; spatio-temporal fusion model

1. Introduction

Surface soil moisture (SSM) of a depth less than 5 cm is an important part of the Earth's water resources and is a key factor controlling the energy and water exchange between the surface and the atmosphere [1,2]. It plays a vital role in the processes of precipitation, runoff, infiltration, evapotranspiration, and agricultural application [3–5]. How to accurately monitor SSM dynamic change on the Earth's surface is a hot topic in geoscience. Due to its characteristics of large coverage of surface changes, long duration, relatively low cost, and real-time dynamic monitoring [6,7], satellite remote sensing has become one of the effective technology approaches for SSM monitoring.

Microwave remote sensing has the advantages of wide range of Earth observation and all-weather monitoring of the real surface situation. Moreover, it is not disturbed by clouds and can penetrate the depth of surface soil to about 5 cm (e.g., C band) [8]. Since the 1970s, a series of active and passive microwave sensors have been utilized to monitor global SSM [3,8]. Based on the observed data of microwave remote sensing, microwave SSM products with varied spatial coverage, varied time coverage, and varied accuracy have been retrieved and released. The widely used global products are as follows: the Soil Moisture Active Passive (SMAP, 3 km, 9 km and 36 km) [9,10] and the Advanced Microwave Scanning Radiometer—Earth Observing System sensor (AMSR-E) datasets (25 km) released by the National Aeronautics and Space Administration (NASA) [11,12], the Soil Moisture and Ocean Salinity (SMOS, 25 km) [13,14] and the Climate Change Initiative (CCI, 0.25°) datasets released by the European Space Agency [3,15], the Advanced Microwave Scanning Radiometer 2 (AMSR2, 0.25°) dataset released by the Japan Aerospace Exploration Agency (JAXA) [16], and the FengYun-3 dataset (25 km) released by the China Meteorological Administration [17,18]. The revisit interval of the above satellites/sensors is 1–3 days. For the radar failure of SMAP, the C band Sentinel-1 data was to substitute the SMAP radar for global scale 3 km and 1 km SSM estimation. However, the temporal resolution degrades from 3 days to 12 days [10]. In addition, the coarse spatial resolution (tens of kilometers in pixel size) of passive microwave SSM products makes it difficult to meet the applications (e.g., drought monitoring and irrigation management) at the regional scale. Therefore, it is urgent to carry out research on passive microwave SSM downscaling, improve its spatial resolution, and upgrade the applications of passive microwave SSM from global scale to regional scale.

At present, passive microwave SSM downscaling mainly relies on high spatial resolution auxiliary data including optical/thermal infrared (OTI), radar, terrain, and other data [19–24]. The empirical and physical models between passive microwave SSM and auxiliary data are built for downscaling and for spatial resolution improvement. OTI data are seriously polluted by clouds [25], which means that the traditional downscaling methods often lack OTI data for spatial resolution improvement in long time series. In other words, the traditional downscaling method only realizes the downscaling of microwave soil moisture on some dates and does not make effective use of the long time series characteristics of microwave data. Therefore, the traditional microwave SSM downscaling methods struggle to estimate OTI-scale SSM with long time series effectively. In general, there is a compromise between temporal and spatial in remote-sensing data. To alleviate the temporal and spatial compromise contradiction faced by remote-sensing data, Gao et al. [26] proposed a spatio-temporal fusion model (STFM) in 2006. It assumes that the temporal change is the same at the varied scales. Once proposed and optimized, the model has been widely used in phenological analysis [27], vegetation monitoring [28], urban heat island monitoring [29,30], and other research fields, because it can estimate the OTI-scale data with long time series of many surface parameters, such as vegetation index [31], surface temperature [32], reflectance [33,34], and evapotranspiration [35,36]. However, STFM for OTI-scale SSM estimation is rarely reported due to the lack of OTI-scale (no more than 1 km in pixel size) SSM.

In the study, a SSM retrieval and fusion model (SMRFM) is proposed to estimate the OTI-scale SSM with long time series by passive microwave SSM downscaling. The proposed SMRFM will reduce the limitations of traditional microwave SSM downscaling method caused by the availability of OTI data. The SMRFM is implemented in two steps: the first step is to build an empirical equation to retrieve the MODerate-resolution Imaging Spectro-radiometer (MODIS) scale SSM using MODIS OTI data; the second step is to construct the SSM STFM using paired AMSR-E and MODIS SSM for the estimation of MODIS SSM (1 km) with long time series. The proposed SMRFM is implemented over the

Naqu, central Tibet Plateau. The estimated MODIS SSM by SMRFM is validated by the in situ data and Noah land surface model 0.01-degree SSM. Moreover, the estimated MODIS SSM not only can be used to understand the coupling process of land water, energy, and carbon cycles on a more precise scale [9], but also can be helpful for the practical application of regional moisture monitoring, crop production status monitoring, and yield estimation. Therefore, it can be considered that the study will provide convenience for the long time series and MODIS SSM estimation at the regional scale and will have important theoretical and practical significance.

2. Materials and Methods

2.1. Materials

2.1.1. Study Area and In Situ Surface Soil Moisture Data

The study area is located in Naqu, central Tibet Plateau (Figure 1a). The in situ SSM data used are from the Soil Moisture and Temperature Monitoring Network (SMTMN) in Naqu, which is deployed by Yang et al. [37]. The SMTMN covers $1° \times 1°$ geographical space (91.5° E–92.5° E, 31° N–32° N), which contains 57 in situ sites. The real time SSM of 0–5, 10, 20, and 40 cm is measured in volumetric water content using the EC-TM and 5TM monitoring equipment, which is manufactured by Decagon. The sensors measure SSM according to the sensitivity of soil dielectric permittivity to liquid soil water with an accuracy of 0.001 m^3/m^3. The SSM data is recorded every 30 min, and each record reflects the average of SSM over the past half-hour. A total of 48 SSM records are collected per day. The in situ data of SMTMN has been shared to the International Soil Moisture Network (https://ismn.geo.tuwien.ac.at/en/networks/?id=CTP_SMTMN, accessed on 15 September 2021), and the time range is from 2008 to 2016. The first layer (0–5 cm) of in situ data was selected in this study. Regarding soil texture of SMTMN, silt and sand are dominant components with a comparable magnitude, while clay content consistently maintains at a low level (less than 10%). The range of altitudes of in situ sites is 4450–5000 m and is grassland-dominated. The relatively homogeneous area is convenient for MODIS SSM retrieval. This area belongs to the sub-frigid climate zone. According to a previous study [38], the period from October to May is defined as the frozen period, as the land surface temperature (LST) is below 0 °C most of the time. Meanwhile, the other period of the year (June to September) is the unfrozen period. Soil freezing in frozen period has an adverse impact on SSM monitoring. To improve the reliability of the study, the proposed SMRFM is only implemented in unfrozen period.

2.1.2. Aqua AMSR-E Soil Moisture

The multi-frequency dual polarization AMSR-E sensor mounted on the Aqua satellite is developed by JAXA and can be used to monitor the changes of SSM [39,40]. The ascending and descending time of AMSR-E are 01:30 PM and 01:30 AM local time, respectively. The spatial resolution of SSM released by AMSR-E is ~25 km. The expected accuracy of AMSR-E is 0.06 m^3/m^3 in low-to-medium vegetation coverage areas [11]. A variety of products have been retrieved and released based on AMSR-E observations [41], the most notable of which are released by NASA and JAXA. The root mean square error (*RMSE*) of JAXA AMSR-E SSM (<0.12 m^3/m^3) is lower than NASA data (>0.16 m^3/m^3) in the study area shown in previous study [12]. Therefore, JAXA AMSR-E SSM is used for SSM downscaling in the study. For the sensor failure, AMSR-E could not continuously observe the Earth and release the SSM product after October 2011 [16]. It was officially retired after nearly ten years of in-orbit operation. The time range of SSM products released by AMSR-E data is from May 2002 to October 2011. The Shizuku satellite equipped with AMSR2 sensor was launched by JAXA for the replacement of AMSR-E in May 2012. It continues to carry out Earth observation and release global ~25 km SSM products [16]. To match the pixel size of MODIS data, AMSR-E data were resampled to 1 km using the cubic convolution interpolation method in ArcGIS software.

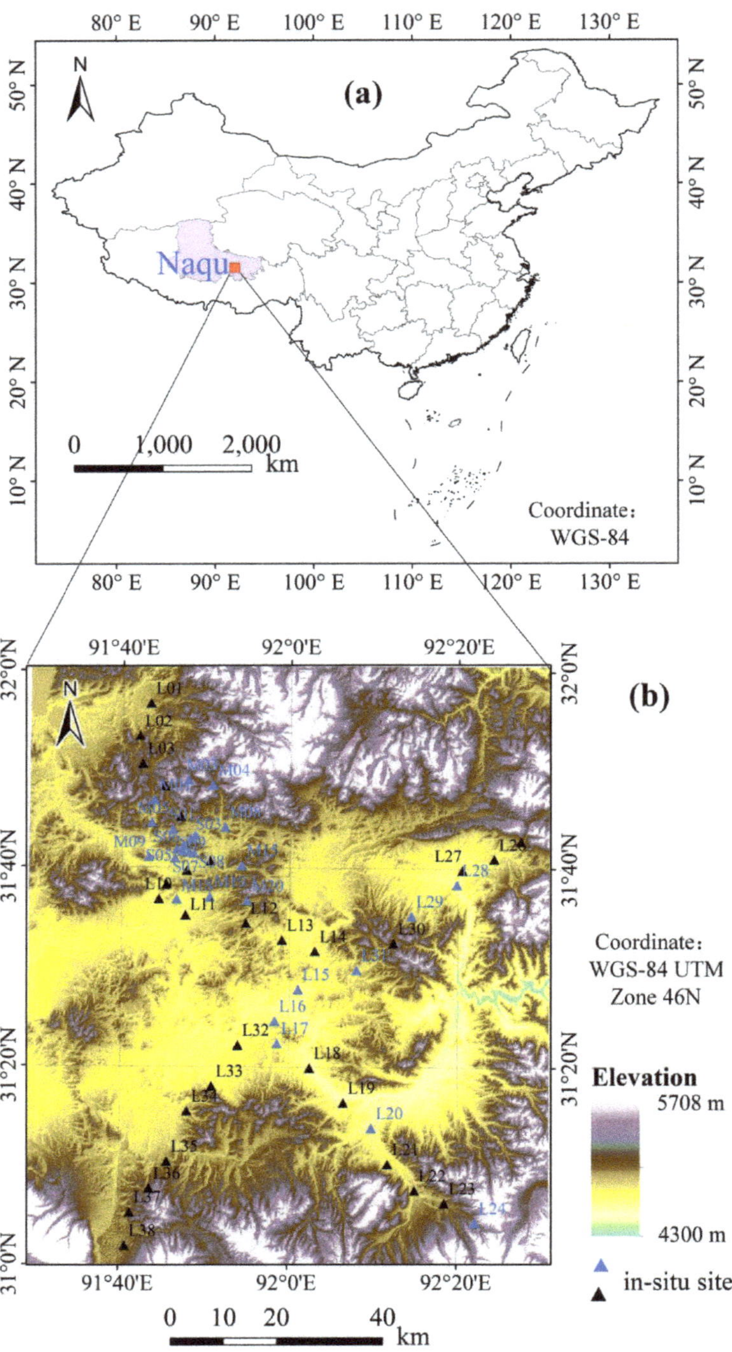

Figure 1. Location of the study area in southwest China (**a**) and the 57 in situ soil moisture sites (**b**). All in situ sites (in black and blue triangle) data is used for MODIS soil moisture retrieval and daily evaluation. The selected 29 in situ sites (in black triangle) data is used for temporal evaluation.

2.1.3. Aqua MODIS Optical and Thermal Infrared Data

MODIS sensors are carried on Terra and Aqua satellites. Aqua MODIS data is selected to reduce the adverse impact of AMSR-E and MODIS on observation time mismatch. The MYD11A1 daily 1 km LST data and the MYD13A1 8-day 500 m composited Normalized Difference Vegetation Index (NDVI) data are used to fit the empirical equation for MODIS SSM retrieval. Because of the composited product of MODIS NDVI, it assumes that the NDVI is constant in the 8-day composited date [42]. The LST gradients are normally reduced at nighttime, which is more beneficial to SSM retrieval [3]. Therefore, the Aqua MODIS LST at nighttime is used to eliminate the observation time difference between the two datasets for improved MODIS SSM retrieval. As the visible light cannot be used at nighttime, the Aqua MODIS NDVI data (visible light data) is used at daytime.

2.1.4. SRTM DEM Data

The SRTM digital elevation model (DEM) data, produced by NASA originally, are a major breakthrough in the digital mapping of the world. The 90 m SRTM DEM (version 4) used was downloaded from https://srtm.csi.cgiar.org/srtmdata/ (accessed on 12 October 2021) in this study. For more information about the used SRTM DEM, refer to [43]. After data mosaicking and clipping, the 90 m STRM DEM were resampled to 1 km pixel size using the cubic convolution interpolation method in ArcGIS software. As altitude (m) and slope (°) extracted from STRM DEM data play an important role in the redistribution of SSM, they were selected for MODIS SSM retrieval in the study.

2.1.5. Noah Land Surface Model L4 Central Asia Daily Soil Moisture

As OTI data are seriously polluted by clouds, it is difficult for traditional microwave SSM downscaling methods to effectively estimate MODIS SSM with long time series. Therefore, it is inappropriate to compare the SMRFM method proposed in the study with the traditional microwave SSM downscaling methods. Alternatively, the SSM data simulated by the land surface model were used as comparative data to verify the MODIS SSM with long time series estimated by SMRFM. The comparative data were acquired from the FLDAS Noah Land Surface Model L4 Central Asia Daily dataset (version 001) [44], which is simulated from the Noah 3.6.1 model in the Famine Early Warning Systems Network Land Data Assimilation System, adapted from Land Information System. This dataset contains a series of land surface parameters in a 0.01-degree spatial resolution over the Central Asia region (30–100° E, 21–56° N) from October 2000 to present. The four layers SSM data were comprised by the daily dataset and the top layer (0–10 cm) SSM in volumetric water content was used as the comparative dataset in the study. The 0.01-degree simulated SSM is resampled and then clipped to 1-km size for matching the pixel of fused MODIS SSM. In November 2020, all FLDAS Noah data were post-processed with the MOD44 MODIS land mask, so the simulated SSM data were missing over inland water in the study.

All data used in this study are shown in Table 1. As the temporal coverage of different data is different, the temporal intersection of AMSR-E and MODIS data in unfrozen period (1 August–31 September 2010 and 1 June–31 September 2011, six months in total) was used as the study period.

2.2. Methods

The MODIS SSM with long time series is estimated by downscaling AMSR-E SSM from coarse scale using proposed SMRFM. A reference MODIS SSM is retrieved from the coupling of in situ data, MODIS OTI data and topographic information. Then, the long time series AMSR-E SSM is downscaled to MODIS scale using the reference MODIS SSM and STFM. Therefore, SMRFM solves the difficulty of MODIS SSM acquisition in STFM and can be taken as an improvement for STFM in SSM.

Table 1. A general description of data used in the study. OTI is optical/thermal infrared, DEM is digital elevation model, AMSR-E is the Advanced Microwave Scanning Radiometer—Earth Observing System sensor, MODIS is the Moderate-resolution Imaging Spectro-radiometer, LST and NDVI is the land surface temporal and Normalized Difference Vegetation Index, SSM is the surface soil moisture.

	Microwave Data	OTI Data	Noah Data	DEM Data	In Situ Data
Sensor	AMSR-E	MODIS	/	SRTM	EC-TM, 5TM
Data Type	JAXA SSM	LST, NDVI	simulated SSM	altitude, slope	in situ SSM
Temporal coverage	May 2002 to 1 October 2011	May, 2002 to now	October 2000 to now	/	August 2010 to September 2016
Spatial resolution	~25 km	1 km/500 m	0.01°	90 m	/
Temporal resolution	1–3 days	daily	daily	Static	daily
unit	m^3/m^3	/	m^3/m^3	m, °	m^3/m^3

2.2.1. Data Pre-Processing

The average daily in situ SSM is used for further analysis. As the in situ SSM is 0–0.6 m^3/m^3 in SMTMN [37,39], AMSR-E data higher than 0.6 m^3/m^3 is excluded in the study. In addition, the pre-processing manners of in situ data are different for different application scenarios in the study. For reference MODIS SSM retrieval and daily evaluation, the daily 57 in situ data (the triangle shown in Figure 1b) is used. For temporal evaluation, the in situ data is selected according to the following four conditions: (1) the data quality of in situ data should be marked "G" (Good); (2) the monitored surface soil depth should be less than 5 cm at first layer; (3) the temporal correlation between in situ SSM and the corresponding AMSR-E SSM should be positive and pass the hypothesis test (p-value is less than 0.05). After the selection, 29 in situ sites (the black triangle shown in Figure 1b) are used for temporal evaluation in the study.

2.2.2. Spatio-Temporal Fusion Model

The SSM STFM (Figure 2) takes the known MODIS SSM and the corresponding AMSR-E SSM as the paired reference data at t_0 date and then again to fuse t_k date AMSR-E SSM for the unknown MODIS SSM estimation at the date. Notably, the date of reference data is taken as the reference date. The estimation needs to excavate the spatio-temporal correlation characteristics between AMSR-E and MODIS SSM without the help of other additional remote sensing auxiliary data. In practice, one or more paired reference datasets can be used to estimate MODIS SSM at t_k date. The more paired reference data that is used, the more restrictive the conditions of the model. As the main aim of the study is to verify the feasibility of the proposed SMRFM, only one paired reference dataset is used in the STFM.

From the reference date t_0 to the prediction date t_k, temporal variation of SSM can be fitted by a linear equation. For AMSR-E SSM (SSM_M), the linear equation is as follows:

$$SSM_M(t_k) = a(x, y, \Delta t) \times SSM_M(t_0) + b(x, y, \Delta t) \quad (1)$$

where $\Delta t = t_k - t_0$, a and b are the regression coefficients of the linear equation that are calculated by the least-squares method. As the model assumes that SSM has the same temporal change at different scales [45], the regression coefficients estimated at AMSR-E scale can be applied to MODIS scale. Thus, the MODIS SSM at date t_k can be estimated using Equation (2).

$$SSM_F(t_k) = a(x, y, \Delta t) \times SSM_F(t_0) + b(x, y, \Delta t) \quad (2)$$

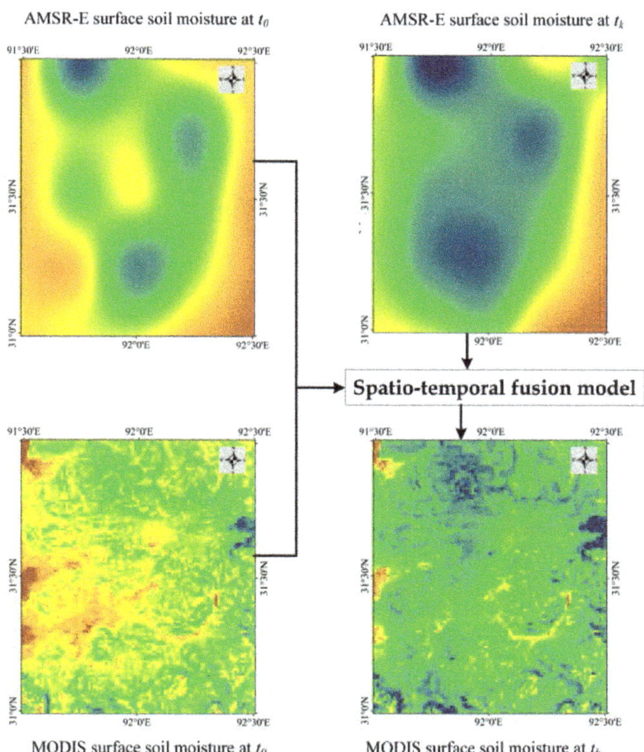

Figure 2. Schematic diagram of surface soil moisture spatio-temporal fusion model.

The change of SSM in each pixel may be different with the change of time. The fixed regression coefficients for the remotely sensed image may have a negative impact on prediction. To make the prediction result more accurate, the information of similar pixels in the neighborhood moving sliding window (e.g., 5 × 5) is used in the model. The prediction expression is as follows,

$$SSM_F(x_{w/2}, y_{w/2}, t_k) = \sum_{i=1}^{l} W(x_i, y_i, t_0) \times [a(x, y, \Delta t) \times SSM_F(t_0) + b(x, y, \Delta t)] \quad (3)$$

where w and $w/2$ are the size and center of moving window, respectively, (x_i, y_i) indicates similar pixels, and l is the number of similar pixels. Thus, the regression coefficients of each pixel may be different for MODIS SSM estimation. For the selection of similar pixels and the calculation of linear regression coefficients in the fusion model, please refer to [33] for details.

From the above equations, it can be seen that the paired AMSR-E and MODIS SSM at reference date t_0 and the AMSR-E SSM at t_k are used to predict the MODIS SSM at t_k. In this process, there is no need to rely on other remote-sensing auxiliary data. However, the MODIS SSM at reference date t_0 is also unknown in most cases. Therefore, the Aqua MODIS LST and NDVI data are used for SSM retrieval in the study and then again to estimate the MODIS SSM at reference date t_0.

2.2.3. Aqua MODIS Surface Soil Moisture Retrieval

OTI data cannot penetrate clouds, vegetation, and soil surface layers, and do not meet the conditions of the remote-sensing radiation transfer equation for SSM retrieval. Therefore, the SSM retrieval from OTI data lacks physical basis. In many cases, the use of OTI data to monitor SSM is mainly based on the correlation between SSM and remote-sensing surface parameters such as vegetation index, surface temperature [46], thermal inertia [47], surface reflectance [48], drought index [49,50]. Then, empirical equations between SSM and the remotely sensed surface parameters are established to retrieve regional SSM. The study develops a MODIS SSM retrieval model using in situ data, OTI data (LST and NDVI), altitude, and surface slope data which are calculated by the DEM data (Equation (4)).

$$SSM_F = a_1 \times LST + a_2 \times NDVI + a_3 \times altitude + a_4 \times Slope + a_5 \quad (4)$$

where a_i $(i = 1, 2, 3, 4, 5)$ are regression coefficients fitted by the least-squares method. To weaken the uncertainty caused by the spatial matching between the in situ SSM data and the remotely sensed pixel data, a 3 × 3 neighborhood average of the pixel corresponding to the in situ site location is taken as the matching value. Moreover, neighborhood average can weaken the information distortion that may exist in the single pixel value corresponding to the in situ site location. This can improve the robustness of MODIS SSM retrieval model (Equation (4)).

To estimate MODIS SSM using Equation (4), MODIS LST should meet the following two conditions: cloud-free and temperature higher than 0 °C (unfrozen soil). In general, the more training samples, the higher probability of the accuracy and stability of the fitting formula. To achieve this goal, the percentage of uncontaminated pixels in daily MODIS LST during the study period was calculated (Figure 3). This showed that the number of days for which the percentage of uncontaminated pixels is greater than 80% does not exceed 21 days. This suggests that number of days is relatively small for SSM estimation using Equation (4) in the study period. After careful screening, it was found that only 5 days of MODIS LST data were 100% uncontaminated. At the same time, the number of in situ sites for the days was counted. It was found that the number of effective in situ sites was the largest on 24 July 2011 among the 5 days, reaching as many as 48. Therefore, the MODIS SSM on 24 July 2011 is retrieved by Equation (4).

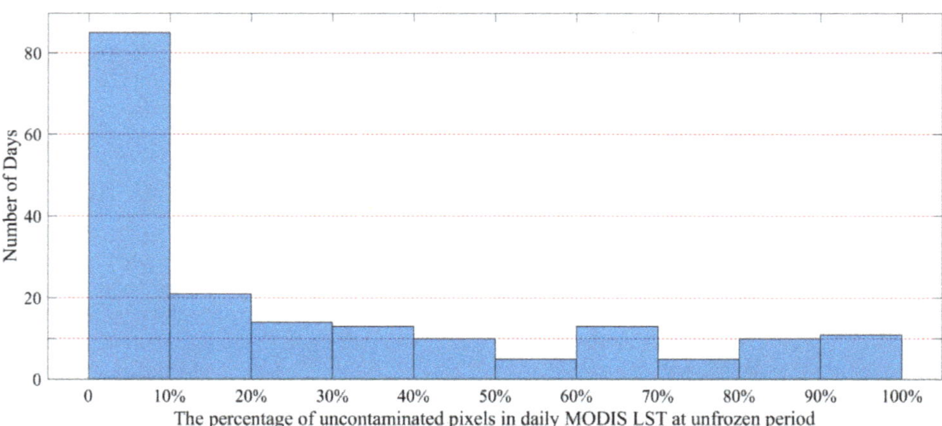

Figure 3. The histogram of percentage of uncontaminated pixels in daily MODIS LST at unfrozen period.

To avoid over-fitting of SSM retrieval equation (Equation (4)) using least-squares method, the 48-sample dataset (in situ data and its corresponding remote sensing data) on 24 July 2011 is sorted by in situ SSM in ascending order. Then, the dataset is divided into five subsets at an interval of five, and the second subset with a sample size of ten is taken as the validation dataset, and the remaining 38 samples are taken as the training dataset.

Theoretically, the average value of regional SSM should not change with the varied scale. AMSR-E SSM has a stronger theoretical basis than that of SSM retrieved from OTI data. Therefore, the regional SSM retrieved by microwave data should be better than that of OTI data in theory. To ensure the same regional average SSM between AMSR-E and MODIS SSM, the AMSR-E SSM is taken as the benchmark and is used to correct retrieved MODIS SSM using Equation (5).

$$SSM_{FC} = \frac{SSM_F}{average(SSM_F)} \times average(SSM_M) \quad (5)$$

where SSM_{FC} is the corrected MODIS SSM, *average ()* is the average of SSM. The corrected MODIS SSM is taken as the reference SSM in the STFM.

2.2.4. Evaluation Methods

The correlation coefficient (*r*), *RMSE*, *bias*, and the unbiased *RMSE* (*µbRMSE*) are used as the indicators for accuracy evaluation.

$$r = \frac{\sum_{i=1}^{n} (SSM_{pixel,i} - \overline{SSM_{pixel}}) \times (SSM_{ref,i} - \overline{SSM_{ref}})}{\sqrt{\sum_{i=1}^{n} (SSM_{pixel,i} - \overline{SSM_{pixel}})^2 \times \sum_{i=1}^{n} (SSM_{ref,i} - \overline{SSM_{ref}})^2}} \quad (6)$$

$$RMSE = \sqrt{\frac{\sum_{i=1}^{n} (SSM_{pixel,i} - SSM_{ref,i})^2}{n}} \quad (7)$$

$$bias = SSM_{pixel} - SSM_{ref} \quad (8)$$

$$\mu bRMSE = \sqrt{RMSE^2 - bias^2} \quad (9)$$

where $SSM_{pixel,i}$ and $\overline{SSM_{pixel}}$ are the pixel SSM and the average pixel SSM, and $SSM_{ref,i}$ and $\overline{SSM_{ref}}$ are the reference SSM and the average reference SSM. The direct and indirect evaluations are implemented to evaluate the accuracy of pixel SSM and to investigate the feasibility of the proposed SMRFM for MODIS SSM estimation.

The in situ data are taken as the reference SSM and then again to directly compare the difference between the in situ SSM and the pixel SSM neglecting the spatial matching difference. This method is often used for accuracy evaluation of the satellite based SSM in previous studies [23,51,52]. The in situ sites for temporal evaluation (the black triangle shown in Figure 1b) are evenly distributed throughout the study area, and their average value can be considered as the SSM at SMTMN scale. Therefore, the fused MODIS SSM, AMSR-E SSM, and Noah SSM are evaluated against in situ SSM at SMTMN scale. It can be used to evaluate the overall temporal accuracy of pixel SSM. To demonstrate the individual difference of pixel SSM at each in situ site, the temporal accuracy of pixel SSM against in situ data is calculated at MODIS scale. Like the evaluation at SMTMN scale, the temporal variation of pixel SSM is used to directly compare in situ data. Instead of using overall average of all selected in situ data, the observed temporal SSM is used at each site. In addition, the daily evaluation of pixel SSM against in situ data is also explored at MODIS scale in the study, so as to display all daily accuracies in pixel SSM. In general, the focus of temporal variation accuracy evaluation and daily accuracy evaluation are different. The former focuses on depicting temporal variation of SSM, and the temporal characteristics are emphasized. Therefore, the evaluation index pays more attention to temporal *r* and *µbRMSE*. Meanwhile, the latter focuses on describing spatial variation of SSM, and the characteristics of absolute value change are emphasized. The evaluation index pays more attention to *RMSE* and *bias*. Therefore, it is more convincing to carry out the evaluations in view of temporal variation and absolute value of pixel SSM at SMTMN scale and MODIS scale.

In fact, there is a spatial matching error between in situ data and pixel SSM, although evaluations based on in situ data are widely used. To eliminate the uncertainty of spatial matching, the triple collocation (TC) method is used for further evaluation. The TC method was proposed by Stoffelen [53] and was used to evaluate wind and wave height observations in oceanography. It was later introduced into remote-sensing SSM observation error estimation. For example, TC method is used to evaluate the global errors for ASCAT, AMSR-E, and ERA reanalysis SSM [54], which has shown that TC method is robust and can generate objective error estimates. There are four assumptions of TC method for temporal r estimation in the case of unknown truth values [55]: (1) there is a linear correlation between the three kinds of SSM and the unknown truth SSM; (2) the error is stable and does not change with temporal variation; (3) the errors of the three kinds of SSM are independent of each other; (4) the errors of the three kinds of SSM are independent of unknown truth values. As fused MODIS SSM and AMSR-E SSM are related to each other, a triplet pattern of in situ, Noah, and remote sensing SSM is built for the TC evaluation in the study. Two kinds of TC triplets are constructed: in situ Noah-fused MODIS SSM (TC1) and in situ Noah-AMSR-E SSM (TC2), so as to compare the temporal accuracy difference of the three kinds of pixel SSM at MODIS scale.

3. Results

3.1. Accuracy Analysis of MODIS Surface Soil Moisture Retrieval

The training and validation accuracy of the equation fitting for the retrieval of MODIS SSM is shown in Table 2.

Table 2. Fitting accuracy of MODIS surface soil moisture retrieval equation.

	RMSE (m^3/m^3)	r
Training accuracy	0.073	0.656
Validation accuracy	0.088	0.669

Table 2 shows that the fitted equation (Equation (4)) has a good robustness for the comparable accuracy of training and validation datasets. The *RMSE* is less than 0.09 m^3/m^3 and the r is higher than 0.65. This indicates that the fitted equation can estimate MODIS SSM well. Then, the fitted equation is applied to retrieve MODIS SSM on 24 July 2011 (Figure 4) in the study.

The spatial distribution of the AMSR-E and MODIS SSM is consistent as a whole, but there are still certain spatial and numerical differences between them (Figure 4). The range of AMSR-E is 0.132–0.548 m^3/m^3, with an average of 0.323 m^3/m^3. The range of retrieved MODIS SSM is 0.036–0.690 m^3/m^3, with an average of 0.365 m^3/m^3. The coefficient of variation for AMSR-E is 0.259, and for retrieved MODIS SSM is 0.189. This suggests that the AMSR-E is more discrete than the retrieved MODIS SSM.

The average values of the AMSR-E and MODIS SSM are different, which is consistent with our expectation. Thus, the corrected SSM is calculated using Equation (5). The spatial distribution of the corrected MODIS SSM (Figure 4c) has not changed obviously when compared to Figure 4b in visual representation. However, the spatial fitness between AMSR-E and MODIS SSM is slightly improved. *RMSE* between them has decreased from 0.112 m^3/m^3 to 0.099 m^3/m^3 after correction. Therefore, the corrected MODIS SSM (Figure 4c) and AMSR-E SSM (Figure 4a) are used to construct the paired reference datasets of STFM in the study.

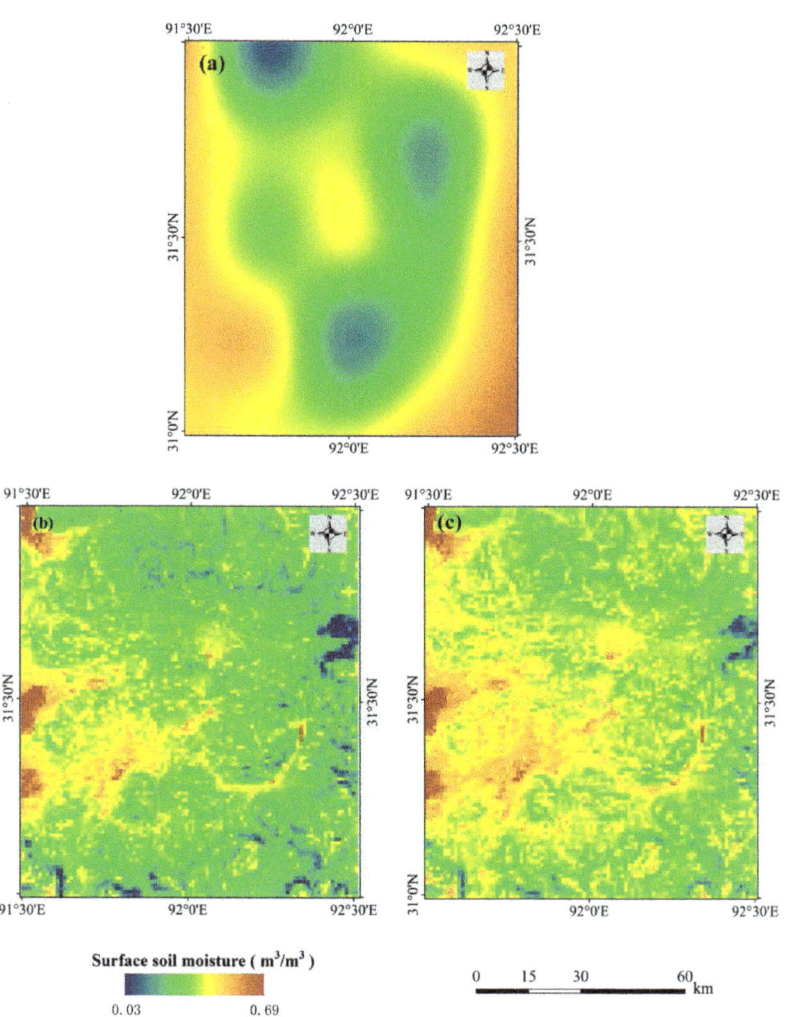

Figure 4. Surface soil moisture on 24 July 2011 in 1 km pixel size: (**a**) AMSR-E data resampled to MODIS scale; (**b**) the retrieved MODIS surface soil moisture; (**c**) the corrected MODIS surface soil moisture.

3.2. Fused MODIS Surface Soil Moisture

The long time series MODIS SSM is fused by SMRFM using one fixed paired reference dataset and the corresponding AMSR-E SSM at the unfrozen period. To validate the spatial downscaling ability of SMRFM, the spatial distribution of fused MODIS SSM at different dates is shown in Figure 5.

More detailed spatial information is presented in the fused MODIS SSM. It suggests that the SMRFM can improve the spatial resolution of AMSR-E SSM from microwave scale to MODIS scale well. The enhanced spatial information of fused MODIS SSM will be beneficial for applications at the regional scale. Meanwhile, the AMSR-E and fused MODIS SSM have relatively good consistency in the spatial distribution, indicating that the STFM can downscale AMSR-E SSM to fine scale from coarse scale well under large spatial resolution differences. For the large difference in spatial resolution of the two kinds of SSM (the paired reference data), there may be some inconsistencies in the fused results. It is mainly that the spatial variation of MODIS SSM in special areas cannot be well represented in AMSR-E SSM.

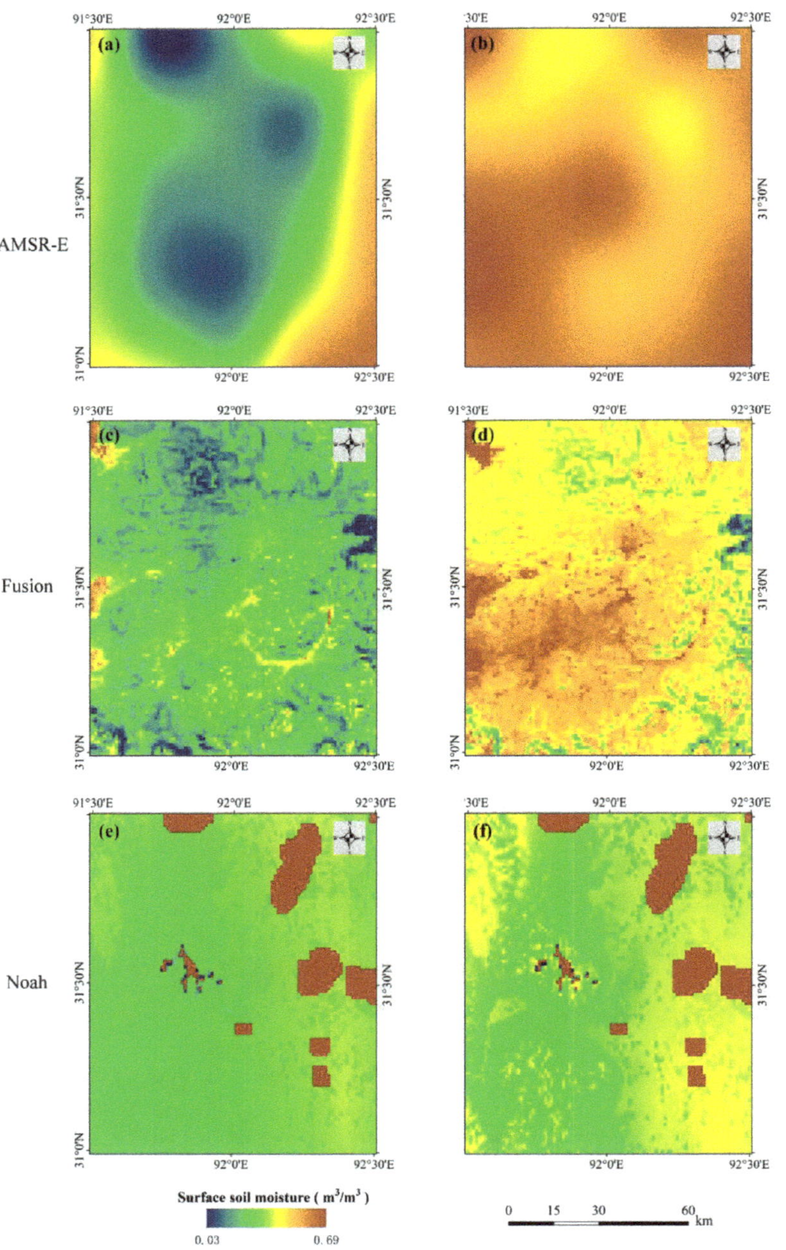

Figure 5. The spatial distribution of AMSR-E SSM (upper), fused MODIS SSM (middle) and Noah SSM at different dates in 1 km pixel size; (**a**,**c**,**e**) are data at 31 August 2010; (**b**,**d**,**f**) are data at 13 September 2011.

There are many spatial void data for Noah SSM, as they are masked by the land surface data, and the other two kinds of SSM are not masked. However, this does not affect the presentation of the results. In addition, the spatial distributions of AMSR-E and fused MODIS SSM are quite different when compared to Noah SSM. Nevertheless, the variation characteristics in temporal are still captured by the three kinds of SSM, but each SSM has some deviation in its depiction. This suggests the three kinds of SSM all have certain uncertainty, consistent with previous studies on satellite-based SSM [10,15,16].

3.3. Evaluations against In Situ Data at SMTMN Scale

The fused MODIS SSM, AMSR-E SSM, Noah SSM, and in situ SSM are aggregated to the SMTMN scale. Then the in situ site-based temporal variation differences between the three kinds of pixel SSM are compared (Figure 6). It shows that the pixel SSM can well monitor the temporal variations of regional SSM and display a good consistency in unfrozen period compared to in situ data. Nevertheless, the temporal variation of the four kinds of SSM differs greatly. The range of in situ data is 0.182–0.403 m^3/m^3, AMSR-E SSM is 0.106–0.601 m^3/m^3, fused MODIS SSM is 0.152–0.557 m^3/m^3 and Noah SSM is 0.238–0.455 m^3/m^3 at SMTMN scale. The ability of the fused MODIS SSM to capture in situ data is between AMSR-E and Noah SSM. As the fused MODIS SSM is downscaled by AMSR-E, there is high consistency in temporal variation curves.

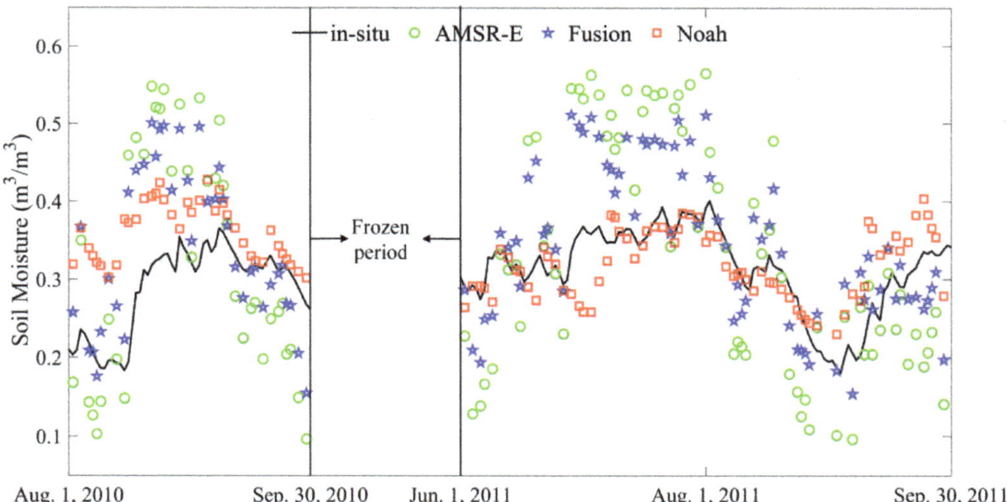

Figure 6. Comparison of the in situ, AMSR-E, Noah, and the fused MODIS SSM at SMTMN scale in unfrozen period. The gap between the two solid black lines is the frozen period. The in situ data are represented by line symbol and the pixel SSM is represented by point symbol.

The quantitative evaluation results (Table 3) show that the fused MODIS SSM is slightly higher than AMSR-E SSM and obviously higher than Noah SSM in terms of temporal *r* (0.673). Meanwhile, it presents lower temporal *μbRMSE* (0.070 m^3/m^3) than AMSR-E SSM against in situ data. Noah data present the lowest *μbRMSE*, however, it also gets the lowest temporal *r*. As the highest temporal *r* and the moderate temporal *μbRMSE* of fused SSM, it shows that the fused data have more advantages than the other pixel data against in situ data. Compared to improving the temporal *r* of AMSR-E, the fused MODIS SSM has more advantages in decreasing the temporal *RMSE* of AMSR-E. This suggests that the fused SSM has higher accuracy than AMSR-E SSM in overall temporal variation at SMTMN scale.

Table 3. Temporal accuracy evaluations of the AMSR-E, Noah, and the fused MODIS SSM against in situ data at SMTMN scale in unfrozen period.

	r	$RMSE$ (m^3/m^3)	$bias$ (m^3/m^3)	$\mu bRMSE$ (m^3/m^3)
AMSR-E	0.661 **	0.112	0.017	0.111
Fused	0.673 **	0.078	0.034	0.070
Noah	0.438 **	0.062	0.030	0.054

Note: ** indicates the temporal r passed the p-value < 0.01.

3.4. Evaluations against In Situ Soil Moisture at MODIS Scale

In terms of overall temporal accuracy at SMTMN scale, it can be considered that the fused MODIS SSM outperforms AMSR-E in describing temporal variation of in situ data. However, the accuracy difference between the pixel SSM is still unclear at the MODIS scale and needs to be further explored.

3.4.1. Daily Accuracy Evaluation

To calculate the daily evaluation of pixel SSM against in situ data effectively, all the available daily pixel SSM and in situ SSM are collected at MODIS scale during unfrozen period. The scatter plots between them are shown in Figure 7. It shows that the fitting line of the scatter between in situ and fused MODIS SSM is closest to the 1:1 line. Noah and in situ SSM present the lowest slope of fitting line. Among quantitative indexes, fused MODIS SSM presents the highest r (0.714) and the lowest $RMSE$ (0.117 m^3/m^3) compared to AMSR-E and Noah SSM. It reveals that the fused MODIS SSM gets the best evaluation indexes at the MODIS scale indicating the advantages of SMRFM.

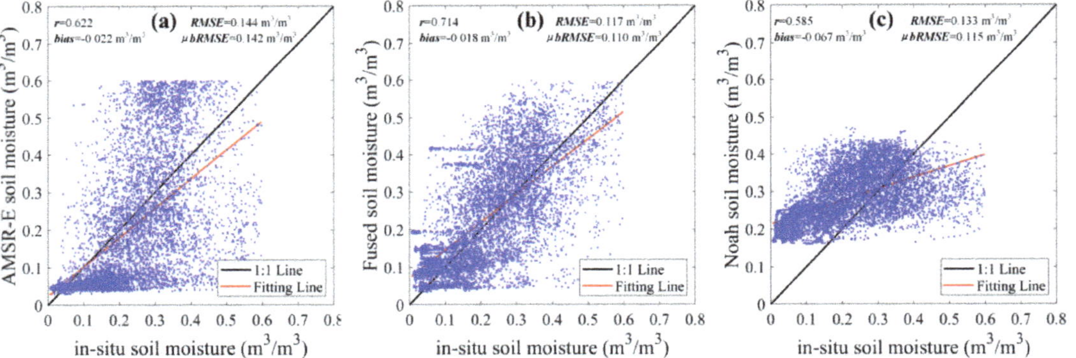

Figure 7. Scatter plots between in situ soil moisture and AMSR-E (**a**), fused MODIS (**b**), and Noah surface soil moisture (**c**).

3.4.2. Temporal Accuracy Evaluation

To further demonstrate the difference between the three kinds of pixel SSM, the temporal accuracy is investigated at MODIS scale. The fused MODIS SSM, AMSR-E SSM, and Noah SSM are extracted based on the selected 29 in situ sites. Then they are directly temporal evaluated against the in situ data. The evaluation results at the 29 in situ sites were obtained in the study (Figure 8), and the average values are shown in Table 4.

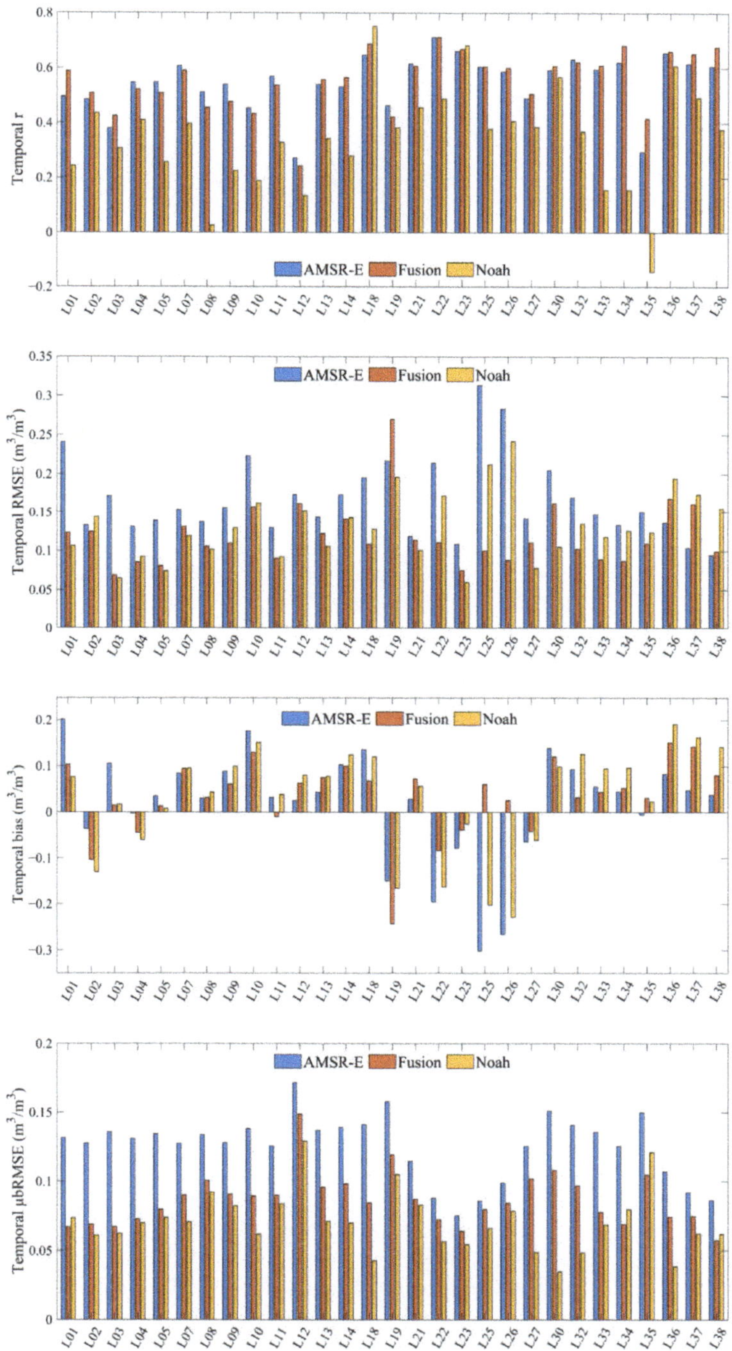

Figure 8. Temporal accuracy evaluation of AMSR-E, Noah, and the fused MODIS SSM against each in situ data at MODIS scale.

Table 4. Average of temporal variation evaluations against in situ data at MODIS scale.

	r (No. of p-Value > 0.05)	$RMSE$ (m^3/m^3)	$bias$ (m^3/m^3)	$\mu bRMSE$ (m^3/m^3)
AMSR-E	0.547 (0)	0.167	0.017	0.126
Fusion	0.557 (0)	0.119	0.035	0.087
Noah	0.348 (5)	0.131	0.031	0.071

In terms of temporal r, the range of fused MODIS SSM is 0.243–0.722 with an average of 0.557, the range of AMSR-E SSM is 0.275–0.728 with an average of 0.547, and the range of Noah SSM is −0.143–0.759 with an average of 0.348. For temporal $RMSE$, the range of fused MODIS SSM is 0.073–0.276 m^3/m^3 with an average of 0.119 m^3/m^3, the range of AMSR-E SSM is 0.097–0.314 m^3/m^3 with an average of 0.167 m^3/m^3, and the temporal RMSE range of Noah SSM is 0.067–0.248 m^3/m^3 with an average of 0.131 m^3/m^3. For temporal *bias*, the range of fused MODIS SSM is −0.243–0.151 m^3/m^3 with an average of 0.035 m^3/m^3, the range of AMSR-E SSM is −0.301–0.204 m^3/m^3 with an average of 0.017 m^3/m^3, and the range of Noah SSM is −0.237–0.191 m^3/m^3 with an average of 0.031 m^3/m^3. For temporal $\mu bRMSE$, the range of fused MODIS SSM is 0.053–0.152 m^3/m^3 with an average of 0.087 m^3/m^3, the range of AMSR-E SSM is 0.074–0.176 m^3/m^3 with an average of 0.126 m^3/m^3, and the range of Noah SSM is 0.037 −0.134 with an average of 0.071 m^3/m^3.

In most cases, the temporal r of Noah SSM is lower than AMSR-E and fused SSM (Figure 8). The negative correlation of Noah SSM at L35 site indicates that it could not describe the temporal variation of in situ data well. The temporal r of fused MODIS SSM is higher than AMSR-E at 17 in situ sites. The temporal *bias* is positive at most sites, indicating that the pixel SSM overestimates the in situ data. For temporal $RMSE$ and $\mu bRMSE$, the similar change characteristics are displayed. The higher $RMSE$ and $\mu bRMSE$ are obtained by AMSR-E at each site. Meanwhile, the difference between fused MODIS SSM and Noah SSM is not very large in the temporal $RMSE$ and $\mu bRMSE$ at each site.

Compared to the temporal evaluation at SMTMN scale, the fused SSM presents better evaluation indexes than the AMSR-E in terms of temporal r, $RMSE$ and $\mu bRMSE$ (Tables 3 and 4). Meanwhile, Noah SSM presents the lowest average temporal r and temporal $\mu bRMSE$ at MODIS scale, which is consistent with the temporal evaluation at SMTMN scale. Notably, the temporal r of five sites failed the hypothesis test (p-value > 0.05) for Noah SSM.

3.5. Evaluations Based on Triple Collocation Method

Referring to previous studies [45,55], the valid number of data points should be greater than 100 for each pixel SSM in the TC triplet. Like the temporal evaluation against in situ data, the TC evaluations are still carried out at the selected 29 in situ sites (the black triangle shown in Figure 1b). The boxplot of TC1 (in situ Noah-Fusion TC triplet) and TC2 (in situ Noah-AMSR-E TC triplet) evaluations is shown in Figure 9.

The average temporal r of in situ SSM is the best in each TC triplet. The ranges of temporal r for in situ data, Noah SSM, and fused SSM are 0.526–0.990, 0.361–0.837, and 0.623–0.991, with averages of 0.762, 0.521, and 0.761 in TC1. Meanwhile, the ranges of temporal r for in situ data, Noah SSM, and AMSR-E SSM are 0.563–0.990, 0.348–0.826, and 0.602–0.991, with averages of 0.766, 0.518, and 0.755 in TC2. The average temporal r of in situ data is comparable in each TC triplet, as is the Noah SSM. Thus, direct comparison can be implemented between the TC temporal r of AMSR-E and the fused MODIS SSM. Therefore, the average temporal r of the four kinds of SSM can be sorted as follows: in situ SSM > the fused MODIS SSM > AMSR-E SSM > Noah SSM. This suggests that the proposed SMRFM can be used to estimate fine-scale SSM with long time series and that the estimated SSM is better than the AMSR-E SSM in temporal variation evaluated by TC method at MODIS scale in the study.

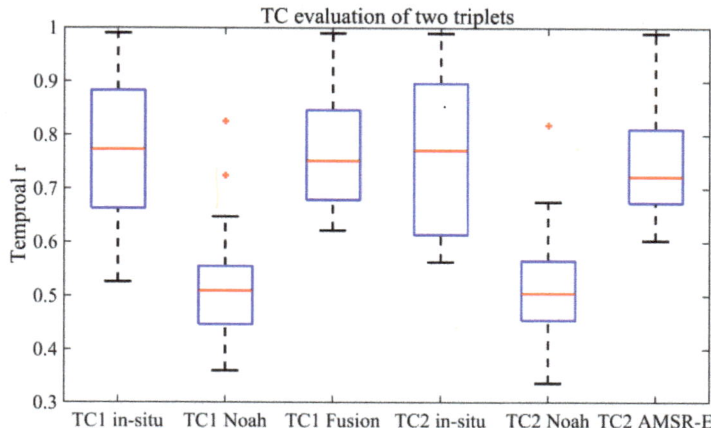

Figure 9. TC evaluations of the two triplets. TC1 indicates in situ Noah-Fusion TC triplet and TC2 indicates in situ Noah-AMSR-E TC triplet.

4. Discussion

The SMRFM is proposed to downscale AMSR-E SSM to MODIS scale with long time series in this study. To evaluate the accuracy of estimated MODIS SSM, the *r*, *RMSE*, *bias*, and *μbRMSE* are used in the study. A higher *r* indicates a higher explained variability and a lower *RMSE* indicates a higher agreement between the pixel SSM and in situ data in absolute value. Positive *bias* indicates that the in situ data are overestimated by pixel SSM. The lower *μbRMSE* indicates a higher agreement between the pixel SSM and in situ data in relative value. Therefore, the high accuracy of pixel SSM indicates the high *r*, the low *RMSE*, and *μbRMSE*.

For the spatial mismatch between pixel and in situ SSM, the direct comparison between them has always been controversial [3,7,9]. To evaluate the pixel SSM better using in situ data, the upscaling methods are developed and used in the previous studies [47,56]. Nevertheless, the direct comparison is still the most basic evaluation for pixel SSM, as the in situ data are first-hand real data and can more directly express the changes of actual SSM. Moreover, the effect of spatial mismatch on absolute value comparison of SSM is higher than that of temporal variation [45]. More importantly, the TC method is used for SSM evaluation under the unknown true data. There are two TC triples for evaluations in the study. Both of them show that the in situ data present the highest temporal *r* (Figure 9). Therefore, it is reasonable to evaluate the pixel SSM using the in situ data in temporal variation. As there is only one in situ site in each MODIS pixel, the daily accuracy evaluations in Section 3.4.1 may be a compromise way to evaluate the absolute SSM in the case of absent true pixel SSM.

There are two keys for MODIS SSM estimation using proposed SMRFM. One is the OTI-data-based fine-scale SSM retrieval, another is the STFM. The training and validation accuracies of Equation (4) are comparable in the study. The *RMSE* of retrieved fine-scale SSM was less than 0.09 m^3/m^3 on 24 July 2011. Meanwhile, the *RMSE* of AMSR-E and Noah were 0.128 m^3/m^3 and 0.122 m^3/m^3 against in situ data on that day. The *RMSE* of AMSR-E is no less than 0.11 m^3/m^3 [39] and the downscaled AMSR-E [39] and SMAP SSM [24] is no less than 0.08 m^3/m^3 at Naqu, central Tibet Plateau. It can be concluded that the *RMSE* of retrieved OTI-based SSM is better than the AMSR-E SSM and is comparable with the downscaled SSM. The slope and altitude information of topographic attributes are used to fit the Equation (4). The impact of topographic changes on soil moisture may not be fully considered in the study. Therefore, the index characterizing the information of topographic wetness [57] for OTI-data-based SSM retrieval will be explored in future research.

The fused MODIS SSM significantly improves the spatial detailed information of AMSR-E SSM. Meanwhile, the evaluation indexes of fused data are better than AMSR-E SSM at SMTMN and MODIS scale. There may be several reasons for this result. The key reason may be that the neighborhood information is used by SMRFM for fine-scale SSM estimation. This is equivalent to denoising remote-sensing images using spatial filtering methods [33], which weakens the outliers in temporal variation of SSM. Thus, the temporal variation of estimated data is much smoother than the AMSR-E SSM (Figure 6). The reference MODIS SSM of SMRFM is estimated by Equation (4), which is fitted by the MODIS OTI data, in situ SSM, and the topographic information. Then, the SMRFM is used to estimate MODIS SSM with long time series using the fixed reference data. Therefore, the estimated MODIS SSM by SMRFM can be considered as coupled with the in situ SSM information. This may be another reason for the high accuracy of fused MODIS SSM. According to the basic principles of STFM [26,33], the smaller the difference in temporal variation, the better that MODIS SSM can be estimated [45]. The dominate land cover type of the study area is grassland, and the implementation of SMRFM should not exceed one-and-a-half years. This indicates that the spatial and temporal pattern of SSM will not change much in a relatively long time under the homogeneous land type. This may be a possible factor in the high accuracy of fused MODIS SSM.

There is an assumption that the temporal change is scale-invariant in STFM. The assumption was proposed in 2006 for surface reflectance estimation [26] and was then applied for other surface parameters estimation [27–32]. It is used as a downscaling method for long time series MODIS SSM estimation in the study. Similar with STFM, the scale-invariant assumption also exists in traditional microwave SSM downscaling, but it refers to scale-invariance of the relation between microwave SSM and other remotely sensed parameters for traditional methods in most cases [22,24]. It has been shown that the downscaling capability of STFM is better than that of the traditional downscaling method [45], although the scale-invariant assumption of temporal change is fitted by a linear equation. This may reveal that the scale-invariant assumption in temporal change is more reasonable than scale-invariance in the relation.

To investigate the relation between surface parameters and SSM, the correlations between LST, NDVI, altitude, and slope are calculated in Table 5. It shows that the correlations between SSM and the first two factors (LST and NDVI) are obviously better than the last two (altitude and slope). The correlation of the factors can be sorted as follows: NDVI > LST > Slope > Altitude. This suggests that the topographic factors may be limited in SSM estimation in this study.

Table 5. The correlations between SSM and LST, NDVI, altitude, and slope.

	LST	NDVI	Altitude	Slope
In Situ	0.714 **	0.725 **	0.072 **	0.145 **
AMSR-E	0.647 **	0.737 **	0.092 **	−0.014
Fusion	0.586 **	0.701 **	0.182 **	0.250 **
Noah	0.715 **	0.676 **	0.007 **	−0.029 *

Note: * and ** indicate the correlation passed p-value < 0.05 and p-value < 0.01, respectively.

Since MODIS SSM is downscaled from AMSR-E data using SMRFM, they have the same temporal resolution. Nevertheless, the effectiveness of proposed SMRFM in estimating SSM in highly heterogeneous areas and longer time series needs to be further explored. Notably, the premise of SMRFM for SSM estimation is to estimate the reference SSM. In the study, the OTI data are used for reference SSM retrieval. As OTI data cannot penetrate the surface, the use of Sentinel-1 and OTI data in SMRFM framework may enhance the accuracy and spatial resolution of estimated SSM. Therefore, the SMRFM has the potential to estimate long time series finer-scale (less than 1 km) SSM with the finer-scale reference SSM provides.

5. Conclusions

Given the difficulty of taking into account the long time series characteristics of current downscaling method, which integrates microwave SSM data and OTI data to estimate fine scale SSM, an SSM retrieval-and-fusion model named SMRFM is proposed to downscale AMSR-E SSM for MODIS SSM with long time series estimation in the study. The method was applied to the SMTMN over Naqu, central Tibet Plateau to obtain the MODIS SSM with long time series characteristics of microwave data. To validate the SMRFM, in situ data and Noah land surface model 0.01-degree SSM were used in the study. The main conclusions of the study are as follows:

(1) A method that integrates in situ data, remote sensing OTI data, and terrain data was developed for MODIS SSM retrieval, and the estimated MODIS SSM by this method obtains an *RMSE* of less than 0.09 m^3/m^3.
(2) The MODIS SSM fused by the SMRFM can well maintain the spatial distribution and temporal variation of AMSR-E data, although there are certain differences in the special distinction between the two kinds of pixel SSM.
(3) Six months of MODIS SSM in unfrozen period were fused by the proposed SMRFM. The evaluations show that the fused MODIS SSM has better temporal accuracy than that of AMSR-E at SMTMN and MODIS scale. Compared to Noah SSM, the fused SSM presents higher temporal r and slightly lower $\mu bRMSE$. In addition, the fused SSM has better daily accuracy than AMSR-E and Noah SSM. Therefore, it can be considered that the proposed SMRFM can be used to estimate fine-scale SSM with long time series and that the estimated SSM is better than AMSR-E SSM in temporal variation. This will promote the development of research and applications with long time series SSM at regional scale.

Author Contributions: Conceptualization, H.J. and S.C.; methodology, H.J. and J.W.; software, J.W.; validation, H.J., J.Z. and L.W.; formal analysis, X.L.; investigation, J.W.; resources, S.C.; data curation, H.J.; writing—original draft preparation, H.J.; writing—review and editing, S.C., X.L. and J.W. All authors have read and agreed to the published version of the manuscript.

Funding: This research was funded by the National Natural Science Foundation of China (Grand no. 42101401), Guangdong Provincial Water Conservancy Science and Technology Innovation Project (Grand no. 2020-30), Scientific research project of Zhongkai University of Agriculture and Engineering (Grand no. KA22016B770).

Data Availability Statement: The study was performed based on public access remote sensing and in situ data. The fused MODIS surface soil moisture with long time series that support the findings of the study are available from the corresponding author and first author upon reasonable request.

Acknowledgments: We would like to thank JAXA for making the AMSR-E soil moisture data publicly available and ISMN for providing the SMTMN in situ soil moisture data. We would also like to thank the anonymous reviewers for their insightful comments on the manuscript.

Conflicts of Interest: The authors declare no conflict of interest.

References

1. Sharma, K.; Irmak, S.; Kukal, M.S. Propagation of soil moisture sensing uncertainty into estimation of total soil water, evapotranspiration and irrigation decision-making. *Agric. Water Manag.* **2021**, *243*, 106454. [CrossRef]
2. Babaeian, E.; Sadeghi, M.; Jones, S.B.; Montzka, C.; Vereecken, H.; Tuller, M. Ground, Proximal, and Satellite Remote Sensing of Soil Moisture. *Rev. Geophys.* **2019**, *57*, 530–616. [CrossRef]
3. Dorigo, W.A.; Gruber, A.; De Jeu, R.A.M.; Wagner, W.; Stacke, T.; Loew, A.; Albergel, C.; Brocca, L.; Chung, D.; Parinussa, R.M.; et al. Evaluation of the ESA CCI soil moisture product using ground-based observations. *Remote Sens. Environ.* **2015**, *162*, 380–395. [CrossRef]
4. Jadidoleslam, N.; Mantilla, R.; Krajewski, W.F.; Goska, R. Investigating the role of antecedent SMAP satellite soil moisture, radar rainfall and MODIS vegetation on runoff production in an agricultural region. *J. Hydrol.* **2019**, *579*, 124210. [CrossRef]
5. Yoon, J.; Leung, L.R. Assessing the relative influence of surface soil moisture and ENSO SST on precipitation predictability over the contiguous United States. *Geophys. Res. Lett.* **2015**, *42*, 5005–5013. [CrossRef]

6. Sabaghy, S.; Walker, J.P.; Renzullo, L.J.; Jackson, T.J. Spatially enhanced passive microwave derived soil moisture: Capabilities and opportunities. *Remote Sens. Environ.* **2018**, *209*, 551–580. [CrossRef]
7. Wigneron, J.P.; Jackson, T.J.; O'Neill, P.; De Lannoy, G.; de Rosnay, P.; Walker, J.P.; Ferrazzoli, P.; Mironov, V.; Bircher, S.; Grant, J.P.; et al. Modelling the passive microwave signature from land surfaces: A review of recent results and application to the L-band SMOS & SMAP soil moisture retrieval algorithms. *Remote Sens. Environ.* **2017**, *192*, 238–262. [CrossRef]
8. Karthikeyan, L.; Pan, M.; Wanders, N.; Kumar, D.N.; Wood, E.F. Four decades of microwave satellite soil moisture observations: Part 1. A review of retrieval algorithms. *Adv. Water Resour.* **2017**, *109*, 106–120. [CrossRef]
9. Chan, S.K.; Bindlish, R.; O'Neill, P.E.; Njoku, E.; Jackson, T.; Colliander, A.; Chen, F.; Burgin, M.; Dunbar, S.; Piepmeier, J.; et al. Assessment of the SMAP Passive Soil Moisture Product. *IEEE Trans. Geosci. Remote Sens.* **2016**, *54*, 4994–5007. [CrossRef]
10. Das, N.N.; Entekhabi, D.; Dunbar, R.S.; Chaubell, M.J.; Colliander, A.; Yueh, S.; Jagdhuber, T.; Chen, F.; Crow, W.; O'Neill, P.E.; et al. The SMAP and Copernicus Sentinel 1A/B microwave active-passive high resolution surface soil moisture product. *Remote Sens. Environ.* **2019**, *233*, 111380. [CrossRef]
11. Njoku, E.G.; Jackson, T.J.; Lakshmi, V.; Chan, T.K.; Nghiem, S.V. Soil moisture retrieval from AMSR-E. *IEEE Trans. Geosci. Remote Sens.* **2003**, *41*, 215–229. [CrossRef]
12. Zeng, J.; Li, Z.; Chen, Q.; Bi, H.; Qiu, J.; Zou, P. Evaluation of remotely sensed and reanalysis soil moisture products over the Tibetan Plateau using in-situ observations. *Remote Sens. Environ.* **2015**, *163*, 91–110. [CrossRef]
13. Kerr, Y.H.; Waldteufel, P.; Richaume, P.; Wigneron, J.P.; Ferrazzoli, P.; Mahmoodi, A.; Al Bitar, A.; Cabot, F.; Gruhier, C.; Juglea, S.E.; et al. The SMOS Soil Moisture Retrieval Algorithm. *IEEE Trans. Geosci. Remote Sens.* **2012**, *50*, 1384–1403. [CrossRef]
14. Jamei, M.; Mousavi Baygi, M.; Oskouei, E.A.; Lopez-Baeza, E. Validation of the SMOS Level 1C Brightness Temperature and Level 2 Soil Moisture Data over the West and Southwest of Iran. *Remote Sens.* **2020**, *12*, 2819. [CrossRef]
15. Wang, Y.; Leng, P.; Peng, J.; Marzahn, P.; Ludwig, R. Global assessments of two blended microwave soil moisture products CCI and SMOPS with in-situ measurements and reanalysis data. *Int. J. Appl. Earth Obs.* **2021**, *94*, 102234. [CrossRef]
16. Kim, S.; Liu, Y.Y.; Johnson, F.M.; Parinussa, R.M.; Sharma, A. A global comparison of alternate AMSR2 soil moisture products: Why do they differ? *Remote Sens. Environ.* **2015**, *161*, 43–62. [CrossRef]
17. Liu, Y.; Zhou, Y.; Lu, N.; Tang, R.; Liu, N.; Li, Y.; Yang, J.; Jing, W.; Zhou, C. Comprehensive assessment of Fengyun-3 satellites derived soil moisture with in-situ measurements across the globe. *J. Hydrol.* **2021**, *594*, 125949. [CrossRef]
18. Song, C.; Jia, L. A Method for Downscaling FengYun-3B Soil Moisture Based on Apparent Thermal Inertia. *Remote Sens.* **2016**, *8*, 703. [CrossRef]
19. Merlin, O.; Walker, J.; Chehbouni, A.; Kerr, Y. Towards deterministic downscaling of SMOS soil moisture using MODIS derived soil evaporative efficiency. *Remote Sens. Environ.* **2008**, *112*, 3935–3946. [CrossRef]
20. Peng, J.; Loew, A.; Merlin, O.; Verhoest, N.E.C. A review of spatial downscaling of satellite remotely sensed soil moisture. *Rev. Geophys.* **2017**, *55*, 341–366. [CrossRef]
21. Piles, M.; Sanchez, N.; Vall-llossera, M.; Camps, A.; Martinez-Fernandez, J.; Martinez, J.; Gonzalez-Gambau, V. A Downscaling Approach for SMOS Land Observations: Evaluation of High-Resolution Soil Moisture Maps over the Iberian Peninsula. *IEEE J. Sel. Top. Appl. Earth Obs. Remote Sens.* **2014**, *7*, 3845–3857. [CrossRef]
22. Senanayake, I.P.; Yeo, I.Y.; Willgoose, G.R.; Hancock, G.R. Disaggregating satellite soil moisture products based on soil thermal inertia: A comparison of a downscaling model built at two spatial scales. *J. Hydrol.* **2021**, *594*, 125894. [CrossRef]
23. Wei, Z.; Meng, Y.; Zhang, W.; Peng, J.; Meng, L. Downscaling SMAP soil moisture estimation with gradient boosting decision tree regression over the Tibetan Plateau. *Remote Sens. Environ.* **2019**, *225*, 30–44. [CrossRef]
24. Nasta, P.; Penna, D.; Brocca, L.; Zuecco, G.; Romano, N. Downscaling near-surface soil moisture from field to plot scale: A comparative analysis under different environmental conditions. *J. Hydrol.* **2018**, *557*, 97–108. [CrossRef]
25. Li, Z.; Shen, H.; Li, H.; Xia, G.; Gamba, P.; Zhang, L. Multi-feature combined cloud and cloud shadow detection in GaoFen-1 wide field of view imagery. *Remote Sens. Environ.* **2017**, *191*, 342–358. [CrossRef]
26. Gao, F.; Masek, J.; Schwaller, M.; Hall, F. On the blending of the Landsat and MODIS surface reflectance: Predicting daily Landsat surface reflectance. *IEEE Trans. Geosci. Remote Sens.* **2006**, *44*, 2207–2218. [CrossRef]
27. Park, S.; Jeong, S.; Park, Y.; Kim, S.; Lee, D.; Mo, Y.; Jang, D.; Park, K. Phenological Analysis of Sub-Alpine Forest on Jeju Island, South Korea, Using Data Fusion of Landsat and MODIS Products. *Forests* **2021**, *12*, 286. [CrossRef]
28. Meng, J.; Du, X.; Wu, B. Generation of high spatial and temporal resolution NDVI and its application in crop biomass estimation. *Int. J. Digit. Earth* **2013**, *6*, 203–218. [CrossRef]
29. Shen, H.; Huang, L.; Zhang, L.; Wu, P.; Zeng, C. Long-term and fine-scale satellite monitoring of the urban heat island effect by the fusion of multi-temporal and multi-sensor remote sensed data: A 26-year case study of the city of Wuhan in China. *Remote Sens. Environ.* **2016**, *172*, 109–125. [CrossRef]
30. Weng, Q.; Fu, P.; Gao, F. Generating daily land surface temperature at Landsat resolution by fusing Landsat and MODIS data. *Remote Sens. Environ.* **2014**, *145*, 55–67. [CrossRef]
31. Qiu, Y.; Zhou, J.; Chen, J.; Chen, X. Spatiotemporal fusion method to simultaneously generate full-length normalized difference vegetation index time series (SSFIT). *Int. J. Appl. Earth Obs.* **2021**, *100*, 102333. [CrossRef]
32. Zhao, W.; Duan, S. Reconstruction of daytime land surface temperatures under cloud-covered conditions using integrated MODIS/Terra land products and MSG geostationary satellite data. *Remote Sens. Environ.* **2020**, *247*, 111931. [CrossRef]

33. Cheng, Q.; Liu, H.; Shen, H.; Wu, P.; Zhang, L. A Spatial and Temporal Nonlocal Filter-Based Data Fusion Method. *IEEE Trans. Geosci. Remote Sens.* **2017**, *55*, 4476–4488. [CrossRef]
34. Emelyanova, I.V.; McVicar, T.R.; Van Niel, T.G.; Li, L.T.; van Dijk, A.I.J.M. Assessing the accuracy of blending Landsat–MODIS surface reflectances in two landscapes with contrasting spatial and temporal dynamics: A framework for algorithm selection. *Remote Sens. Environ.* **2013**, *133*, 193–209. [CrossRef]
35. Bhattarai, N.; Quackenbush, L.J.; Dougherty, M.; Marzen, L.J. A simple Landsat-MODIS fusion approach for monitoring seasonal evapotranspiration at 30 m spatial resolution. *Int. J. Remote Sens.* **2015**, *36*, 115–143. [CrossRef]
36. Pieri, M.; Chiesi, M.; Battista, P.; Fibbi, L.; Gardin, L.; Rapi, B.; Romani, M.; Sabatini, F.; Angeli, L.; Cantini, C.; et al. Estimation of Actual Evapotranspiration in Fragmented Mediterranean Areas by the Spatio-Temporal Fusion of NDVI Data. *IEEE J. Sel. Top. Appl. Earth Obs. Remote Sens.* **2019**, *12*, 5108–5117. [CrossRef]
37. Yang, K.; Qin, J.; Zhao, L.; Chen, Y.; Tang, W. A Multiscale Soil Moisture and Freeze–Thaw Monitoring Network on the Third Pole. *Bull. Am. Meteorol. Soc.* **2013**, *94*, 1907–1916. [CrossRef]
38. Zhao, L.; Yang, K.; Qin, J.; Chen, Y.; Tang, W.; Lu, H.; Yang, Z. The scale-dependence of SMOS soil moisture accuracy and its improvement through land data assimilation in the central Tibetan Plateau. *Remote Sens. Environ.* **2014**, *152*, 345–355. [CrossRef]
39. Jiang, H.; Shen, H.; Li, H.; Lei, F.; Gan, W.; Zhang, L. Evaluation of Multiple Downscaled Microwave Soil Moisture Products over the Central Tibetan Plateau. *Remote Sens.* **2017**, *9*, 402. [CrossRef]
40. Njoku, E.G.; Chan, S.K. Vegetation and surface roughness effects on AMSR-E land observations. *Remote Sens. Environ.* **2006**, *100*, 190–199. [CrossRef]
41. Lu, H.T.U.J.; Koike, T.; Fujii, H.; Ohta, T.; Tamagawa, K. Development of a physically-based soil moisture retrieval algorithm for spaceborne passive microwave radiometers and its application to AMSR-E. *J. Remote Sens. Soc. Jpn.* **2009**, *29*, 253–262. [CrossRef]
42. Spruce, J.P.; Sader, S.; Ryan, R.E.; Smoot, J.; Kuper, P.; Ross, K.; Prados, D.; Russell, J.; Gasser, G.; McKellip, R. Assessment of MODIS NDVI time series data products for detecting forest defoliation by gypsy moth outbreaks. *Remote Sens. Environ.* **2011**, *115*, 427–437. [CrossRef]
43. Jarvis, A.; Reuter, H.I.; Nelson, A.; Guevara, E. 2008, Hole-Filled SRTM for the Globe Version 4, Available from the CGIAR-CSI SRTM 90 m Database. Available online: http://srtm.csi.cgiar.org (accessed on 12 October 2021).
44. Jacob, J.; Slinski, K. FLDAS Noah Land Surface Model L4 Central Asia Daily 0.01 × 0.01 degree. *Goddard Earth Sci. Data Inf. Serv. Cent. (GES DISC)* **2021**. [CrossRef]
45. Hongtao, J.; Huanfeng, S.; Xinghua, L.; Chao, Z.; Huiqin, L.; Fangni, L. Extending the SMAP 9-km soil moisture product using a spatio-temporal fusion model. *Remote Sens. Environ.* **2019**, *231*, 111224. [CrossRef]
46. Zhao, W.; Li, A.; Zhao, T. Potential of Estimating Surface Soil Moisture With the Triangle-Based Empirical Relationship Model. *IEEE Trans. Geosci. Remote Sens.* **2017**, *55*, 6494–6504. [CrossRef]
47. Qin, J.; Yang, K.; Lu, N.; Chen, Y.; Zhao, L.; Han, M. Spatial upscaling of in-situ soil moisture measurements based on MODIS-derived apparent thermal inertia. *Remote Sens. Environ.* **2013**, *138*, 1–9. [CrossRef]
48. Zhan, Z.; Qin, Q.; Ghulan, A.; Wang, D. NIR-red spectral space based new method for soil moisture monitoring. *Sci. China. Ser. D Earth Sci.* **2007**, *50*, 283–289. [CrossRef]
49. AghaKouchak, A.; Farahmand, A.; Melton, F.S.; Teixeira, J.; Anderson, M.C.; Wardlow, B.D.; Hain, C.R. Remote sensing of drought: Progress, challenges and opportunities. *Rev. Geophys.* **2015**, *53*, 452–480. [CrossRef]
50. Wang, J.; Ling, Z.; Wang, Y.; Zeng, H. Improving spatial representation of soil moisture by integration of microwave observations and the temperature–vegetation–drought index derived from MODIS products. *ISPRS J. Photogramm. Remote Sens.* **2016**, *113*, 144–154. [CrossRef]
51. Ma, H.; Zeng, J.; Chen, N.; Zhang, X.; Cosh, M.H.; Wang, W. Satellite surface soil moisture from SMAP, SMOS, AMSR2 and ESA CCI: A comprehensive assessment using global ground-based observations. *Remote Sens. Environ.* **2019**, *231*, 111215. [CrossRef]
52. Abowarda, A.S.; Bai, L.; Zhang, C.; Long, D.; Li, X.; Huang, Q.; Sun, Z. Generating surface soil moisture at 30 m spatial resolution using both data fusion and machine learning toward better water resources management at the field scale. *Remote Sens. Environ.* **2021**, *255*, 112301. [CrossRef]
53. Stoffelen, A. Toward the true near-surface wind speed: Error modeling and calibration using triple collocation. *J. Geophys. Res.* **1988**, *103*, 7755–7766. [CrossRef]
54. Scipal, K.; Holmes, T.; de Jeu, R.; Naeimi, V.; Wagner, W. A possible solution for the problem of estimating the error structure of global soil moisture data sets. *Geophys. Res. Lett.* **2008**, *35*, L24403. [CrossRef]
55. McColl, K.A.; Vogelzang, J.; Konings, A.G.; Entekhabi, D.; Piles, M.; Stoffelen, A. Extended triple collocation: Estimating errors and correlation coefficients with respect to an unknown target. *Geophys. Res. Lett.* **2014**, *41*, 6229–6236. [CrossRef]
56. Greifeneder, F.; Notarnicola, C.; Bertoldi, G.; Niedrist, G.; Wagner, W. From point to pixel scale; an upscaling approach for in situ soil moisture measurements. *Vadose Zone J.* **2016**, *15*. [CrossRef]
57. Schönbrodt-Stitt, S.; Ahmadian, N.; Kurtenbach, M.; Conrad, C.; Romano, N.; Bogena, H.R.; Vereecken, H.; Nasta, P. Statistical Exploration of SENTINEL-1 Data, Terrain Parameters, and in-situ Data for Estimating the Near-Surface Soil Moisture in a Mediterranean Agroecosystem. *Front. Water* **2021**, *3*, 75. [CrossRef]

Article

Hybrid Methodology Using Sentinel-1/Sentinel-2 for Soil Moisture Estimation

Simon Nativel [1], Emna Ayari [1,2,†], Nemesio Rodriguez-Fernandez [1], Nicolas Baghdadi [3], Remi Madelon [1], Clement Albergel [4] and Mehrez Zribi [1,*]

1 CESBIO, CNES/CNRS/INRAE/IRD/UPS, Université de Toulouse, 18 Av. Edouard Belin, Bpi 2801, CEDEX 9, 31401 Toulouse, France; simon.nativel@univ-tlse3.fr (S.N.); emna.ayari@inat.u-carthage.tn (E.A.); nemesio.rodriguez-fernandez@univ-tlse3.fr (N.R.-F.); remi.madelon@univ-tlse3.fr (R.M.)
2 National Agronomic Institute of Tunisia, Carthage University, Tunis 1082, Tunisia
3 CIRAD, CNRS, INRAE, TETIS, University of Montpellier, AgroParisTech, CEDEX 5, 34093 Montpellier, France; nicolas.baghdadi@teledetection.fr
4 European Space Agency Climate Office, ECSAT, Harwell Campus, Oxforshire, Didcot OX11 0FD, UK; Clement.albergel@esa.int
* Correspondence: mehrez.zribi@ird.fr; Tel.: +33-56155-8505
† LR17AGR01 InteGRatEd Management of Natural Resources: remoTE Sensing, Spatial Analysis and Modeling (GREEN-TEAM).

Abstract: Soil moisture is an essential parameter for a better understanding of water processes in the soil–vegetation–atmosphere continuum. Satellite synthetic aperture radar (SAR) is well suited for monitoring water content at fine spatial resolutions on the order of 1 km or higher. Several methodologies are often considered in the inversion of SAR signals: machine learning techniques, such as neural networks, empirical models and change detection methods. In this study, we propose two hybrid methodologies by improving a change detection approach with vegetation consideration or by combining a change detection approach together with a neural network algorithm. The methodology is based on Sentinel-1 and Sentinel-2 data with the use of numerous metrics, including vertical–vertical (VV) and vertical–horizontal (VH) polarization radar signals, the classical change detection surface soil moisture (SSM) index I_{SSM}, radar incidence angle, normalized difference vegetation index (*NDVI*) optical index, and the VH/VV ratio. Those approaches are tested using in situ data from the ISMN (International Soil Moisture Network) with observations covering different climatic contexts. The results show an improvement in soil moisture estimations using the hybrid algorithms, in particular the change detection with the neural network one, for which the correlation increases by 54% and 33% with respect to that of the neural network or change detection alone, respectively.

Keywords: soil moisture; Sentinel-1; Sentinel-2; change detection; artificial neural network

Citation: Nativel, S.; Ayari, E.; Rodriguez-Fernandez, N.; Baghdadi, N.; Madelon, R.; Albergel, C.; Zribi, M. Hybrid Methodology Using Sentinel-1/Sentinel-2 for Soil Moisture Estimation. *Remote Sens.* **2022**, *14*, 2434. https://doi.org/10.3390/rs14102434

Academic Editors: Wei Zhao, Jian Peng, Hongliang Ma, Chunfeng Ma and Jiangyuan Zeng

Received: 7 April 2022
Accepted: 17 May 2022
Published: 19 May 2022

1. Introduction

Soil moisture is a key parameter for understanding different processes related to the transfer of the soil–vegetation–atmosphere flux [1–3]. It is also an essential parameter in the management of water resources, particularly for optimizing irrigation [4,5]. In this context, remote sensing has greatly contributed to allowing the spatial and temporal monitoring of this parameter at different spatial scales from global to local [6,7].

Most of the currently available operational surface soil moisture products are on a global scale with spatial resolutions of several kilometers. They are essentially based on active and passive microwave measurements [8–12]. In passive microwaves, these are mainly products based on SMOS [8] and SMAP [9] missions dedicated to monitoring soil moisture with L-Band measurements and other non-dedicated sensors using higher frequency bands. In active microwaves, these are measurements based on acquisitions with a scatterometer, particularly data acquired by the ASCAT/METOP satellite series [13]. The European Space Agency (ESA) Climate Change Initiative (CCI) soil moisture project also provides long time series by merging soil moisture estimations from active and passive sensors [14].

For soil moisture estimation at high spatial resolution, we identify products with an average resolution at approximately 1 km or at the plot scale [15–24]. There have been various studies that have developed methodologies based on low-resolution data disaggregation techniques, notably with measurements acquired in thermal infrared (MODIS) [25] or, more recently, data acquired by SAR sensors. The Synthetic Aperture Radar (SAR) technique offers a high spatial resolution estimate of the radar signal adapted to applications at agricultural field scale. The measured signal is dependent on the radar configurations (frequency, incidence angle, and polarization) and the dielectric and geometric properties of the surface. After numerous demonstration space missions (ERS, ASAR/ENVISAT, RADARSAT, etc.), the arrival of Sentinel-1 constellation [26] in the context of the Copernicus program has enabled exponential growth in the use of these signals for monitoring soil moisture and the dynamics of the vegetation cover. Other soil moisture products are then offered only based on Sentinel-1 data, with three types of methodologies: one based on the direct inversion of physical or semiempirical models [27–29]; one based on the application of machine learning approaches and particularly neural networks [30–32]; and one based on the change detection technique [33–35]. For example, at plot scale, El Hajj et al. [31] presented an Artificial Neural Network (ANN) approach with training using the coupling of the Integral Equation Model (IEM) and the Water Cloud Model (WCM) to provide an estimate of soil moisture at the scale of the agricultural plot. Gao et al. [35] also proposed an approach at the plot scale with greater consideration of the vegetation cover and its effect on the temporal variation of the radar signal. Bauer-Marschallinger et al. [33] proposed a change detection approach very close to the initial approach proposed with data from ASCAT scatterometers [13] at a 1 km scale.

For these products, which are highly useful for regional hydrology, the validation of existing products, despite the very interesting potential, still shows some limitations in different contexts, in particular that of dense vegetation covers but also in relatively complex contexts with strong heterogeneities in terms of land use and topography [36].

In this context, this study proposes to test hybrid approaches to soil moisture retrieval at a 1 km scale with the objective of improving the estimation accuracy of soil moisture. The approaches consider hybrid methodologies with a combination of a change detection approach with empirical modeling or machine learning.

Section 2 presents in the first subsection the database used in this study in terms of soil moisture data and satellite measurements. The second subsection presents the methodologies tested and proposed in this study. Section 3 illustrates the results. Section 4 includes the discussion of the proposed applications. The conclusions are presented in Section 5.

2. Materials and Methods

2.1. Database

2.1.1. ISMN Soil Moisture Data

The training and validation of the proposed methods are conducted based on data from the International Soil Moisture Network (ISMN) [37]. The data are available in conjunction with additional datasets of Koppen–Geiger climate classes, ESA's CCI land cover, and soil characteristics. The upper soil layer (0–10 cm) moisture measurements are harmonized as fractional volumetric soil moisture (m^3/m^3) and converted into Coordinated Universal Time (UTC). After data quality verification, some ISMN networks suffer from a lack of measurements. Therefore, we considered 21 networks among a total of 71 spatially distributed as shown in Figure 1. The data of each station should cover a period of two years with at least 20 dates between the start and the end date of acquisitions—1 January 2015 and 19 August 2021, respectively. Consequently, in the same network, we retain only stations with valid dataset as detailed in Table 1.

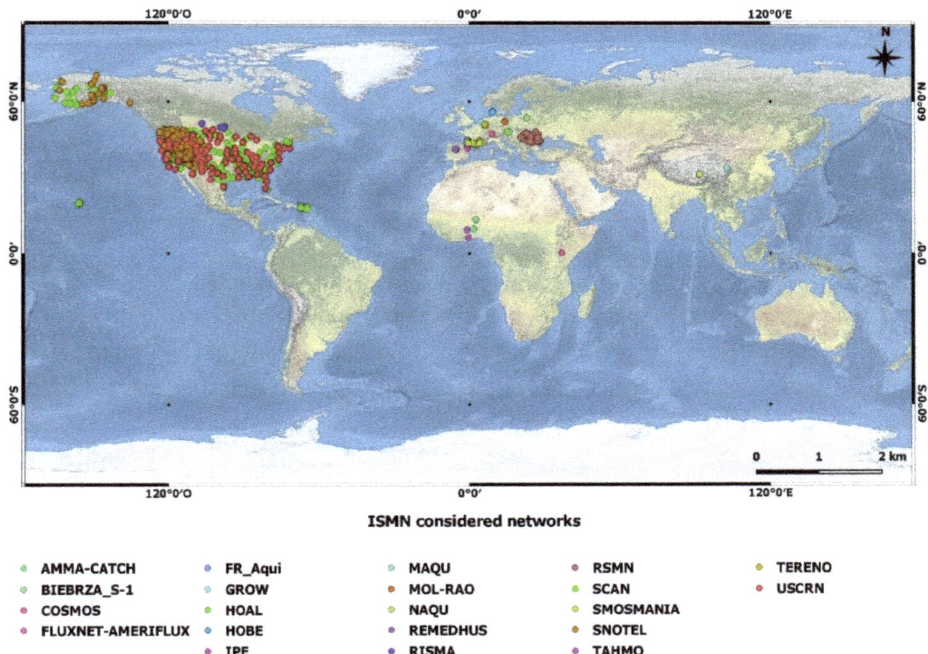

Figure 1. The global distribution of the International Soil Moisture Network (ISMN).

Table 1. Overview of the considered ISMN networks.

Network	Country	Number of Selected Stations	SM Sensors	References
AMMA-CATCH	Benin, Niger	7	CS616	Cappelaere et al. [38]; De Rosnay et al. [39]; Lebel et al. [40]; Mougin et al. [41]; Pellarin et al. [42]; Galle et al. [43].
BIEBRZA-S-1	Poland	8	GS-3	Musial et al. 2016 [44]
COSMOS	USA	2	Cosmic-ray-Probe	Zreda et al. [45]; Zreda et al. [46]
HOBE	Denmark	3	Decagon-5TE	Bircher et al. [47]; Jensen et al. [48]
FLUXNET-AMERIFLUX	USA	4	CS655, ThetaProbe-ML3 ThetaProbe-ML2X,	
FR-Aqui	France	3	ThetaProbe ML2X	Al-Yaari et al. [49]; Wigneron et al. [50]
GROW	UK	20	Flower-Power	Zappa et al. [51]; Xaver et al. [52]; Zappa et al. 2020 [53]
HOAL	Austria	32	SPADE-Time-Domain-Transmissivity	Vreugdenhil M. et al. [54]; Blöschl, Günter, et al. [55]
IPE	Spain	2	CS655, ThetaProbe-ML2X	Alday et al. [56]

The main objective of introducing the change as a function of time is to reduce the dependency to other variables affecting the radar signal such as soil roughness, which can be very important, particularly in the context of strong topography or even important spatial changes in microtopography, that change little with time for a given site, in contrast to soil moisture. Two vegetation-related quantities were tested for the $V1$ parameter: the optical vegetation index $NDVI$ estimated from Sentinel-2 data, as illustrated in Section 2.3, and the VH/VV ratio, considered to be strongly linked to the dynamics of the vegetation cover. This second option could be particularly interesting in the context of a humid climate with limited optical data.

2.2.3. Artificial Neural Network Hybrid Approach

The multilayer perceptron (MLP), which is a multilayer feed-forward ANN, is one of the most widely used ANNs, mainly in the field of water resources [79,80]. A multilayer perceptron has one or more hidden layers between its input and output layers. The neurons are organized in layers such that neurons of the same layer are not interconnected and that the connections are directed from lower to upper layers. Each neuron returns an output based on a weighted sum of all inputs and according to a nonlinear function called the transfer or activation function. The input layer, made up of different metrics from Sentinel-1 and Sentinel-2 data, is connected to the hidden layer(s), which is made up of hidden neurons. The final estimates of the ANN are given by an activation function associated with the final layer called the output layer, using a sum of the weighted outputs of the hidden neurons.

The ANN model architecture consists of three hidden layers of 20 neurons with a rectified linear function (ReLu) as activation functions and an output layer with a single neuron with a linear activation function. The mean square error was used as the loss function and the gradient backpropagation was carried out using a first order stochastic gradient-based optimizer (Adam).

Different predictors based on Sentinel-1 and Sentinel-2 were tested to estimate soil moisture: VV, VH, incidence angle, VH/VV, $NDVI$, and I_{SSM}.

1. The VV and VH signals are identified for their high sensitivity to soil moisture.
2. The classical change detection SSM index I_{SSM} is calculated as a function of radar backscattering coefficients in VV polarization to use it for soil moisture estimation.
3. The incidence angle has an effect on the contribution of soil and vegetation components on the radar signal.
4. The $NDVI$ index is identified to take into account the effect of vegetation cover on the backscattering signal.
5. The VH/VV ratio is identified to take into account the effect of vegetation cover on the backscattering signal [81].
6. SSM_t estimated from the classic change detection approach described in Section 3.1, Equation (2) is also considered as input.

The ANN models were trained using in situ soil moisture measurements retrieved from the ISMN as target. The training of the ANN models was conducted using 70% of the data samples. Thirty percent were kept for validation.

2.3. *Statistical Parameters for Accuracy Assessment*

Datasets are randomly subdivided into two parts: 70% of the database for model calibration and 30% for validation. The training data are used to calculate the different parameters to be estimated in the empirical and semiempirical models.

The *Bias*, root mean square error (*RMSE*) and Pearson's correlation (*R*) are considered to estimate the precision of the models.

$$Bias = P_i^{estimated} - P_i^{measured} \tag{8}$$

$$RMSE = \sqrt{\frac{1}{N}\sum_{i=1}^{N}\left(P_i^{estimated} - P_i^{measured}\right)^2} \quad (9)$$

where N is the number of data samples, $P_i^{estimated}$ is the estimated value of sample i, and $P_i^{measured}$ is the measured value of sample i.

$$R = \frac{\sum_{i=1}^{N}(x_i - \overline{x})(y_i - \overline{y})}{\sqrt{\sum_{i=1}^{N}(x_i - \overline{x})^2}\sqrt{\sum_{i=1}^{N}(y_i - \overline{y})^2}} \quad (10)$$

where x_i and y_i are individual samples taken at points indexed with the variable i.

3. Results

3.1. Improved Change Detection Approach

The empirical improved change detection approach has a double objective, taking into account the effect of vegetation and limiting the effect of surface geometry. The calibration of α and β parameters were conducted by using 70% of the dataset selected randomly (23869 samples). For the validation, the remaining 10,229 samples of the dataset were used.

Figure 2 illustrate the validations of the different algorithms described in Section 3.2 tested with ISMN data. The proposed results show an improvement in accuracy by considering the effect of vegetation cover in the tested relationships. The *RMSE* (*R*) values decrease (increase) from 0.074 m^3/m^3 (0.58) from the change detection approach (Equation (2), Figure 2a) to 0.073 m^3/m^3 (0.59) for the improved change detection approach (Equation (7)) when the VH/VV ratio is used as the V1 parameter and to 0.068 m^3/m^3 (0.63) when NDVI is used as the *V1* (Figure 2b).

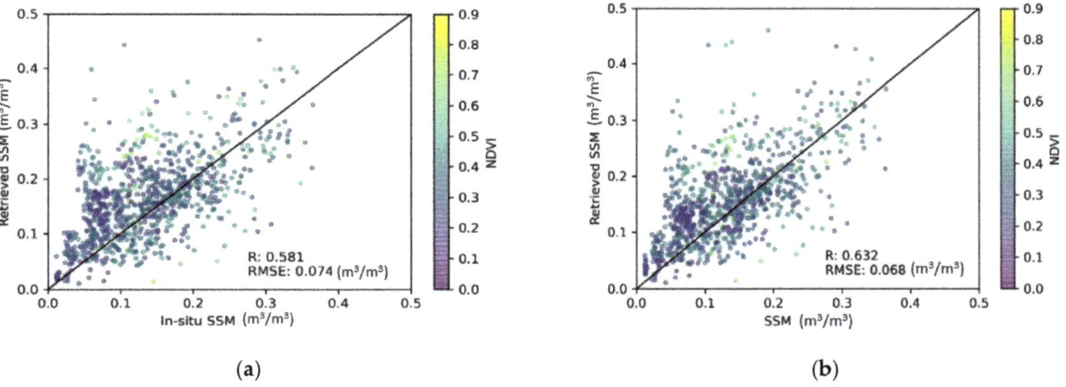

Figure 2. Scatterplots of the retrieved surface soil moisture (SSM) as a function of in situ SSM measurements colored according to NDVI value variation using two change detection approaches: (**a**) classic approach and (**b**) new approach expressed in Equation (7), where *V1* is the *NDVI*.

3.2. Neural Network Hybrid Approach

The different combinations of input metrics are tested to estimate soil moisture. Figure 3 illustrates the results of validations applied for 30% of the database, for each case of combination with the statistical parameters *RMSE* and *R*.

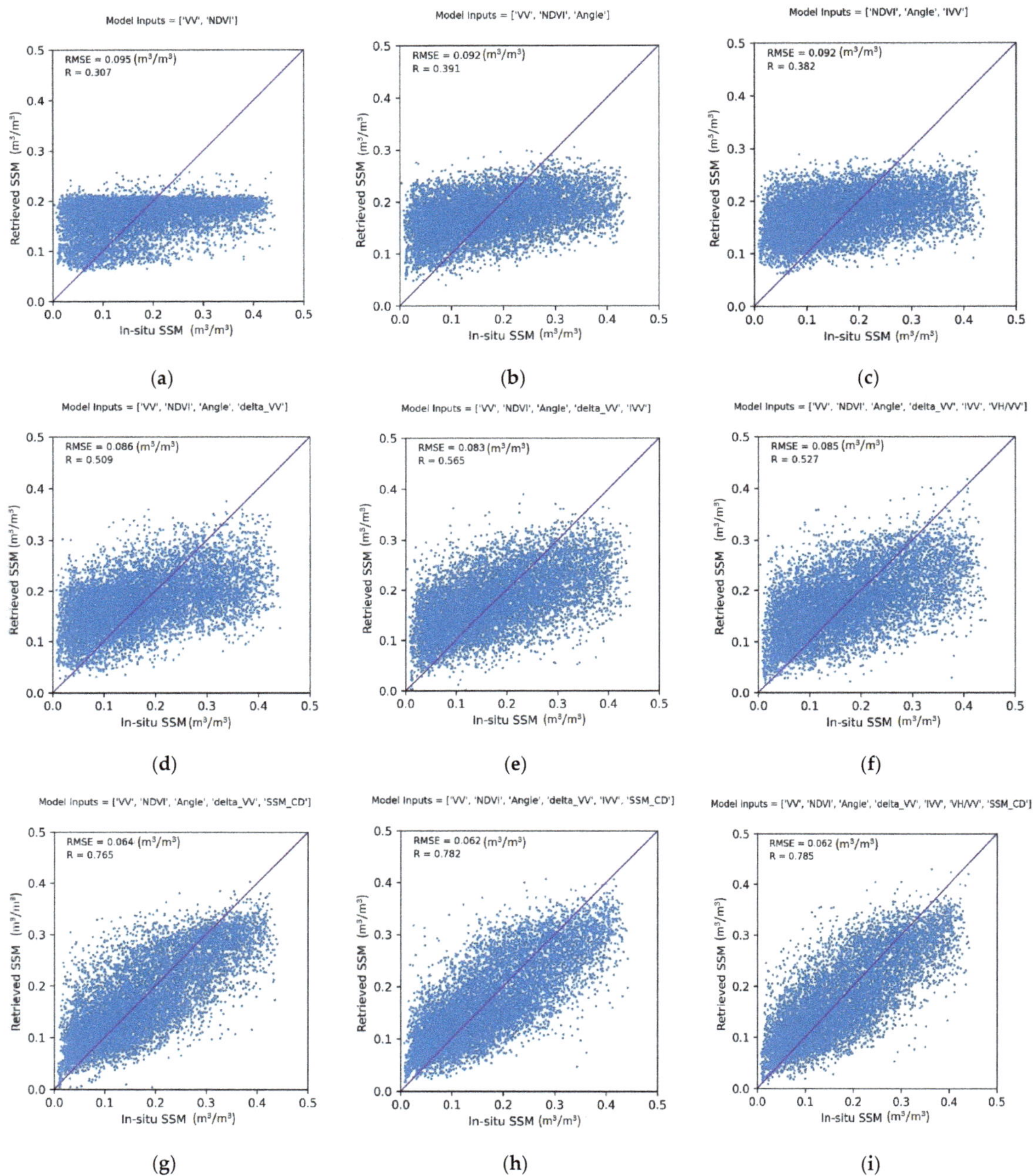

Figure 3. Scatterplots of the retrieved SSM as a function of in situ SSM measurements using the ANN approach using multiple combinations of features: (**a**) VV, *NDVI*, (**b**) VV, *NDVI*, the incidence angle, (**c**) *NDVI*, the incidence angle, I_{SSM}, (**d**) VV, *NDVI*, the incidence angle, Δ_{VV}, (**e**) VV, *NDVI*, the incidence angle, Δ_{VV}, I_{SSM}, (**f**) VV, NDVI, the incidence angle, Δ_{VV}, I_{SSM}, VH/VV ratio, (**g**) VV, *NDVI*, the incidence angle, Δ_{VV}, SSM_t, (**h**) VV, *NDVI*, the incidence angle, Δ_{VV}, I_{SSM}, SSM_t, (**i**) VV, *NDVI*, the incidence angle, Δ_{VV}, I_{SSM}, VH/VV ratio, SSM_t.

For the first six predictor combinations (Figure 3a–f), we observe relatively close precision with *RMSE* values in the range of 0.095 m^3/m^3 and 0.083 m^3/m^3 and *R* of 0.3–0.6. The introduction of moisture estimated by the classic change detection algorithm (Equation (3)) as input to ANN allows a strong improvement in the accuracy of soil moisture estimation with an *RMSE* equal to 0.063 m^3/m^3 and *R* = 0.76 when we consider the predictors: VV, *NDVI*, the incidence angle, Δ_{VV}, and SSM_t. By adding the VH/VV ratio and I_{SSM}, the *RMSE* value decreases to 0.062 m^3/m^3, and the correlation coefficient reaches a value of approximately 0.79. This result confirms the contribution of the hybrid approach to estimating soil moisture. This first estimated soil moisture strongly contributes to a better estimate of soil moisture by the ANN.

Figure 4 illustrates the accuracy of intercomparisons between in situ measurements and satellite estimates for the optimal case for different tested networks, where we represent the *RMSE* and *R* parameters by blue and orange boxes. The *RMSE* values vary from 0.03 m^3/m^3 to 0.09 m^3/m^3, and *R*-values fluctuate between 0.37 and 0.84.

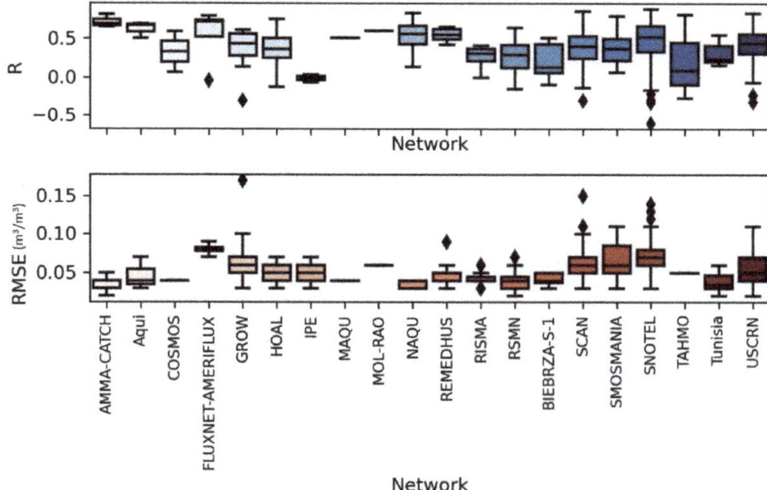

Figure 4. Boxplots of statistical parameters (*R* and *RMSE*) of soil moisture retrieval as a function of ISMN-considered networks using the hybrid methodology of change detection and ANN.

Good consistency is generally observed for networks such as AMMA-CATCH, COSMOS, MAQU, RSMN, HOAL, HOBE, IPE, BIEBRZA S-1, TAHMO, REMEDHUS and RISMA, and *RMSE* values are under or equal to 0.05 m^3/m^3. The NAQU network is characterized by the lowest *RMSE* value of 0.03 m^3/m^3 and *R* value of 0.77.

Within the same soil moisture in situ network, the accuracy of soil moisture retrieval varies from one station to another. For the REMEHDUS case characterized by an *RMSE* equal to 0.05 m^3/m^3, *RMSE* values per station range between 0.03 m^3/m^3 and 0.09 m^3/m^3, and R values vary between 0.34 and 0.69, as represented in Figure 5.

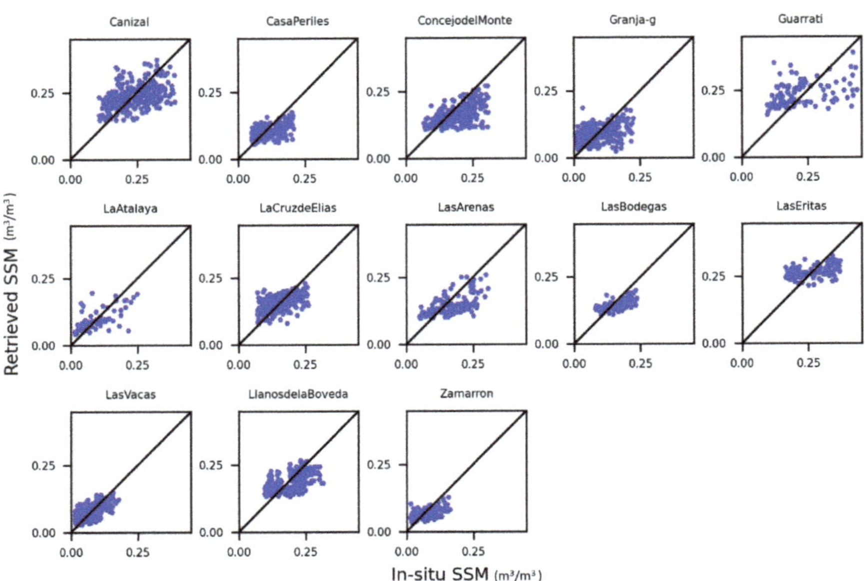

Figure 5. Scatterplots of the estimated soil moisture as a function of ISMN measurements in the REMEDHUS network per considered station.

4. Discussion

The proposed hybrid approaches have allowed more or less strong improvements compared to the initial estimates based on change detection or a separate ANN approach. With an improved change detection method, we observe a negligible contribution of the considered vegetation cover compared to a basic approach directly linking the radar signal to soil moisture. This can be explained by the highly diversified context at the scale of many soil moisture stations with very varied landscape contexts (crops, trees, bare soils, etc.) and different vegetation densities, which can generate significant noise in the modeling of the scattered signal that is difficult to take into account without a more precise description in terms of land use. This noise is particularly observed with the VH/VV index, which is very sensitive to the dynamics of the vegetation cover in a homogeneous context [81], but it could also mix different effects and particularly those of soil roughness [82].

To better analyze proposed results, we examined the time series of the in situ and retrieved soil moisture per station and network. Figure 6 displays the time series of the radar signal (VV), *NDVI*, and soil moisture SSM_t. The in situ soil moisture measurements are illustrated in blue, and the hybrid approach results are drawn in red. The intercomparison between the proposed approach performance within the LasBodegas and Canizal stations reveals *RMSE* values of 0.04 m^3/m^3 and 0.07 m^3/m^3, respectively. The two stations belong to the same climatic region of the arid steppe and characterize a clay fraction interval of approximately 35%. The performance difference may be induced by the land cover, where the Canizal station is occupied by shrubs and the LasBodegas station is covered by trees. The aforementioned land cover may impact the accuracy of soil moisture retrieval due to the vegetation volume impact on the radar signal in the C-band. Additionally, the measured soil moisture values are lower than 0.3 m^3/m^3 at the LasBodegas station, and higher values reach 0.4 m^3/m^3 at the Canizal station. Hence, the soil water content retrieval is more accurate in the first case due to the saturation of the C-band signal at high values of soil moisture.

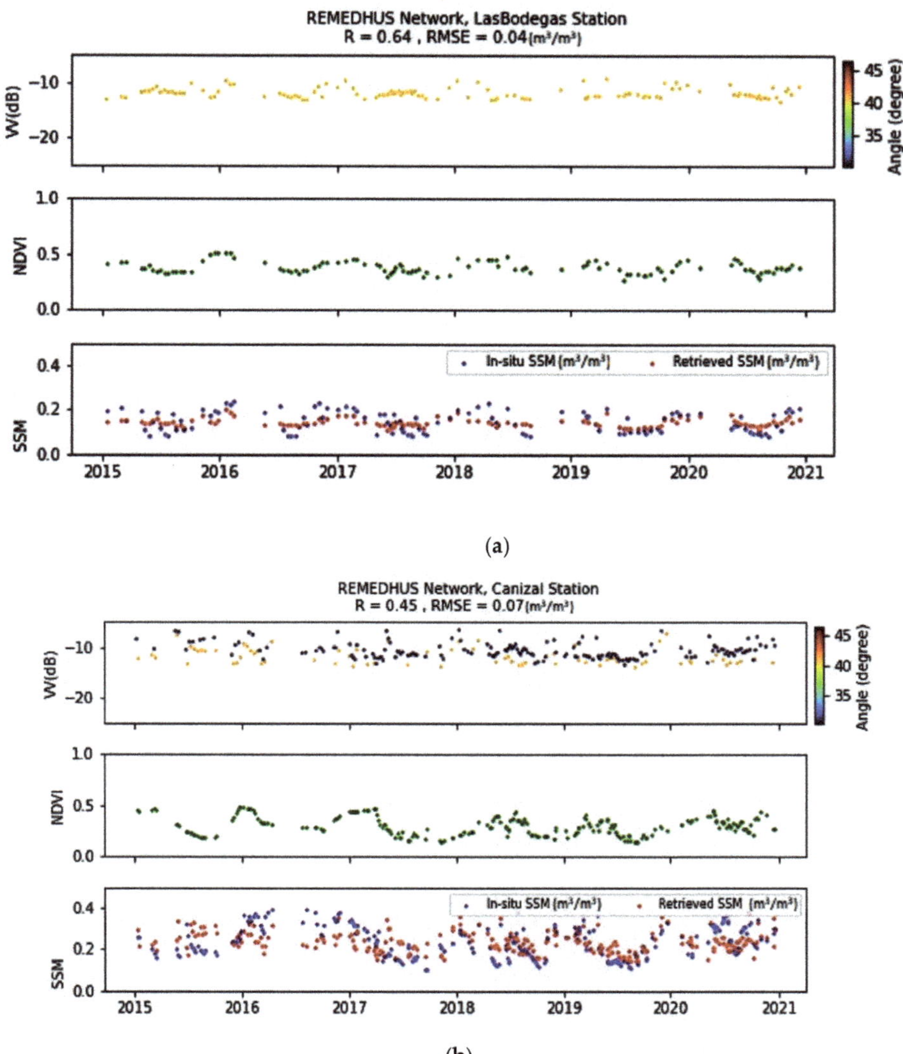

Figure 6. Scatterplots of the temporal evolution of radar signals in VV polarization, NDVI, and the predicted and in situ measurements of soil moisture using the hybrid methods within two stations of the REMEHDUS network: (**a**) LasBodegas station, (**b**) Canizal station.

However, the approach has difficulties for certain stations, as shown for FLUXNET-AMERFLUX, GROW, SNOTAL, and SMOSMANIA networks, where the *RMSE* values reach a maximum of 0.09 m^3/m^3. The analysis of these cases generally leads to contexts of dense vegetation cover that can induce a low sensitivity of the radar signal to soil moisture.

In Figure 7, we scatterplot the statistical parameters as a function of *NDVI* values. According to Figure 7a,b, we observe the increase of *RMSE* and *Bias* values as a function of the increase of *NDVI* values where *RMSE* can reach 0.10 m^3/m^3. The vegetation development may induce a *Bias* between -0.06 m^3/m^3 and 0.04 m^3/m^3, where *NDVI* values exceed 0.5. This behavior may be explained by the C-band potential which is otherwise limited in dense canopies where *NDVI* values are higher than 0.5.

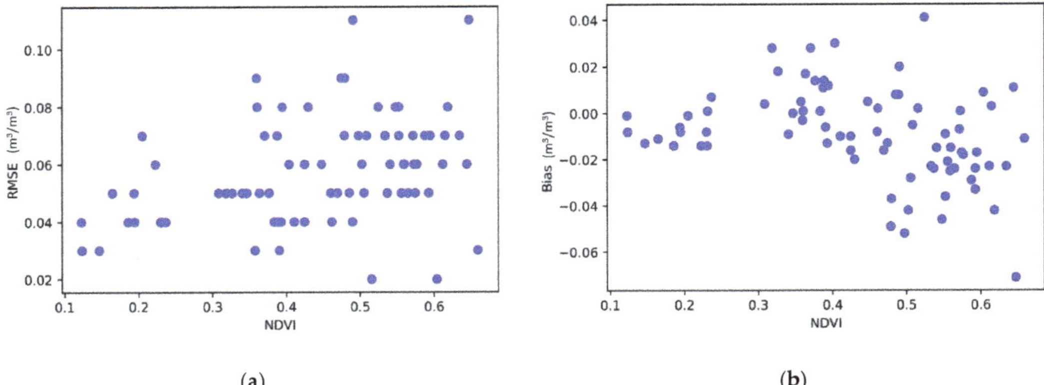

(a) (b)

Figure 7. Scatterplots of statistical parameters of the hybrid approach performance to retrieve SSM as a function of *NDVI* values stations: (**a**) *RMSE*, (**b**) *Bias*.

This is difficult to take into account in a general approach based on a neural network trained on stations with different types of surface conditions, such as the case of the FLUXNET-AMERFLUX network. The station land covers are a mixture of grasslands, temporary crops followed by harvest and bare soil periods, and woody savanna characterized by forest canopy cover between 30% and 60% and vegetation height exceeding 2 m. In this land cover context, the vegetation volume impacts the radar signal and complicates the soil moisture retrieval. We observe the vegetation impact within many stations in the SMOSMANIA network, such as the Mazan-Abbaye, Cabriers Avignon, and Ville Vieille stations occupied by trees or shrubs.

Furthermore, the use of *NDVI* as a vegetation descriptor may induce other limits, such as the availability of data in regions with temperate climates. The presence of clouds contaminates the surface reflectance, which damages the radiometric information. As a result, many time series suffer from gaps and lack data, which complicates the training and validation of the proposed model, such as the case of some stations of the USCRN network, where the mean *RMSE* value is equal to 0.06 m^3/m^3.

By considering the GROW network data, the *RMSE* is equal to 0.07 m^3/m^3. This relatively low accuracy in retrieving soil moisture may be linked to the predominant cold climate of the considered stations. This low-temperature climate may impact the radar signal, especially with the freeze–thaw phenomenon. This change in the physical state of the soil water content generates a fast variability in the Sentinel-1 signal, as discussed for agricultural plots in metropolitan France by Baghdadi et al. [83] and Fayad et al. [84].

5. Conclusions

Different approaches have been proposed for SSM estimation from space. The goal is to improve estimates by combining change detection logic with empirical or other approaches based on an ANN. The study is based on Sentinel-1 and Sentinel-2 data tested on the ISMN moisture network.

Relationships between temporal changes in radar signals and temporal changes in soil moisture are tested. Improved change detection relationships combine these effects with the contribution of vegetation through two optical and radar indices (*NDVI* and VH/VV ratio). The integration of the effect of vegetation slightly improves the precision with an *RMSE* that decreases slightly from 0.074 m^3/m^3 to 0.073 m^3/m^3 and 0.068 m^3/m^3 for VH/VV and *NDVI*, respectively.

Testing an ANN approach through numerous metrics based on radar and optical (VV, VH, VH/VV, *NDVI*, ΔVV, incidence angle, etc.) time series illustrates precision within a 0.08 m^3/m^3–0.09 (m^3/m^3) range. These results are greatly improved with the integration as input of soil moisture estimated from the change detection approach. Thus, we move on to precision below the bar of 0.07 m^3/m^3 for the different possible combinations of metrics. Thus, it seems highly useful to propose this combination to improve the precision of the estimated soil moisture. Despite this improvement, there are some limitations at some stations, particularly related to the vegetation density and presence of forests or extreme climates with cold conditions. In the future, it would be very useful to propose a spatialization of this approach by considering auxiliary information of soil properties and land use for a better application of the proposed algorithms and improvement of proposed precision. In fact, this allows us to distinguish effects due more precisely to vegetation for which volume and attenuation scattering are different from one cover to another. For a high-resolution scale, this aspect, which is generally not considered for a low-resolution scale, seems important.

Author Contributions: S.N., E.A. and M.Z. developed methods and analyzed the data; all authors contributed to the materials/analysis tools; and M.Z. and E.A. wrote the paper. All authors have read and agreed to the published version of the manuscript.

Funding: This study was funded by Projects; ESA No. 4000126684/19/I-NB "ESA CCI+" and TAPAS TOSCA/CNES.

Data Availability Statement: The data presented in this study are available in International Soil Moisture Network (ISMN) https://ismn.geo.tuwien.ac.at/en/ and https://scihub.copernicus.eu/sites (accessed on 16 May 2022).

Acknowledgments: The authors thank the International Soil Moisture Network (ISMN) and the supporting networks for the availability of soil moisture data.

Conflicts of Interest: The authors declare no conflict of interest.

References

1. Koster, R.D.; Dirmeyer, P.A.; Guo, Z.; Bonan, G.; Chan, E.; Cox, P.; Gordon, C.T.; Kanae, S.; Kowalczyk, E.; Lawrence, D.; et al. Regions of Strong Coupling Between Soil Moisture and Precipitation. *Science* **2004**, *305*, 1138–1140. [CrossRef] [PubMed]
2. Anguela, T.P.; Zribi, M.; Hasenauer, S.; Habets, F.; Loumagne, C. Analysis of surface and root-zone soil moisture dynamics with ERS scatterometer and the hydrometeorological model SAFRAN-ISBA-MODCOU at Grand Morin watershed (France). *Hydrol. Earth Syst. Sci.* **2008**, *12*, 1415–1424. [CrossRef]
3. Albergel, C.; Zakharova, E.; Calvet, J.-C.; Zribi, M.; Pardé, M.; Wigneron, J.-P.; Novello, N.; Kerr, Y.; Mialon, A.; Fritz, N.-E. A first assessment of the SMOS data in southwestern France using in situ and airborne soil moisture estimates: The CAROLS airborne campaign. *Remote Sens. Environ.* **2011**, *115*, 2718–2728. [CrossRef]
4. Brocca, L.; Ciabatta, L.; Moramarco, T.; Ponziani, F.; Berni, N.; Wagner, W. Use of Satellite Soil Moisture Products for the Operational Mitigation of Landslides Risk in Central Italy. In *Satellite Soil Moisture Retrieval*; Elsevier: New York, NY, USA, 2016; Volume 7, pp. 231–247. [CrossRef]
5. Le Page, M.; Jarlan, L.; El Hajj, M.M.; Zribi, M.; Baghdadi, N.; Boone, A. Potential for the Detection of Irrigation Events on Maize Plots Using Sentinel-1 Soil Moisture Products. *Remote Sens.* **2020**, *12*, 1621. [CrossRef]
6. Ulaby, F.T.; Bradley, G.A.; Dobson, M.C. Microwave Backscatter Dependence on Surface Roughness, Soil Moisture, and Soil Texture: Part II-Vegetation-Covered Soil. *IEEE Trans. Geosci. Electron.* **1979**, *17*, 33–40. [CrossRef]
7. Jackson, T.J.; Cosh, M.H.; Bindlish, R.; Starks, P.J.; Bosch, D.D.; Seyfried, M.; Goodrich, D.C.; Moran, M.S.; Du, J.Y. Validation of Advanced Microwave Scanning Radiometer Soil Moisture Products. *IEEE Trans. Geosci. Remote Sens. Dec.* **2010**, *48*, 4256–4272. [CrossRef]
8. Kerr, Y.H.; Waldteufel, P.; Richaume, P.; Wigneron, J.-P.; Ferrazzoli, P.; Mahmoodi, A.; Al Bitar, A.; Cabot, F.; Gruhier, C.; Juglea, S.E.; et al. The SMOS Soil Moisture Retrieval Algorithm. *IEEE Trans. Geosci. Remote Sens.* **2012**, *50*, 1384–1403. [CrossRef]
9. Entekhabi, D.; Njoku, E.G.; O'Neill, P.E.; Kellogg, K.H.; Crow, W.T.; Edelstein, W.N.; Entin, J.K.; Goodman, S.D.; Jackson, T.J.; Johnson, J.; et al. The Soil Moisture Active Passive (SMAP) Mission. *Proc. IEEE* **2010**, *98*, 704–716. [CrossRef]
10. Kim, H.; Parinussa, R.; Konings, A.G.; Wagner, W.; Cosh, M.H.; Lakshmi, V.; Zohaib, M.; Choi, M. Global-scale assessment, and combination of SMAP with ASCAT (active) and AMSR2 (passive) soil moisture products. *Remote Sens. Environ.* **2018**, *204*, 260–275. [CrossRef]
11. Motte, E.; Zribi, M.; Fanise, P.; Egido, A.; Darrozes, J.; Al-Yaari, A.; Baghdadi, N.; Baup, F.; Dayau, S.; Fieuzal, R.; et al. GLORI: A GNSS-R Dual Polarization Airborne Instrument for Land Surface Monitoring. *Sensors* **2016**, *16*, 732. [CrossRef]
12. Colliander, A.; Reichle, R.; Crow, W.; Cosh, M.H.; Chen, F.; Chan, S.K.; Das, N.N.; Bindlish, R.; Chaubell, M.J.; Kim, S.; et al. Validation of Soil Moisture Data Products From the NASA SMAP Mission. *IEEE J. Sel. Top. Appl. Earth Obs. Remote Sens.* **2021**, *15*, 364–392. [CrossRef]

13. Wagner, W.; Bloeschl, G.; Pamaloni, P.; Calvet, J.C. Operational readiness of microwave remote sensing of soil moisture for hydrologic applications. *Nord. Hydrol.* **2007**, *38*, 1–20. [CrossRef]
14. Dorigo, W.; Wagner, W.; Albergel, C.; Albrecht, F.; Balsamo, G.; Brocca, L.; Chung, D.; Ertl, M.; Forkel, M.; Gruber, A.; et al. ESA CCI Soil Moisture for improved Earth system understanding: State-of-the art and future directions. *Remote Sens. Environ.* **2017**, *203*, 185–215. [CrossRef]
15. Moran, M.S.; Hymer, D.C.; Qi, J.; Sano, E.E. Soil moisture evaluation using multi-temporal synthetic aperture radar (SAR) in semiarid rangeland. *Agric. For. Meteorol.* **2000**, *105*, 69–80. [CrossRef]
16. Pierdicca, N.; Pulvirenti, L.; Bignami, C. Soil moisture estimation over vegetated terrains using multitemporal remote sensing data. *Remote Sens. Environ.* **2010**, *114*, 440–448. [CrossRef]
17. Bousbih, S.; Zribi, M.; Lili-Chabaane, Z.; Baghdadi, N.; El Hajj, M.; Gao, Q.; Mougenot, B. Potential of Sentinel-1 Radar Data for the Assessment of Soil and Cereal Cover Parameters. *Sensors* **2017**, *17*, 2617. [CrossRef]
18. Şekertekin, A.; Marangoz, A.M.; Abdikan, S. Soil Moisture Mapping Using Sentinel-1A Synthetic Aperture Radar Data. *Int. J. Environ. Geoinform.* **2018**, *5*, 178–188. [CrossRef]
19. Hajj, M.E.; Baghdadi, N.; Belaud, G.; Zribi, M.; Cheviron, B.; Courault, D.; Hagolle, O.; Charron, F. Irrigated Grassland Monitoring Using a Time Series of TerraSAR-X and COSMO-SkyMed X-Band SAR Data. *Remote Sens.* **2014**, *6*, 10002–10032. [CrossRef]
20. Srivastava, H.S.; Patel, P.; Sharma, Y.; Navalgund, R.R. Large-Area Soil Moisture Estimation Using Multi-Incidence-Angle RADARSAT-1 SAR Data. *IEEE Trans. Geosci. Remote Sens.* **2009**, *47*, 2528–2535. [CrossRef]
21. Balenzano, A.; Mattia, F.; Satalino, G.; Davidson, M.W.J. Dense Temporal Series of C- and L-band SAR Data for Soil Moisture Retrieval Over Agricultural Crops. *IEEE J. Sel. Top. Appl. Earth Obs. Remote Sens.* **2010**, *4*, 439–450. [CrossRef]
22. Ma, C.; Li, X.; McCabe, M.F. Retrieval of High-Resolution Soil Moisture through Combination of Sentinel-1 and Sentinel-2 Data. *Remote Sens.* **2020**, *12*, 2303. [CrossRef]
23. Wang, H.; Magagi, R.; Goita, K.; Jagdhuber, T. Refining a Polarimetric Decomposition of Multi-Angular UAVSAR Time Series for Soil Moisture Retrieval Over Low and High Vegetated Agricultural Fields. *IEEE J. Sel. Top. Appl. Earth Obs. Remote Sens.* **2019**, *12*, 1431–1450. [CrossRef]
24. Wang, H.; Magagi, R.; Goita, K.; Jagdhuber, T.; Hajnsek, I. Evaluation of Simplified Polarimetric Decomposition for Soil Moisture Retrieval over Vegetated Agricultural Fields. *Remote Sens.* **2016**, *8*, 142. [CrossRef]
25. Molero, B.; Merlin, O.; Malbeteau, Y.; Al Bitar, A.; Cabot, F.; Stefan, V.; Kerr, Y.; Bacon, S.; Cosh, M.; Bindlish, R.; et al. SMOS disaggregated soil moisture product at 1 km resolution: Processor overview and first validation results. *Remote Sens. Environ.* **2016**, *180*, 361–376. [CrossRef]
26. Kim, S.-B.; Moghaddam, M.; Tsang, L.; Burgin, M.; Xu, X.; Njoku, E.G. Models of L-Band Radar Backscattering Coefficients Over Global Terrain for Soil Moisture Retrieval. *IEEE Trans. Geosci. Remote Sens.* **2013**, *52*, 1381–1396. [CrossRef]
27. Kim, S.-B.; Van Zyl, J.J.; Johnson, J.T.; Moghaddam, M.; Tsang, L.; Colliander, A.; Dunbar, R.S.; Jackson, T.J.; Jaruwatanadilok, S.; West, R.; et al. Surface Soil Moisture Retrieval Using the L-Band Synthetic Aperture Radar Onboard the Soil Moisture Active–Passive Satellite and Evaluation at Core Validation Sites. *IEEE Trans. Geosci. Remote Sens.* **2017**, *55*, 1897–1914. [CrossRef]
28. Bousbih, S.; Zribi, M.; El Hajj, M.; Baghdadi, N.; Lili-Chabaane, Z.; Gao, Q.; Fanise, P. Soil Moisture and Irrigation Mapping in A Semi-Arid Region, Based on the Synergetic Use of Sentinel-1 and Sentinel-2 Data. *Remote Sens.* **2018**, *10*, 1953. [CrossRef]
29. Ezzahar, J.; Ouaadi, N.; Zribi, M.; Elfarkh, J.; Aouade, G.; Khabba, S.; Er-Raki, S.; Chehbouni, A.; Jarlan, L. Evaluation of Backscattering Models and Support Vector Machine for the Retrieval of Bare Soil Moisture from Sentinel-1 Data. *Remote Sens.* **2019**, *12*, 72. [CrossRef]
30. Notarnicola, C.; Angiulli, M.; Posa, F. Soil moisture retrieval from remotely sensed data: Neural network approach versus Bayesian method. *IEEE Trans. Geosci. Remote Sens.* **2008**, *46*, 547–557. [CrossRef]
31. El Hajj, M.; Baghdadi, N.; Zribi, M.; Bazzi, H. Synergic Use of Sentinel-1 and Sentinel-2 Images for Operational Soil Moisture Mapping at High Spatial Resolution over Agricultural Areas. *Remote Sens.* **2017**, *9*, 1292. [CrossRef]
32. Zribi, M.; Kotti, F.; Amri, R.; Wagner, W.; Shabou, M.; Chabaane, Z.L.; Baghdadi, N. Soil moisture mapping in a semiarid region, based on ASAR/Wide Swath satellite data. *Water Resour. Res.* **2014**, *50*, 823–835. [CrossRef]
33. Bauer-Marschallinger, B.; Freeman, V.; Cao, S.; Paulik, C.; Schaufler, S.; Stachl, T.; Modanesi, S.; Massari, C.; Ciabatta, L.; Brocca, L.; et al. Toward Global Soil Moisture Monitoring With Sentinel-1: Harnessing Assets and Overcoming Obstacles. *IEEE Trans. Geosci. Remote Sens.* **2018**, *57*, 520–539. [CrossRef]
34. Foucras, M.; Zribi, M.; Albergel, C.; Baghdadi, N.; Calvet, J.-C.; Pellarin, T. Estimating 500-m Resolution Soil Moisture Using Sentinel-1 and Optical Data Synergy. *Water* **2020**, *12*, 866. [CrossRef]
35. Gao, Q.; Zribi, M.; Escorihuela, M.J.; Baghdadi, N. Synergetic Use of Sentinel-1 and Sentinel-2 Data for Soil Moisture Mapping at 100 m Resolution. *Sensors* **2017**, *17*, 1966. [CrossRef]
36. Bazzi, H.; Baghdadi, N.; El Hajj, M.; Zribi, M.; Belhouchette, H. A Comparison of Two Soil Moisture Products S2MP and Copernicus-SSM over Southern France. *IEEE J. Sel. Top. Appl. Earth Obs. Remote Sens.* **2019**, *100*, 10–18. [CrossRef]
37. Dorigo, W.A.; Wagner, W.; Hohensinn, R.; Hahn, S.; Paulik, C.; Xaver, A.; Gruber, A.; Drusch, M.; Mecklenburg, S.; van Oevelen, P.; et al. The International Soil Moisture Network: A data hosting facility for global in situ soil moisture measurements. *Hydrol. Earth Syst. Sci.* **2011**, *15*, 1675–1698. [CrossRef]

38. Cappelaere, C.; Descroix, L.; Lebel, T.; Boulain, N.; Ramier, D.; Laurent, J.-P.; Le Breton, E.; Boubkraoui, S.; Bouzou Moussa, I.; Quantin, G.; et al. The AMMA Catch observing system in the cultivated Sahel of Southwest Niger- Strategy, Implementation and Site conditions. *J. Hydrol.* **2009**, *375*, 34–51. [CrossRef]
39. de Rosnay, P.; Gruhier, C.; Timouk, F.; Baup, F.; Mougin, E.; Hiernaux, P.; Kergoat, L.; LeDantec, V. Multi-scale soil moisture measurements at the Gourma meso-scale site in Mali. *J. Hydrol.* **2009**, *375*, 241–252. [CrossRef]
40. Lebel, T.; Cappelaere, B.; Galle, S.; Hanan, N.; Kergoat, L.; Levis, S.; Vieux, B.; Descroix, L.; Gosset, M.; Mougin, E.; et al. AMMA-CATCH studies in the Sahelian region of West-Africa: An overview. *J. Hydrol.* **2009**, *375*, 3–13. [CrossRef]
41. Galle, S.; Grippa, M.; Peugeot, C.; Moussa, I.B.; Cappelaere, B.; Demarty, J.; Mougin, E.; Panthou, G.; Adjomayi, P.; Agbossou, E.; et al. AMMA-CATCH, a Critical Zone Observatory in West Africa Monitoring a Region in Transition. *Vadose Zone J.* **2018**, *17*, 1–24. [CrossRef]
42. Mougin, E.; Hiernaux, P.; Kergoat, L.; Grippa, M.; de Rosnay, P.; Timouk, F.; Le Dantec, V.; Demarez, V.; Lavenu, F.; Arjounin, M.; et al. The AMMA-CATCH Gourma observatory site in Mali: Relating climatic variations to changes in vegetation, surface hydrology, fluxes and natural resources. *J. Hydrol.* **2009**, *375*, 14–33. [CrossRef]
43. Pellarin, T.; Laurent, J.; Cappelaere, B.; Decharme, B.; Descroix, L.; Ramier, D. Hydrological modelling and associated microwave emission of a semi-arid region in South-western Niger. *J. Hydrol.* **2009**, *375*, 262–272. [CrossRef]
44. Musial, J.P.; Dabrowska-Zielinska, K.; Kiryla, W.; Oleszczuk, R.; Gnatowski, T.; Jaszczynski, J. Derivation and validation of the high-resolution satellite soil moisture products: A case study of the biebrza sentinel-1 validation sites. *Geoinf. Issues* **2016**, *8*, 37–53.
45. Zreda, M.; Desilets, D.; Ferré Ty, P.A.; Scott, R.L. Measuring soil moisture content non-invasively at intermediate spatial scale using cosmic-ray neutrons. *Geophys. Res. Lett.* **2008**, *35*, 1–5. [CrossRef]
46. Zreda, M.; Shuttleworth, W.J.; Zeng, X.; Zweck, C.; Desilets, D.; Franz, T.; Rosolem, R. COSMOS: The COsmic-ray Soil Moisture Observing System. *Hydrol. Earth Syst. Sci.* **2012**, *16*, 4079–4099. [CrossRef]
47. Jensen, K.H.; Refsgaard, J.C. HOBE: The Danish Hydrological Observatory. *Vadose Zone J.* **2018**, *17*, 1–24. [CrossRef]
48. Bircher, S.; Skou, N.; Jensen, K.H.; Walker, J.P.; Rasmussen, L. A soil moisture and temperature network for SMOS validation in Western Denmark. *Hydrol. Earth Syst. Sci.* **2012**, *16*, 1445–1463. [CrossRef]
49. Al-Yaari, A.; Dayau, S.; Chipeaux, C.; Aluome, C.; Kruszewski, A.; Loustau, D.; Wigneron, J.-P. The AQUI Soil Moisture Network for Satellite Microwave Remote Sensing Validation in South-Western France. *Remote Sens.* **2018**, *10*, 1839. [CrossRef]
50. Wigneron, J.-P.; Dayan, S.; Kruszewski, A.; Aluome, C.; Al-Yaari, A.; Fan, L.; Guven, S.; Chipeaux, C.; Moisy, C.; Guyon, D.; et al. The aqui network: Soil moisture sites in the "les landes" forest and graves vineyards (Bordeaux Aquitaine region, France). In Proceedings of the IGARSS 2018-2018 IEEE International Geoscience and Remote Sensing Symposium, Valencia, Spain, 22–27 July 2018; pp. 3739–3742.
51. Zappa, L.; Forkel, M.; Xaver, A.; Dorigo, W. Deriving Field Scale Soil Moisture from Satellite Observations and Ground Measurements in a Hilly Agricultural Region. *Remote Sens.* **2019**, *11*, 2596. [CrossRef]
52. Xaver, A.; Zappa, L.; Rab, G.; Pfeil, I.; Vreugdenhil, M.; Hemment, D.; Dorigo, W.A. Evaluating the suitability of the consumer low-cost Parrot Flower Power soil moisture sensor for scientific environmental applications. *Geosci. Instrum. Methods Data Syst.* **2020**, *9*, 117–139. [CrossRef]
53. Zappa, L.; Woods, M.; Hemment, D.; Xaver, A.; Dorigo, W. Evaluation of remotely sensed soil moisture products using crowdsourced measurements. In Proceedings of the Eighth international conference on remote sensing and geoinformation of the environment (RSCy2020), Paphos, Cyprus, 26 August 2020; Volume 11524, p. 115241U. [CrossRef]
54. Vreugdenhil, M.; Dorigo, W.; Broer, M.; Haas, P.; Eder, A.; Hogan, P.; Blöschl, G.; Wagner, W. Towards a high-density soil moisture network for the validation of SMAP in Petzenkirchen, Austria. In Proceedings of the 2013 IEEE International Geoscience and Remote Sensing Symposium-IGARSS, Melbourne, VIC, Australia, 21–26 July 2013; pp. 1865–1868. [CrossRef]
55. Blöschl, G.; Blaschke, A.P.; Broer, M.; Bucher, C.; Carr, G.; Chen, X.; Eder, A.; Exner-Kittridge, M.; Farnleitner, A.; Flores-Orozco, A.; et al. The Hydrological Open Air Laboratory (HOAL) in Petzenkirchen: A hypothesis-driven observatory. *Hydrol. Earth Syst. Sci.* **2016**, *20*, 227–255. [CrossRef]
56. Alday, J.G.; Camarero, J.J.; Revilla, J.; De Dios, V.R. Similar diurnal, seasonal and annual rhythms in radial root expansion across two coexisting Mediterranean oak species. *Tree Physiol.* **2020**, *40*, 956–968. [CrossRef] [PubMed]
57. Su, Z.; Wen, J.; Dente, L.; van der Velde, R.; Wang, L.; Ma, Y.; Yang, K.; Hu, Z. The Tibetan Plateau observatory of plateau scale soil moisture and soil temperature (Tibet-Obs) for quantifying uncertainties in coarse resolution satellite and model products. *Hydrol. Earth Syst. Sci.* **2011**, *15*, 2303–2316. [CrossRef]
58. Dente, L.; Su, Z.; Wen, J. Validation of SMOS Soil Moisture Products over the Maqu and Twente Regions. *Sensors* **2012**, *12*, 9965–9986. [CrossRef] [PubMed]
59. Beyrich, F.; Adam, W.K. Site and Data Report for the Lindenberg Reference Site in CEOP—Phase 1. *Berichte des Deutschen Wetterdienstes* **2007**, *230*. Offenbach am Main.
60. Su, Z.; de Rosnay, P.; Wen, J.; Wang, L.; Zeng, Y. Evaluation of ECMWF's soil moisture analyses using observationson the Tibetan Plateau. *J. Geophys. Res. Atmos.* **2013**, *118*, 5304–5318. [CrossRef]
61. Canisius, F. *Calibration of Casselman, Ontario Soil Moisture Monitoring Network*; Agriculture and Agri-Food: Ottawa, ON, Canada, 2011; 37p.
62. L'Heureux, J. Installation Report for AAFC-SAGES Soil Moisture Stations in Kenaston, SK. In *Calibration of Casselman, Ontario Soil Moisture Monitoring Network*; Agriculture and Agri-Food: Ottawa, ON, Canada, 2011; 37p.

63. Ojo, E.R.; Bullock, P.R.; L'Heureux, J.; Powers, J.; McNairn, H.; Pacheco, A. Calibration and Evaluation of a Frequency Domain Reflectometry Sensor for Real-Time Soil Moisture Monitoring. *Vadose Zone J.* **2015**, *14*. [CrossRef]
64. Gonzalez-Zamora, A.; Sanchez, N.; Pablos, M.; Martinez-Fernandez, J. Cci soil moisture assessment with SMOS soil moisture and in situ data under different environmental conditions and spatial scales in Spain. *Remote Sens. Environ.* **2018**, *225*, 469–482. [CrossRef]
65. Schaefer, G.; Cosh, M.; Jackson, T. The usda natural resources conservation service soil climate analysis network (scan). *J. Atmos. Ocean. Technol.* **2007**, *24*, 2073–2077. [CrossRef]
66. Calvet, J.-C.; Fritz, N.; Froissard, F.; Suquia, D.; Petitpa, A.; Piguet, B. In situ soil moisture observations for the CAL/VAL of SMOS: The SMOSMANIA network. In Proceedings of the 2007 IEEE International Geoscience and Remote Sensing Symposium, Barcelona, Spain, 23–28 July 2007; pp. 1196–1199. [CrossRef]
67. Albergel, C.; Rüdiger, C.; Pellarin, T.; Calvet, J.-C.; Fritz, N.; Froissard, F.; Suquia, D.; Petitpa, A.; Piguet, B.; Martin, E. From near-surface to root-zone soil moisture using an exponential filter: An assessment of the method based on in situ observations and model simulations. *Hydrol. Earth Syst. Sci.* **2008**, *12*, 1323–1337. [CrossRef]
68. Calvet, J.-C.; Fritz, N.; Berne, C.; Piguet, B.; Maurel, W.; Meurey, C. Deriving pedotransfer functions for soil quartz fraction in southern france from reverse modeling. *Soil* **2016**, *2*, 615–629. [CrossRef]
69. Leavesley, G.H.; David, O.; Garen, D.C.; Lea, J.; Marron, J.K.; Pagano, T.C.; Strobel, M.L. A modeling framework for improved agricultural water supply forecasting. In *AGU Fall Meeting Abstracts*; American Geophysical Union: Washington, DC, USA, 2008; Volume 2008, pp. C21A–0497.
70. Zacharias, S.; Bogena, H.; Samaniego, L.; Mauder, M.; Fuß, R.; Pütz, T.; Frenzel, M.; Schwank, M.; Baessler, C.; Butterbach-Bahl, K.; et al. A Network of Terrestrial Environmental Observatories in Germany. *Vadose Zone J.* **2011**, *10*, 955–973. [CrossRef]
71. Bogena, H.; Kunkel, R.; Pütz, T.; Vereecken, H.; Kruger, E.; Zacharias, S.; Dietrich, P.; Wollschläger, U.; Kunstmann, H.; Papen, H.; et al. Tereno-long-term monitoring network for terrestrial environmental research. *Hydrol. Und Wasserbewirtsch.* **2012**, *56*, 138–143.
72. Bogena, H.R. Tereno: German network of terrestrial environmental observatories. *J. Large-Scale Res. Facil. JLSRF* **2016**, *2*, 52. [CrossRef]
73. Bell, J.E.; Palecki, M.A.; Baker, C.B.; Collins, W.G.; Lawrimore, J.H.; Leeper, R.; Hall, M.E.; Kochendorfer, J.; Meyers, T.P.; Wilson, T.; et al. U.S. Climate Reference Network Soil Moisture and Temperature Observations. *J. Hydrometeorol.* **2013**, *14*, 977–988. [CrossRef]
74. Schwerdt, M.; Schmidt, K.; Tous Ramon, N.; Klenk, P.; Yague-Martinez, N.; Prats-Iraola, P.; Zink, M.; Geudtner, D. Independent system calibration of Sentinel-1B. *Remote Sens.* **2017**, *9*, 511. [CrossRef]
75. Hagolle, O.; Huc, M.; Villa Pascual, D.; Dedieu, G. A Multi-Temporal and Multi-Spectral Method to Estimate Aerosol Optical Thickness over Land, for the Atmospheric Correction of FormoSat-2, LandSat, VENμS and Sentinel-2 Images. *Remote Sens.* **2015**, *7*, 2668–2691. [CrossRef]
76. Wagner, W.; Lemoine, G.; Rott, H. A method for estimating soil moisture from ERS Scatterometer and soil data. *Remote Sens. Environ.* **1999**, *70*, 191–207. [CrossRef]
77. Pellarin, T.; Calvet, J.C.; Wagner, W. Evaluation of ERS scatterometer soil moisture products over a half-degree region in southwestern France. *Geophys. Res. Lett.* **2006**, *33*, 1–6. [CrossRef]
78. Zribi, M.; André, C.; Decharme, B. A method for soil moisture estimation in Western Africa based on ERS Scatter meter. *IEEE Trans. Geosci. Remote Sens.* **2008**, *46*, 438–448. [CrossRef]
79. ASCE Task Committee on Application of Artificial Neural Networks in Hydrology. Artificial neural networks in hydrology. I: Preliminary concepts. *J. Hydrol. Eng.* **2000**, *5*, 115–123. [CrossRef]
80. Tanty, T.S.D.R.; Desmukh, T.; Bhopal, M. Application of Artificial Neural Network in Hydrology—A Review. *Int. J. Eng. Res.* **2015**, *V4*, 184–188. [CrossRef]
81. Veloso, A.; Mermoz, S.; Bouvet, A.; Le Toan, T.; Planells, M.; Dejoux, J.F.; Ceschia, E. Understanding the temporal behavior of crops using Sentinel-1 and Sentinel-2-like data for agricultural applications. *Remote Sens. Environ.* **2017**, *199*, 415–426. [CrossRef]
82. Baghdadi, N.; Zribi, M. *Land Surface Remote Sensing in Continental Hydrology*; ISTE Press: London, UK; Elsevier: Oxford, UK, 2016; ISBN 9781785481048.
83. Baghdadi, N.; Bazzi, H.; El Hajj, M.; Zribi, M. Detection of Frozen Soil Using Sentinel-1 SAR Data. *Remote Sens.* **2018**, *10*, 1182. [CrossRef]
84. Fayad, I.; Baghdadi, N.; Bazzi, H.; Zribi, M. Near Real-Time Freeze Detection over Agricultural Plots Using Sentinel-1 Data. *Remote Sens.* **2020**, *12*, 1976. [CrossRef]

Review

Recent Progress on Modeling Land Emission and Retrieving Soil Moisture on the Tibetan Plateau Based on L-Band Passive Microwave Remote Sensing

Xiaojing Wu [1,*] and Jun Wen [2]

[1] Key Laboratory of Ecosystem Network Observation and Modeling, Institute of Geographic Sciences and Natural Resources Research, Chinese Academy of Sciences, Beijing 100101, China
[2] Plateau Atmosphere and Environment Key Laboratory of Sichuan Province, College of Atmospheric Sciences, Chengdu University of Information Technology, Chengdu 610225, China
* Correspondence: wuxj@igsnrr.ac.cn

Abstract: L-band passive microwave remote sensing (RS) is an important tool for monitoring global soil moisture (SM) and freeze/thaw state. In recent years, progress has been made in its in-depth application and development in the Tibetan Plateau (TP) which has a complex natural environment. This paper systematically reviews and summarizes the research progress and the main applications of L-band passive microwave RS observations and associated SM retrievals on the TP. The progress of observing and simulating L-band emission based on ground-, aircraft-based and spaceborne platforms, developing regional-scale SM observation networks, as well as validating satellite-based SM products and developing SM retrieval algorithms are reviewed. On this basis, current problems of L-band emission simulation and SM retrieval on the TP are outlined, such as the fact that current evaluations of SM products are limited to a short-term period, and evaluation and improvement of the forward land emission model and SM retrieval algorithm are limited to the site or grid scale. Accordingly, relevant suggestions and prospects for addressing the abovementioned existing problems are finally put forward. For future work, we suggest (i) sorting out the in situ observations and conducting long-term trend evaluation and analysis of current L-band SM products, (ii) extending current progress made at the site/grid scale to improve the L-band emission simulation and SM retrieval algorithms and products for both frozen and thawed ground at the plateau scale, and (iii) enhancing the application of L-band satellite-based SM products on the TP by implementing methods such as data assimilation to improve the understanding of plateau-scale water cycle and energy balance.

Keywords: L-band; passive microwave RS; land emission model; SM retrieval; TP

Citation: Wu, X.; Wen, J. Recent Progress on Modeling Land Emission and Retrieving Soil Moisture on the Tibetan Plateau Based on L-Band Passive Microwave Remote Sensing. *Remote Sens.* **2022**, *14*, 4191. https://doi.org/10.3390/rs14174191

Academic Editors: José Darrozes and Christopher R. Hain

Received: 17 June 2022
Accepted: 23 August 2022
Published: 25 August 2022

1. Introduction

As an essential climate variable, soil moisture (SM) is an important state variable for quantifying water, energy, and carbon exchange processes in the soil–vegetation–atmosphere system [1–4]. It plays an important role in regulating processes such as the partitioning of surface sensible and latent heat flux, surface water budget, and vegetation transpiration [5–8]. This further affects the dynamical and thermal processes in the planetary boundary layer, which in turn impacts the atmospheric state and climate change [9]. SM is also an important factor affecting the growth of vegetation and an important indicator of crop drought, and the effective monitoring of SM can help to accurately implement irrigation measures on farmland [10–12]. Due to its important role in the whole Earth system, SM information is important for a wide range of applications, including climatic modeling, hydrologic modeling, and agriculture growth and drought monitoring. Therefore, effective and large-scale monitoring of SM is important for accurate forecasting of weather and guidance of farming-related measures in the agricultural sector.

The Tibetan Plateau (TP), known as the Third Pole of the World, is one of the most sensitive areas to global climate change due to its special topographic and climatic characteristics. The thermal and dynamic effects of the TP have a very important impact on regulating the weather and climate around the plateau, in Asia, and in the Northern Hemisphere [13]. The TP is also known as the Water Tower of Asia, where the Yellow River, Yangtze River, and Lancang River originate, and its water retention capacity is inextricably linked to the maintenance of ecosystems around the plateau and in Asia. SM, as an important component of the water cycle, is important for understanding and studying the water cycle on the TP [6–8]. In addition, the TP is a typical alpine region with extensive permafrost distribution, and the coexistence of ice and unfrozen water in permafrost can greatly change the soil's hydraulic and thermal properties, thus affecting the regional water and heat exchange and runoff processes [14,15]. Therefore, monitoring SM and freeze–thaw changes on the TP is of great significance for the in-depth understanding of the plateau moisture cycle and energy balance processes.

A ground-based observation network consisting of multiple SM observation sites can provide accurate and long-term SM observations, but its spatial representativeness is limited. At present, several regional-scale SM observation networks have been built on the TP [16–18]. However, due to the complex climate and topographic characteristics of the TP, the SM presents strong spatial heterogeneity, and the regional scale observation networks are insufficient to completely characterize the spatial and temporal distribution of SM across the whole TP.

Since the 1970s, the development of satellite observation technology has provided a new way to monitor SM on a large scale. At present, the technologies commonly used for SM monitoring include visible optical satellites, thermal infrared satellites, and microwave satellites. Research shows that the visible optical remote sensing (RS) and the thermal infrared RS are more frequently influenced by the atmosphere, clouds, and vegetation when retrieving the SM, and the detection depth is only within a few millimeters of the surface soil. On the contrary, the microwave RS not only has the advantage of all-weather and all-day observation capacity but also shows a stronger penetration ability to clouds, rain, snow, and vegetation, which is thus more sensitive to the SM dynamics. Therefore, microwave RS is often treated as the more suitable method to monitor large-scale and long-term SM variations [19,20].

The commonly used microwave RS bands include L-(1–2 GHz), C-(4–8 GHz), and X-band (8–12 GHz). Compared to the C- and X-band, the L-band has a longer wavelength and stronger penetration ability that is more sensitive to SM changes. Therefore, the L-band is usually considered the best band for monitoring global surface SM [20]. In recent years, several L-band microwave RS satellites have been launched worldwide, such as the Soil Moisture and Ocean Salinity (SMOS) satellite of the European Space Agency (ESA) [21], as well as NASA's Aquarius satellite [22] and Soil Moisture Active Passive (SMAP) satellite [23]. In addition, the Global Water Cycle Observation (WCOM) satellite program proposed by Chinese scientists [24] will be expected to achieve continuous observation of L-band microwave RS and to provide higher accuracy and long time series of SM and freeze/thaw state datasets.

Based on satellite observations, researchers around the world have developed a series of L-band microwave emission models and SM retrieval algorithms, which have gone through the process from ground-based validation of theoretical models/algorithms to calibration and validation of satellite observations to global operational monitoring of SM [20]. For the validation of L-band microwave RS observations and products, numerous ground-based and airborne experiments have been conducted in a variety of land conditions and climatic regions around the world, such as the MELBEX III experiment at a Vineyard site in Valencia, Spain [25], the SMOS airborne validation experiment in the Jehol and Erfurt river basins, Germany [26], and the SMAPEx [27] and SMAPVEX15 [28] experiments in Australia and the United States, respectively, which have contributed to the evaluation and improvement of SMOS and SMAP satellite products [20]. Similar experiments have been conducted in China, such as the Heihe Watershed Allied Telemetry Experimental Research (HiWATER) [29] in the Heihe River basins and the L-band SM active-passive thematic experiment in the Luan River basins [30]. In addition, the SMAP satellite team has selected several ground-based core validation networks in various vegetation types and climate regions across the world to calibrate and validate the performance of its products, including the Maqu SM observation network located on the TP [31].

Complex topographic characteristics, the extensive distribution of lakes, the existence of frozen ground with distinct seasonal freeze–thaw transitions, and the lack of accurate soil data have posed many challenges to SM retrievals on the TP. In addition, the impact of Radio Frequency Interference (RFI), topographic relief, and field of view blending has led to the poor quality of satellite observations such as SMOS [32]. To further improve SMOS and SMAP satellite products, Zheng et al. [33] set up an L-band microwave radiometer, i.e., ELBARA-III, in the Maqu SM observation network, which has collected more than five years of consecutive ground-based bright temperature (T_B^p) observations up to now [34,35]. Currently, many studies have been conducted to evaluate L-band satellite-based SM products and retrieval algorithms based on several SM observation networks on the TP. For example, Dente et al. [36] and Chen et al. [37] evaluated the applicability of SMOS and SMAP SM products on the TP, respectively. Zheng et al. [38,39] evaluated and improved the vegetation and surface roughness parameterizations implemented in the current SMAP SM retrieval algorithm and developed a new algorithm for retrieving unfrozen (liquid) water content in the frozen ground. These research efforts related to product validation and algorithm improvement have further promoted the development of L-band microwave RS and the application of L-band satellite products on the TP.

This paper systematically reviews and summarizes the research progress and main applications of L-band passive microwave RS and associated SM retrieval algorithms and products on the TP in recent years. On this basis, the current problems of L-band emission simulation and SM retrieval on the TP are outlined, and relevant suggestions and prospects for addressing the existing problems are finally put forward. Section 2 introduces the study area. In Section 3, we introduce the airborne and ground-based L-band microwave passive RS experiments carried out on the TP and the preliminary validation of satellite-based L-band observations and summarize the research progress in simulating microwave emission on the TP; Section 4 presents the existing SM observation networks on the TP and summarizes the research progress of evaluation and improvement of SM products and retrieval algorithms based on the L-band microwave RS on the TP. On this basis, Section 5 summarizes the main problems of SM retrieval research on the TP and provides related outlooks.

2. Study Area

Known as the Third Pole of the World, the TP is the highest plateau in the world, with an average elevation of over 4000 m. The mountain ranges of the TP extend across Afghanistan, Pakistan, India, China, Bhutan, Myanmar, and Nepal, and more than 4,000,000 km^2 is mainly composed of high-elevation rugged terrain. It is generally high in the northwest and low in the southeast. Grasslands are widely distributed and dominate the vegetation type on the TP. The climate is humid in the southeast and arid in the northwest. In addition, the TP has strong solar insolation and sufficient sunshine, but the overall temperature is low, and the diurnal amplitude of temperature is large. The TP is often regarded as the Asian water tower since more than 10 of the largest rivers in Asia originate from this region, including the Yellow River, the Yangtze River, the Mekong river, the Brahmaputra river, and the Indus river, providing freshwater supply for more than a fifth of the world's population.

3. Progress of L-Band Microwave Emission Observation and Simulation on the TP

In recent years, researchers have validated the satellite observations using L-band observations collected from airborne and ground-based platforms on the TP and have conducted studies related to L-band emission simulation. This section will introduce in detail L-band microwave observation experiments on the TP, including airborne and ground-based experiments and evaluation of satellite observations, and summarize the current forward land emission model adopted by the L-band satellite missions and their applications and improvements in the TP.

3.1. L-Band Microwave Emission Observation

3.1.1. Airborne and Ground-Based Observation Experiments Conducted in the TP

In order to promote observational studies of L-band microwave emission on the TP, airborne and ground-based experiments were carried out in the Heihe River Basin in the northeastern part of the TP and the Maqu area in the southeastern part of the Yellow River source region, respectively [29,33]. In order to improve the observation capability of hydrological and ecological processes at the watershed scale and to establish a leading watershed observation system around the world, an ecohydrological remote sensing experiment, i.e., Heihe Watershed Allied Telemetry Experimental Research (HiWATER), was carried out via combining ground-based, airborne remote sensing, and satellite observation methods [29,40]. Among them, in order to develop passive microwave RS-based SM retrieval products at the watershed scale, several airborne PLMR (Polarimetric L-band Multibeam Radiometer) radiometer-based observations were carried out from 29 June to 2 August 2012 to collect multi-angle dual-polarized T_B^p data in the middle and upper regions of the Heihe River basin. The flight altitude of the airborne experiment was 0.3–3 km, corresponding to a ground resolution of 0.1–1 km, and the incidence angles of the radiometer were $\pm 7°$, $\pm 21.5°$, and $\pm 38.5°$, respectively, with a center frequency of 1.41 GHz.

To validate the SMOS and SMAP satellite T_B^p observations and develop microwave emission models as well as SM retrieval algorithms, Zheng et al. [33] deployed an L-band microwave radiometer (i.e., ELBARA-III) in the Maqu SM observation network at the beginning of 2016. The radiometer was mounted on a 4.8 m height tower with the antenna centered at approximately 6.5 m above the ground, and the antenna beam was generally oriented to the south. The T_B^p observations at both horizontal (T_B^H) and vertical (T_B^V) polarizations were collected every 30 min in steps of 5° from 40° to 70° scanning angles [33,35]. Micro-meteorological observations were also set up near the radiometer to measure a variety of micro-meteorological elements. In late 2016, a rain gauge and an eddy covariance observation system were installed near the radiometer. In addition, vertical SM profile observation probes were added in August 2016 to automatically collect SM observations at 20 soil depths ranging from 2.5 to 100 cm every 15 min [35]. Based on the ELBARA-III microwave radiometer observations, Zheng et al. [33–35,41] conducted several studies on the L-band microwave passive RS of soil freeze–thaw transitions, including the development of a microwave emission model for frozen ground, a new retrieval algorithm for retrieving unfrozen (liquid) soil water content in frozen ground, and a new finding that the sampling depth of L-band microwave radiometry is about 2.5 cm for both frozen and thawed soil conditions.

3.1.2. Satellite Observations and Accuracy Assessment

After a long period of development, three satellites carrying L-band microwave radiometers were successfully launched worldwide, including ESA's SMOS, NASA's Aquarius, and SMAP. The main information about these three satellites is shown in Table 1.

Table 1. Basic information of SMOS, Aquarius, and SMAP satellites.

Satellite Missions	Space Agency	Launched Time	Instruments	Incidence Angle	Overpass Time (d)	Spatial Resolution (km)
SMOS	ESA	2009.11	L-band Radiometry	0–55°	1–3	35–50
Aquarius	NASA	2011.06	L-band Radiometry and Scatterometer	28.7°/37.8°/45.6°	7	76 × 94/84 × 120/96 × 156
SMAP		2015.01	L-band Radiometry and SAR	40°	2–3	40

The SMOS satellite is the world's first L-band passive microwave RS satellite, and one of its main objectives is to provide global surface SM products with an accuracy of about 0.04 m^3 m^{-3} [21,42]. The SMOS satellite carries an L-band microwave radiometer (1.41 GHz) in a sun-synchronous orbit at a mean altitude of 757 km, providing ascending and descending data corresponding to passages through the equator at 6:00 and 18:00 of local solar time, respectively. The microwave radiometer uses a Y-shaped antenna that provides T_B^p observations at incidence angles of 0–55°. SM is retrieved using multi-angular and dual-polarization SMOS T_B^p observations via inverting the L-MEB model in combination with an iterative inversion algorithm [42].

The Aquarius/SAC-D is an ocean observation satellite mission aiming to provide data such as monthly ocean surface salinity for the study of ocean circulation, coupling between global water cycle and climate, and others [22,43]. The observation system consists of three dual-polarized L-band radiometers (1.41 GHz) and one fully polarized L-band scatterometer (1.26 GHz). The orbit of the Aquarius is a sun-synchronous orbit at 657 km, which passes the equator at 6:00 (descending orbit) and 18:00 (ascending orbit) local solar time and covers the globe every 7 days. The Aquarius mission was terminated on 8 June 2015 due to a failure of the power supply and altitude control system. SM is retrieved using Aquarius T_B^p observations at the horizontal polarization using the single channel retrieval algorithm [43].

The SMAP satellite aims to provide high precision and high resolution of SM and freeze/thaw state data on a global scale [23,44]. It carries an L-band microwave radiometer (1.41 GHz) and a synthetic aperture radar (SAR) (1.26 GHz) to obtain simultaneous measurements of T_B^p and backscatter coefficients. The SMAP satellite orbit is in a sun-synchronous orbit at 685 km and passes through the equator at 6:00 (descending orbit) and 18:00 (ascending orbit) local solar time. On 7 July 2015, the SMAP radar stopped working due to a malfunction, and so far, the SMAP radiometer is still working stably. SM is retrieved using SMAP T_B^p observations at vertical polarization using the single channel retrieval algorithm [44].

To validate the accuracy of satellite-based L-band T_B^p observations in the TP, the SMAP and SMOS T_B^p observations are compared to the in situ ELBARA-III observations in the Maqu SM observation network. Figure 1 show the comparison of SMAP, SMOS, and ELBARA-III T_B^p observations from August 2016 to July 2017 for the evening overpass. It can be found that the T_B^p is significantly correlated with soil dryness and wetness and freeze–thaw transitions. For example, the T_B^p increases during the soil freezing period (November to February) and then decreases as the unfrozen (liquid) soil water increases with soil thawing. As shown in the figure, the variations of SMOS and SMAP T_B^p observations are generally consistent with the ELBARA-III measured trends, whereby the SMAP observations are more consistent with the ELBARA-III observations. The correlation coefficients between SMAP and ELBARA-III T_B^p observations are greater than 0.87, and the RMSE and ubRMSE for the T_B^V observations are smaller than these of T_B^H. Good performance of SMAP observations was also reported in ref. [39,45]. Compared to the performance of SMAP T_B^p data, the SMOS data show degraded accuracy and larger fluctuating, which may be related to the influence of RFI and the stability of the radiometer [36]. To further investigate the impact of RFI on the SMOS T_B^p observation, Figure 2 provide the root-mean-square error (RMSE) computed between the SMOS T_B^p observations and simulations produced by the CMEM model for both descending and ascending overpasses performed by the authors. From the figure, it can be found that the RMSE for the SMOS T_B^p observation in the TP is as high as 10–20 K, indicating that the SMOS satellite may be seriously affected by RFI in the TP. A similar finding was also reported by Dente et al. [36].

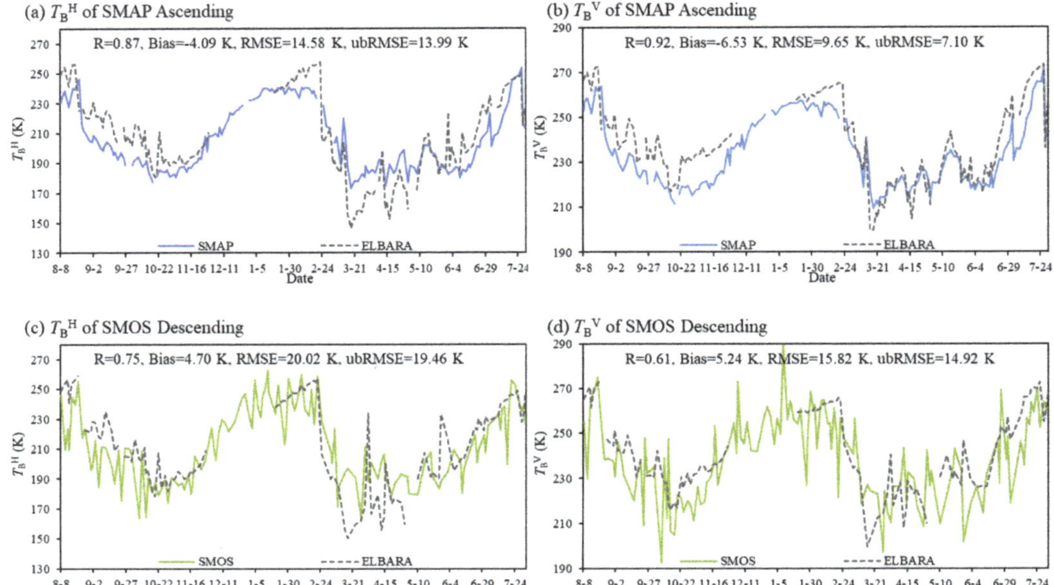

Figure 1. Time series of SMAP and ELBARA-III measured (**a**) T_B^H and (**b**) T_B^V, and SMOS and ELBARA-III measured (**c**) T_B^H and (**d**) T_B^p during the evening overpasses between August 2016 and July 2017. (**a**,**b**) are modified from Zheng et al. [39].

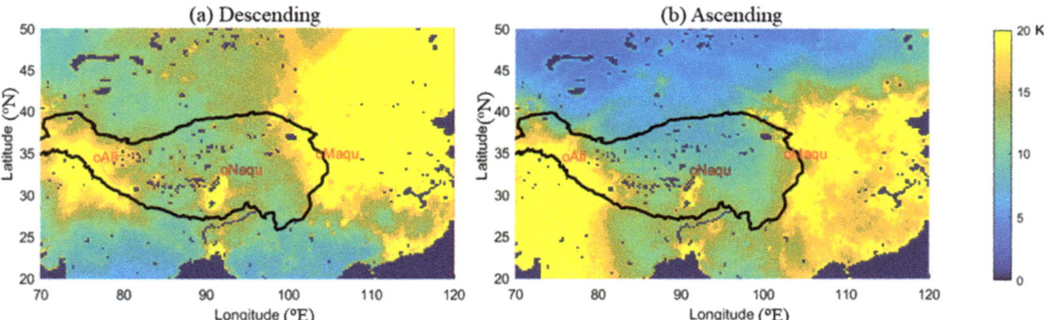

Figure 2. RMSE computed between the SMOS T_B^p observations and simulations produced by the CMEM model for both (**a**) ascending and (**b**) descending overpasses.

3.2. L-Band Microwave Emission Simulation

3.2.1. Forward Land Emission Model Adopted by Current Satellite Missions

The current three L-band satellite missions, i.e., SMAP, SMOS, and Aquarius, all use the zero-order forward microwave emission model, i.e., τ-ω model, developed by Mo et al. [46] for T_B^p simulations. T_B^p generally consists of three components: (1) direct upwelling vegetation emission; (2) downwelling vegetation emission reflected by the soil and attenuated by the canopy layer; (3) upwelling soil emission attenuated by the canopy [38,46]. The model is expressed as follows:

$$T_B^p = (1-\omega^p)(1-\gamma^p)T_C + (1-\omega^p)(1-\gamma^p)\gamma^p r^p T_C + (1-r^p)\gamma^p T_G, \quad (1)$$

$$\gamma^p = exp(-\tau^p/cos(\psi)), \quad (2)$$

where the superscript p represents the polarization (p = V for vertical polarization and p = H for horizontal polarization), ω^p, γ^p, and τ^p are the single scattering albedo, transmittance, and optical depth of vegetation, respectively, T_C and T_G are the effective temperatures of vegetation and soil, respectively, r^p is the reflectivity of rough surface, and ψ is the satellite observation angle.

Table 2 summarize the main parameterizations used in the current forward land emission models for SMOS, Aquarius, and SMAP satellite missions, including the simulation of rough surface reflectivity r^p, soil permittivity ε_s, effective soil temperature T_G, vegetation temperature T_C, single scattering albedo ω, and vegetation optical depth τ^p. Usually, the vegetation single scattering albedo ω is determined by the specific vegetation type that is independent of the polarization. For instance, $\omega = 0$ for sparse vegetation and ω = 0.06–0.08 for forest in the SMOS mission. For the simulation of vegetation optical depth τ^p, which is a function of the leaf area index (LAI) in the SMOS mission (see Table 2) [20], whereby the parameters b′ and b″ depend on the structures of the specific vegetation type. For the Aquarius and SMAP missions, the τ^p is linearly related to the vegetation water content (VWC) [47], whereby the VWC is determined by the normalized vegetation difference index (NDVI) and vegetation type.

The h-Q-N model is adopted by the three satellite missions to simulate the rough surface reflectivity r^p as [48,49]:

$$r^p = \left[(1-Q){r_s}^p + Q r_s^q\right] exp\left(-hcos^N(\psi)\right), \tag{3}$$

where ${r_s}^p$ and r_s^q (p = H, V; q = V, H) are the smooth surface reflectivity, which is related to the soil permittivity ε_s and can be obtained by the Fresnel equation. Parameter h is the roughness height parameter, which is related to the type of land cover, e.g., h = 0.1 for sparsely vegetated subsurface and h = 0.3 for forested subsurface in the SMOS mission. In the Aquarius mission, h is taken as a constant value of 0.1. Parameter Q denotes the polarization mixing factor, which is usually assumed as 0 at L-band. Parameter N represents the angular effect of observation angle, which is introduced to better account for multi-angle and dual-polarization measurements. In the SMOS mission, N is related to the polarization, while it is taken as a constant value of 2 in both Aquarius and SMAP missions.

Various soil dielectric constant models have been developed for passive microwave remote sensing, such as the Dobson model [50], the Wang and Schmugge model [51], and the Mironov model [52]. Currently, the Mironov model [52] is implemented by both SMOS and SMAP satellite missions, and the Wang and Schmugge model is adopted for the Aquarius satellite mission. However, these models are only applicable to unfrozen soil conditions, resulting in the inability of current satellite missions to retrieve the unfrozen (liquid) soil water content under frozen soil conditions [38].

Table 2. Parameterizations adopted by the SMOS, Aquarius, and SMAP satellite missions for key parameters in the forward land emission model.

Parameters	SMOS (L2 and L3)	Aquarius (L2)	SMAP (L2)
r^p	h-Q-N model		
	h = 0.1 for sparse vegetation, and h = 0.3 for forest	h = 0.1	h = f(IGBP)
	Q = 0; N^V = 0, N^H = 2	Q = 0; N^p = 2	Q = 0; N^p = 2
ε_s	Mironov model [52]	Wang and Schmugge model [51]	Mironov model [52]
	ε_s = f(SM, T_G, % clay)		
T_G	T_G = f(T_{soil_surf}, T_{soil_deep})		
	$C_T = (SM/W_0)^{b0}$	C_T = 0.246	
T_C	Skin temperature from ECMWF land surface model	$T_C = T_G$	
ω	ω = 0 for sparse vegetation, and ω = 0.06–0.08 for forest	ω = 0.05	ω = f(IGBP)
τ^p	τ^p = b′·LAI + b″	τ^p = b·VWC, VWC = f(NDVI, IGBP)	
		b = 0.8	b = f(IGBP)

The estimation of T_G is related to the profile soil temperature, which can be estimated as [53]:

$$T_G = T_{soil\_surf} - \left(T_{soil\_surf} - T_{soil\_deep}\right)C_T, \quad (4)$$

where T_{soil_surf} and T_{soil_deep} are the soil temperatures at the surface (~5 cm) and deep layers (~50 cm), respectively. Currently, the SMOS satellite mission uses the soil temperature simulations of the first and third soil layers obtained from the land surface model of the European Centre of Medium Range Weather Forecasting (ECMWF) as the T_{soil_surf} and T_{soil_deep}, and both Aquarius and SMAP missions use the soil temperature simulations of the first and second layers obtained from NASA GEOS-5 (Goddard Earth Observation System Model Version 5) as the T_{soil_surf} and T_{soil_deep} [20]. In addition, C_T is a fitting parameter, which is related to SM and parameters W_0 and b_0 in the SMOS satellite mission, where the standard values of parameters W_0 and b_0 are taken as 0.3 m^3 m^{-3} and 0.3, respectively. The value of C_T is taken as 0.246 for both Aquarius and SMAP missions [20]. Both Aquarius and SMAP satellite missions assume that the atmosphere, vegetation, and near-surface soil are in thermal equilibrium during the satellite overpasses, then the T_C is approximately equal to the T_G, while the SMOS mission uses the surface temperature output from the ECMWF land surface model as the T_C [20].

3.2.2. Progress of L-Band Microwave Emission Simulation on the TP

In the past few years, researchers have used a combination of airborne, ground-based, and satellite-based L-band microwave observations to evaluate the applicability of the widely used τ-ω model and its parameterizations on the TP. Based on this, new parameterizations for surface roughness, vegetation optical depth, and soil permittivity have been developed specifically for the TP conditions, improving microwave emission simulations across different climatic and land conditions of the TP. Zheng et al. [38,54] used the SMAP T_B^p observations to evaluate the applicability of the forward land emission model adopted by the SMAP satellite mission to the desert (Ngari SM observation network) and grassland (Maqu SM observation network) conditions. The results showed that the default SMAP land emission model tends to underestimate the effect of surface roughness and overestimate the effect of vegetation, resulting in the underestimation of year-round T_B^p in the Ngari area. Overestimation of T_B^p during the warm season and underestimation of T_B^V during the cold season in the Maqu area was also found. Based on this, Zheng et al. [38,54] used the surface roughness parameterizations developed by Wigneron et al. [55] to improve the underestimation of T_B^p in both the Ngari and Magu regions. A new vegetation parameterization based on simulations produced by a discrete microwave radiative transfer model was further developed to reduce the simulation bias in the Maqu region. The newly developed surface roughness and vegetation parameterizations were adopted by Wu et al. [45] to implement the two-stream microwave emission model developed by Schwank et al. [56] to simulate the T_B^p in the TP. In comparison to the τ-ω model, the two-stream microwave emission model presents comparable simulations, which consider multiple scattering and reflection and remove the assumption of a "soft layer" that is physically more correct than the τ-ω model [45,56]. In addition, Wu and Zheng [57] firstly investigated the impact of surface roughness on multi-angular T_B^p simulation using the in situ ELBARA-III T_B^p observations conducted in the Maqu SM observation network. The results showed that the multi-angular T_B^p simulation could be improved via site-specific calibration of the h-Q-N model, leading to a nonzero value for the parameter Q. As such, the noncoherent emission contribution to cross-polarization mixing can be accounted for by the h-Q-N model. This indicates that consideration of polarization mixing is necessary for L-band T_B^p simulation [57].

In addition to the currently widely used τ-ω model, researchers have conducted a lot of research on L-band microwave emission simulation on the TP based on the physically based discrete microwave radiative transfer model developed at the Tor Vergata University of Rome (hereafter "Tor Vergata model") [33,35,41,52–58]. Wang et al. [58] used the Tor Vergata model to simulate the active and passive observation signals of the Aquarius mission in the Magu SM observation network. They found that the correlation coefficients computed between the simulated T_B^p and backscatter coefficients produced by the Tor Vergata model and the corresponding Aquarius satellite observations are about 0.86 and 0.68, demonstrating the applicability of the Tor Vergata model in the Magu region. Bai et al. [59] simultaneously simulated the SMAP observed T_B^p and backscatter coefficients using the calibrated Tor Vergata model considering the sensitive parameters and found that the simulation results of the combined active–passive model are in good agreement with the SMAP observations. For the T_B^p simulation under frozen soil conditions, Zheng et al. [33] introduced a four-phase dielectric mixing model [60] to the Tor Vergata model. The results showed that the developed model simulates the ε_s and T_B^p for both frozen and thawed soil conditions well, extending the application of the Tor Vergata model on the TP. Recently, Zheng et al. [41] used the Tor Vergata model in combination with the four-phase dielectric mixing model to explore the active and passive microwave characteristics of diurnal soil freeze–thaw transitions. The results further confirmed the ability of the improved Tor Vergata model to reproduce diurnal variations of ground-based observed T_B^p and backscatter coefficients as well as to quantify their relationships at different observation angles and frequencies. To further explore the impact of SM and soil temperature (SMST) profile dynamics on the diurnal L-band T_B^p observation signatures of frozen soil, an integrated land emission model was developed by Zheng et al. [35]. The model was developed by combining the improved Tor Vergata model with a multilayer soil scattering model developed based on integrating the Wilheit [61] and the advanced integral equation method (AIEM) [62]. The results showed that the Fresnel simulations with a sampling depth of 2.5 cm fit best with the multilayer Wilheit results, indicating that the diurnal L-band T_B^p observation signatures of frozen soil are mainly dominated by the SMST dynamics at the surface layer. A similar finding was recently reported by Wu et al. [63].

In summary, two distinguishing features can be drawn related to the L-band microwave emission simulation on the TP. One is that the polarization mixing effect should be considered in simulating the L-band T_B^p observations on the TP. Figure 3 show the angular dependence of averaged ELBARA-III T_B^p observations and corresponding simulations produced by the h-Q-N model with/without a zero Q value as well as the parameterized model developed by Shi et al. [64] based on the IEM simulations. Overestimations are noted for the simulations produced by the h-Q-N model with a zero Q value and the parameterized model, especially at the vertical polarization, which also becomes larger with increasing incidence angles. The above deficiency is largely addressed by the calibrated h-Q-N model with a nonzero Q value, indicating the necessity to consider the polarization mixing effect in L-band emission modeling on the TP. The other feature is that the diurnal L-band T_B^p observation signatures of both frozen and thawed soil conditions are primary dominated by the SMST dynamics at the surface layer of around 2.5 cm. Figure 4 show the comparisons between both T_B^H and T_B^V simulations produced by the τ-ω model configured either with the multilayer Wilheit [61] model or with the single Fresnel model considering three depths of SMST profile at 2.5 (Sim1), 5 (Sim2) and 10 cm (Sim3) for both warm (from 7 August to 30 September) and cold (from 1 January to 15 March) periods. The Fresnel simulations with input of SMST at 2.5 cm (i.e., Sim1) fit best with the multilayer Wilheit simulations at both polarizations for both periods, indicating that the sampling depth of L-band radiometry is close to 2.5 cm for both frozen and thawed soil conditions on the TP. A similar finding was also reported by Zheng et al. [34].

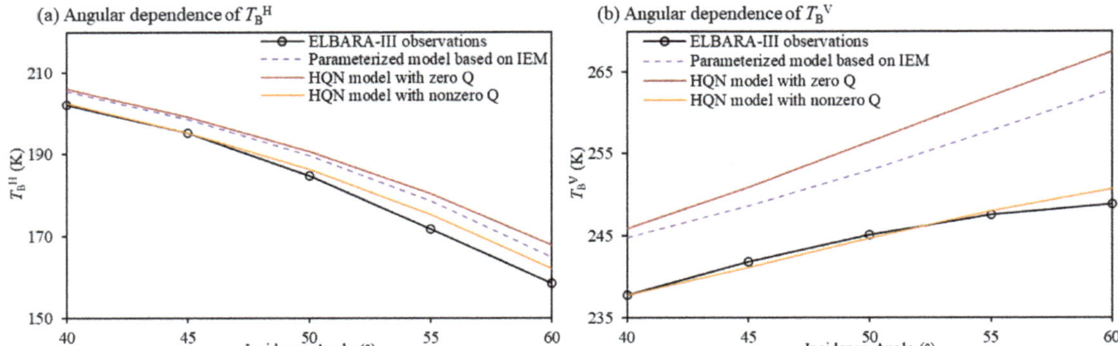

Figure 3. Angular dependence of ELBARA-III T_B^p observations and corresponding simulations produced by the h-Q-N model with/without a zero Q value as well as the parameterized model developed by Shi et al. [57] based on the IEM simulations. The figure is modified from Wu and Zheng [57].

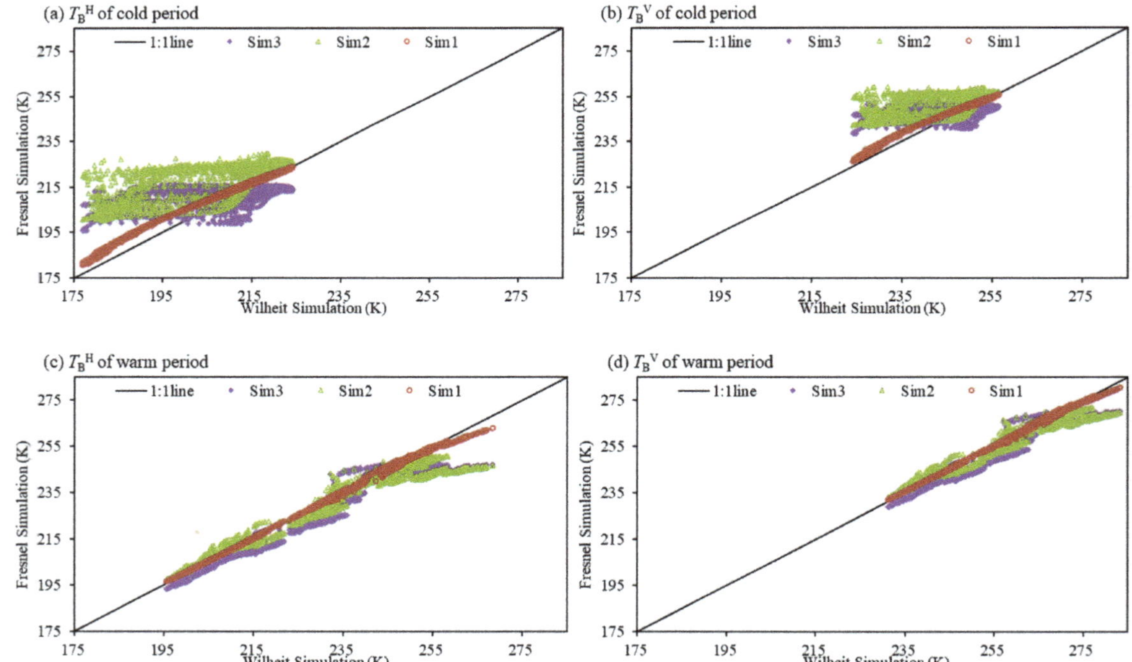

Figure 4. Comparisons of T_B^H (**a**,**c**) and T_B^V (**b**,**d**) simulations produced by the τ-ω model configured either with the multilayer Wilheit [12] model or with the single Fresnel model considering three depths of SMST profile at 2.5 (Sim1), 5 (Sim2), and 10 cm (Sim3) for (**a**) cold (from 1 January to 15 March) and (**b**) warm (from 7 August to 30 September) periods. The figure is modified from Wu [63].

4. Progress of SM Observation and Retrieval Using L-Band Passive Microwave RS on the TP

SM retrieval algorithms for L-band microwave RS show certain errors and limitations for their applications to the TP. In order to obtain higher accuracy of SM products for the TP, further evaluation and improvement of satellite-based SM products and retrieval algorithms are necessary. Among them, in situ data collected by multiple SM observation networks established on the TP are the key basis for the evaluation of L-band SM products and the improvement of retrieval algorithms. In this section, the details of SM observation networks in the TP and the research progress of validating the L-band SM products and improving the corresponding retrieval algorithms are reviewed.

4.1. SM Observation Networks on the TP

Due to the high spatial variability of SM and the large error in using a single station observation to represent the true value of regional-scale SM, several regional-scale SM observation networks have been established on the TP, including the upper Heihe River Basin, Maqu, Naqu, Pali, and Ngari observation networks [16–18,40,65–68] (Figure 5). Dense SM observation stations are distributed within these networks to provide SM data of different soil layers. In addition, by means of soil sampling and laboratory measurements, these observation networks also provide information on soil texture and organic carbon content across the observation stations. Table 3 summarize the basic information of the five SM observation networks on the TP, such as the number of stations deployed, climate type, land cover type, the temporal resolution of observation, and observation depth for each network. A brief description of the five observation networks is provided below.

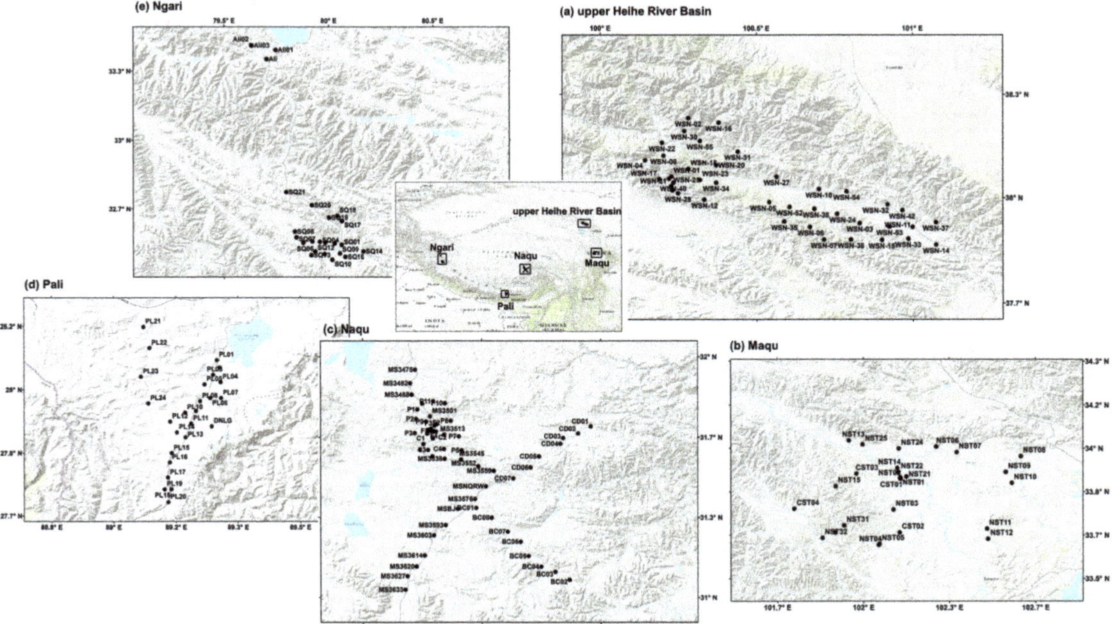

Figure 5. Locations of (**a**) upper Heihe River Basin, (**b**) Maqu, (**c**) Naqu, (**d**) Pali, and (**e**) Ngari SM observations networks and corresponding deployed SM observation stations on the TP.

Table 3. Basic information of SM observation networks on the TP.

Network	Establish Time	Station Number	Climate	Land Cover	Temporal Resolution	Observation Depth (cm)	Reference
upper Heihe River Basin	2012	40	Humid	Alpine Meadow	5 min	4, 10, 20	[66]
Maqu	2008	20 + 6 *			15 min	5, 10, 20, 40, 80	[16,17]
Naqu	2010	56	Semi-Arid	Alpine Steppe	30 min	5, 10, 20, 40	[18,37]
Pali	2015	25					
Ngari	2010	20 + 5 *	Arid	Desert	15 min	5, 10, 20, 40, 60	[16,17]

The number with * indicates the newly established stations.

The Heihe River is the second largest inland river in China, and a variety of land cover types such as oasis, desert, and grassland are distributed across the river basin [29]. The upper reaches of the Heihe River basin have an average elevation of 4869 m, which belong to a humid climate with precipitation mainly falling from May to September. The area is widely covered by permafrost and seasonally frozen ground, and the main land cover is alpine meadows [66]. In 2012, 40 wireless SM observation stations were set up within the framework of the HiWATER experiment [29]. At each station, sensors were installed at soil depths of 4, 10, and 20 cm to collect SM data every 5 min. The relevant data were published on the website of the HiWATER experiment (http://westdc.westgis.ac.cn/data/df372e4a-7da8-4c9d-8479-75cafb44007f (accessed on 22 August 2022)).

The Maqu SM observation network [16,17] is located in the source area of the Yellow River in the northeastern part of the TP, with altitudes ranging from 3400 to 3800 m. The climate type is characterized as cold and humid with rainy summers and cold, dry winters. The average annual temperature is about 1.2 °C, and the annual precipitation is about 600 mm. The main land cover type is alpine meadows. In 2008, 20 observations were originally installed, which a covered area of about 40 × 80 km^2. In 2014, six new stations were installed due to the damage to several old monitoring sites caused by local people or animals [16]. Decagon 5TM ECH2O probes were used to measure SM at depths of 5, 10, 20, 40, and 80 cm with a temporal resolution of 15 min. The relevant data were published by the National Tibetan Plateau Data Center (http://www.tpdc.ac.cn/en/data/d323f0b2-dada-4ed5-aa00-57564da788d2/ (accessed on 22 August 2022)).

The Naqu SM observation network [18,37] is located in the central part of the TP with an average altitude of 4650 m. The climate type is characterized as cold and semi-arid, and the main land cover type is alpine meadows with low vegetation coverage. The soil includes high soil organic carbon content. The mean annual precipitation in the Naqu region is around 500 mm, and 75% of the precipitation is concentrated between May and October due to the impact of South Asian monsoons. There are 56 stations established in the observation network, with 38, 22, and 9 stations distributed in the spatial grids of 1.0°, 0.3°, and 0.1°, respectively, to provide an observational basis for the study of SM upscaling and downscaling. The stations are also equipped with Decagon 5TM ECH2O probes at observation depths of 5, 10, 20, and 40 cm with a temporal resolution of 30 min. The relevant data were published by the National Tibetan Plateau Data Center (https://www.tpdc.ac.cn/en/data/ef949bb0-26d4-4cb6-acc2-3385413b91ee/ (accessed on 22 August 2022)).

The Pali SM observation network [37] is located in the southern part of the TP that is near the northern slope of the Himalayas, with an average altitude of 4486 m. The climate type is characterized as semi-arid, and the main land cover types are sparse grassland and bare soil. The average annual precipitation in the Pali region is less than 400 mm, and about 85% of the precipitation is concentrated between May and October due to the impact of South Asian monsoons. The Pali SM observation network consists of 25 stations with Decagon 5TM ECH2O probes installed at depths of 5, 10, 20, and 40 cm to collect SM data at a temporal resolution of 30 min. The relevant data were published by the National Tibetan

Plateau Data Center (https://www.tpdc.ac.cn/en/data/ef949bb0-26d4-4cb6-acc2-3385413b91ee/ (accessed on 22 August 2022)).

The Ngari SM observation network [16,17] is located in the western part of the TP with an average elevation of 4869 m. The climate is characterized as cold and arid, and the land cover is bare soil and desert. Twenty SM observation stations were established in June 2010 in the Ngari area, of which four stations were set up in the desert area, and the rest were located near the city of Shiquanhe. In 2016, five new stations were installed due to the damage to several old monitoring sites caused by local people or animals [16]. Each station was equipped with Decagon 5TM ECH2O probes at depths of 5, 10, 20, 40, and 60 cm to collect SM observations with a temporal resolution of 15 min. The relevant data were published by the National Tibetan Plateau Data Center (http://www.tpdc.ac.cn/en/data/d323f0b2-dada-4ed5-aa00-57564da788d2/ (accessed on 22 August 2022)).

4.2. Validation of SM Products Retrieved from the L-Band Passive RS on the TP

Due to the impact of different instruments, operational modes, and retrieval algorithms adopted by the three different L-band satellite missions (i.e., SMOS, Aquarius and SMAP, see Tables 1 and 2), the performances of SM products retrieved using the T_B^p observations collected from these three satellites present distinct characteristics for different climate and land cover conditions on the TP. Therefore, it is necessary to validate the performance of these satellite-based SM products on the TP using SM measurements collected from the five in situ SM observation networks (see Figure 5 and Table 3). Table 4 summarize the error statistics for the validations of L-band satellite-based SM products performed on the TP in recent years, which mainly include correlation coefficient (R), bias, and RMSE.

For SMOS SM products, Su et al. [17] firstly made a preliminary evaluation of L2 SM products using measurements collected from the Maqu network and found that the correlation coefficient can reach 0.72 and the RMSE is about 0.09 $m^3\ m^{-3}$. Zhao et al. [69] further evaluated their performances using measurements collected from the Naqu network and found that the L2 and L3 SM products show greater uncertainty at the SMOS original grid (15, 25 km), and the correlation coefficient between SM products and observations can be improved through averaging the values of SM products to the spatial resolution of 100 km. Zeng et al. [70] thoroughly evaluated the performance of L3 SM products using SM measurements collected from both the Maqu and Naqu networks and found that SMOS products show large noise and bias, especially at the descending overpass. They further pointed out that the presence of RFI can be an important factor causing bias. In addition, it was found that the performance of SMOS products in the Naqu network is better than that of the Maqu network covered by denser vegetation. A similar finding was also reported by Chen et al. [37], who found that the L3 SM product performs well in the Naqu network with correlation coefficients of about 0.67 and 0.73 for the ascending and descending overpasses, respectively. Recently, Liu et al. [71] thoroughly evaluated the performance of multiple satellite-based SM products using data from the five in situ SM observation networks for the first time. They found that the SMOS-IC products were affected by RFI with a slight underestimation. Liu et al. [72] further evaluated the performance of SMOS-IC products using the three-corned hat method and also found that it is strongly influenced by the presence of RFI. In general, SMOS SM products can reflect SM conditions across the TP to some extent, but the performance is inconsistent in different areas of the TP. In addition, there is a slight dry bias in most areas, and the uncertainty of SM products is high due to the influence of RFI presence.

Relatively less work has been carried out to validate the SM products of the Aquarius satellite mission. Li et al. [73] used data from the Naqu network to evaluate the Aquarius L3 SM product and found that the correlation coefficient could reach 0.77 with an RMSE of about 0.08 $m^3\ m^{-3}$. It was also shown that the Aquarius SM product could generally reflect the spatial and temporal variations of SM. It is worth noting that the revisit period of the Aquarius satellite is 7 days, resulting in a limited number of SM retrievals within the study time frame.

Regarding SMAP SM products, Chen et al. [37] evaluated the performance of L3 passive SM products using data from both Naqu and Pali networks and found that the products could capture the amplitude and temporal variation of SM observations well. Liu et al. [71] thoroughly evaluated the L3 passive SM products using data from the five in situ SM observation networks on the TP and found that the SMAP product correlates well with SM observations with smaller RMSE and bias in comparison to other products. They also showed that the SMAP product shows higher accuracy in relatively sparsely vegetated areas. A similar finding was also reported by Zeng et al. [74]. Li et al. [75] further evaluated the performance of both L3 original and enhanced passive SM products using data from the Naqu and Magu networks and found that both products capture the temporal variability and spatial distribution characteristics of SM observations with strong correlation. They also showed that the enhanced product presents a higher correlation and provides more details of SM variability. Ma et al. [76] thoroughly evaluated the performance of passive, active, and combined active–passive SM products with resolutions of 3, 9, and 36 km using data from the upper and middle reaches of the Heihe River basin. They found that SMAP products are able to capture spatial and temporal variability of SM observations and typical precipitation events in most of the study areas, with passive SM products performing best. In addition, it was found that SMAP SM products perform better in bare soil areas than the vegetated areas. In general, SMAP SM products can better reflect the spatial and temporal variations of SM in multiple observation network areas of the TP with relatively high accuracy in comparison to other products.

Three distinct features can be drawn from the summary of validating the three L-band SM products on the TP (see Table 4): (1) the applicability of the three satellite-based SM products varies in different climatic and land cover regions, while in most cases they can capture the amplitude and temporal changes of SM observations; (2) through comprehensive analysis, it is found that the SMAP satellite products perform the best, and the SMOS retrieval results have large deviation and relatively high uncertainty due to the presence of RFI; (3) different vegetation cover types show different degrees of influence on the satellite-based soil moisture retrievals, and generally speaking, the accuracy of the products in bare soil areas is better than that in vegetation cover areas.

4.3. *Improvement and Development of SM Retrieval Algorithms Using the L-Band Passive RS on the TP*

Based on the τ-ω model, researchers have developed many SM retrieval algorithms for the L-band passive microwave RS, including the iterative inversion algorithm based on the L-MEB forward model [42], Single Channel Algorithm (SCA) [44], Dual Channel Algorithm (DCA) [44], and Land Parameter Retrieval Model (LPRM) [77]. The SMOS satellite uses the iterative inversion algorithm based on the L-MEB forward model as the default algorithm. This method takes into account a priori information on the retrieved parameters and minimizes the cost function by a generalized least squares iterative algorithm to retrieve both SM and τ [42]. Currently, the default algorithms implemented by the Aquarius and SMAP satellite missions are based on the SCA using the T_B^H (i.e., SCA-H) and T_B^V (i.e., SCA-V) observations, respectively. The SCA firstly converts the T_B^p observation into emissivity using the effective soil temperature and then removes the impact of vegetation and surface roughness based on certain parameterizations to obtain soil emissivity, which finally uses the Fresnel equation in combination with a soil dielectric constant model to obtain SM [20,44]. In general, the errors of satellite-based SM products are mainly sourced from adopted forward land emission models and input parameters [20,71,74]. Our review of the progress of L-band microwave emission simulation on the TP (see Section 3.2) reveals that the forward land emission models adopted by current L-band satellite missions still show deficiencies in their applications to the TP, such as underestimation of effective soil temperature and surface roughness effects, overestimation of vegetation effects, and the inapplicability of the adopted dielectric constant models for frozen soil conditions [33,35,45,54,57], etc. Based on this, researchers have improved the relevant parameterizations adopted in the current SM retrieval algorithm, as well as developed a new SM retrieval algorithm to obtain high accuracy of SM retrievals for the TP environment.

Table 4. Summary of error statistics for the validations of L-band satellite-based SM products performed on the TP.

Satellite	SM Product	Spatial Resolution	SM Network	Error Statistics *			Reference
				R	Bias (m^3 m^{-3})	RMSE (m^3 m^{-3})	
SMOS	L2_SM	25 km	Maqu	0.72	-	0.09	Su et al. [17]
	L2_SM	15 km	Naqu	0.41 [a]/0.41 [d]	−0.02 [a]/0.00 [d]	-	Zhao et al. [69]
	L3_SM	25 km		0.26 [a]/0.17 [d]	−0.06 [a]/0.03 [d]	-	
	L3_SM	25 km	Maqu	0.24 [a]/0.20 [d]	−0.03 [a]/0.25 [d]	0.14 [a]/0.37 [d]	Zeng et al. [70]
			Naqu	0.54 [a]/0.43 [d]	−0.07 [a]/0.00 [d]	0.10 [a]/0.14 [d]	
	L3_SM	25 km	Naqu	0.67 [a]/0.73 [d]	−0.02 [a]/−0.01 [d]	0.07 [a]/0.06 [d]	Chen et al. [37]
			Pali	0.31 [a]/0.37 [d]	−0.02 [a]/−0.04 [d]	0.09 [a]/0.08 [d]	
	SMOS-IC	25 km	Heihe	0.18 [a]/0.30 [d]	−0.04 [a]/−0.12 [d]	0.12 [a]/0.14 [d]	Liu et al. [71]
			Naqu	0.43 [a]/0.47 [d]	−0.13 [a]/−0.05 [d]	0.18 [a]/0.14 [d]	
			Pali	0.60 [a]/0.52 [d]	−0.06 [a]/−0.03 [d]	0.07 [a]/0.09 [d]	
			Maqu	0.49 [a]/0.64 [d]	−0.01 [a]/−0.07 [d]	0.08 [a]/0.11 [d]	
			Ngari	0.12 [a]/0.10 [d]	−0.02 [a]/0.00 [d]	0.09 [a]/0.12 [d]	
Aquarius	L3_SM	1°	Naqu	0.77	−0.07	0.08	Li et al. [73]
SMAP	L3_SM_P	36 km	Naqu	0.87 [d]	−0.03 [d]	0.06 [d]	Chen et al. [37]
			Pali	0.67 [d]	−0.03 [d]	0.04 [d]	
	L3_SM_P	36 km	Heihe	0.64 [a]/0.78 [d]	−0.11 [a]/−0.10 [d]	0.11 [a]/0.11 [d]	Liu et al. [71]
			Naqu	0.84 [a]/0.82 [d]	−0.00 [a]/−0.02 [d]	0.08 [a]/0.07 [d]	
			Pali	0.67 [a]/0.62 [d]	−0.03 [a]/−0.05 [d]	0.05 [a]/0.06 [d]	
			Maqu	0.72 [a]/0.81 [d]	−0.07 [a]/−0.07 [d]	0.09 [a]/0.08 [d]	
			Ngari	0.57 [a]/0.34 [d]	−0.04 [a]/−0.05 [d]	0.05 [a]/0.05 [d]	
	L3_SM_P_E	9 km	Naqu	0.88	0.00	0.06	Li et al. [75]
			Maqu	0.65	0.11	0.13	
	L3_SM_P	36 km	Naqu	0.88	0.00	0.06	
			Maqu	0.64	0.12	0.13	
	L3_SM_P	36 km	Maqu	0.55 [d]	0.07 [d]	0.12 [d]	Zeng et al. [74]
			Naqu	0.78 [d]	−0.01 [d]	0.06 [d]	
			Pali	0.73 [d]	−0.05 [d]	0.06 [d]	
	L2_SM_A	3 km	Heihe	0.21~0.78	−0.12~0.09	0.03~0.17	Ma et al. [76]
	L2_SM_P	36 km		0.55~0.78	−0.00~0.09	0.03~0.09	
	L2_SM_AP	9 km		0.39~0.81	−0.20~0.03	0.04~0.81	

* The superscripts [a] and [d] represent the SM products retrieved using the T_B^p observations collected during the ascending and descending overpasses.

For instance, current commonly used soil dielectric constant models (e.g., Dobson model [50], Wang and Schmugge model [51], and Mironov model [52]) are unable to simulate the dielectric constant of frozen soils, leading to the failure of retrieving unfrozen (liquid) water content for frozen ground based on current L-band satellite missions. Zheng et al. [38,39] validated the applicability of the four-phase dielectric mixing model for estimating the soil permittivity of frozen ground on the TP, which divides the components of wet soil into the air, ice, matrix, and liquid water and is able to simulate the dielectric constants of soils under both frozen and thawed conditions [56]. Later on, Zheng et al. [35] compared the performance of three dielectric mixing models that are suitable for both frozen and thawed soil conditions on the TP, i.e., the four-phase dielectric mixing model and another two models developed by Zhang et al. [78] and Mironov [79]. The results showed that the four-phase dielectric mixing model is more suitable for the TP condition. On this basis, Zheng et al. [38,54] further improved the underestimation of the surface roughness effect in the SMAP SM retrieval algorithm by adopting a new surface roughness parameterization, thus improving the accuracy of SM retrievals in desert areas (e.g., Ngari network) and in vegetated areas during the freezing period (e.g., Maqu network) on the TP. Furthermore, Zheng et al. [38,39] introduced a new vegetation parameterization and found that the SM retrieval in the Maqu network can be further improved with ubRMSE reduced by more than 40%. Recently, Wu et al. [45] introduced the four-phase dielectric mixing model and the new parameterizations of surface roughness and vegetation developed by Zheng et al. [38,39] to the two-stream microwave emission model that is physically more correct than the τ-ω model [56]. The improved two-stream microwave emission model was further adopted to replace the τ-ω model adopted by the SMAP default SM retrieval algorithm to improve the SM retrievals on the TP. Figure 6 show the time series of θ_{liq} measurements and retrievals obtained by the SCA-V and DCA based on the improved two-stream microwave emission model using the SMAP T_B^p measurements during the descending and ascending overpasses for the period from August 2016 to July 2017. The SMAP SM products are also shown for comparison purposes, which are only available for the warm season due to the fact that the Mironov model [52] adopted by the current SMAP SM retrieval algorithm (see Table 2) is only suitable for thawed soil conditions. On the contrary, the SCA-V and DCA developed based on the improved two-stream microwave emission model with the implementation of a four-phase dielectric mixing model are able to retrieve unfrozen (liquid) water content θ_{liq} under both frozen and thawed soil conditions. The two methods are generally comparable to each other and are better than the SMAP product, whereby the latter tends to underestimate the θ_{liq}. Therefore, usage of the improved two-stream microwave emission model configured with the four-phase dielectric mixing model to replace the τ-ω model implemented by the current SMAP SM retrieval algorithm has improved the accuracy of SM retrievals and extended the retrieval algorithm to the frozen ground that widely covers the TP. In summary, three distinct features can be drawn from the abovementioned efforts made to improve the accuracy of SM retrievals on the TP using the algorithms implemented by current L-band satellite missions. First, a new soil dielectric mixing model was introduced and validated to fill the gap in retrieving unfrozen water content in frozen soil. Second, surface roughness and vegetation parameterizations embedded with default algorithms of current satellite missions were updated for the TP environment, which leads to better SM retrievals. Third, a more physical-based forward land emission model was implemented to release the assumptions made by the current widely used τ-ω model, providing the potential to retrieve SM from more complex land conditions.

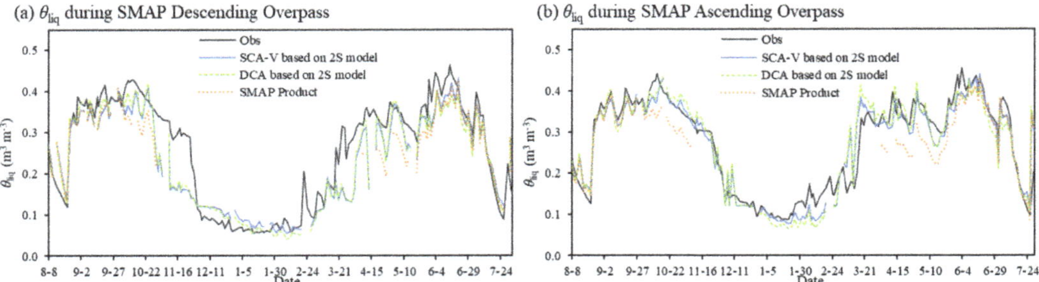

Figure 6. Time series of θ_{liq} measurements and retrievals obtained by the SCA-V and DCA based on the improved two-stream (2S) microwave emission model using the SMAP T_B^p measurements during the (**a**) descending and (**b**) ascending overpasses. The values derived from the SMAP SM products are also shown. The figure is modified from Wu [45].

In addition to improving the default retrieval algorithms implemented by current satellite missions, researchers also improved the current SM retrieval accuracy by developing new retrieval algorithms. Wang et al. [58,80] developed a new SM retrieval algorithm based on the physical-based Tor Vergata model to retrieve SM in the Maqu network based on the combination of Aquarius active and passive observations. The obtained SM retrievals were found to be able to reflect SM variations in the study area, providing a new way for the simultaneous use of active and passive observations to retrieve SM. Recently, Zeng et al. [81] developed a physical-based SM Index (SMI), which was shown to be able to reproduce measured θ_{liq} dynamics for both frozen and thawed conditions in the Naqu and Pali networks. The developed SMI shows great potential to produce better θ_{liq} retrievals on the TP based on the SMAP T_B^p measurements.

5. Conclusions

L-band passive microwave RS observation is an important tool for monitoring global SM and its freeze/thaw state, which can provide large-scale and long time series SM products for the TP in a complex natural environment. In recent years, researchers conducted ground-based and airborne L-band microwave radiometry observation experiments and established regional-scale in situ SM observation networks on the TP. In addition, a lot of work has been carried out to evaluate and improve the accuracy of current forward land emission models and SM retrieval algorithms to further improve the applicability of L-band satellite-based SM products to the TP condition. Progress related to L-band microwave emission modeling on the TP have highlighted the necessity to consider the impact of polarization mixing. For the first time, it was reported that the diurnal T_B^p observation signatures of both frozen and thawed soil conditions are primarily dominated by SMST dynamics at the surface layer around 2.5 cm. To further address the deficiencies in retrieving SM on the TP, such as lack of product under frozen ground, new parameterizations of soil permittivity, surface roughness, and vegetation are developed or introduced, which largely improve the accuracy of current SM retrievals. Moreover, to overcome the deficiency of the current widely used τ-ω model, more physical-based models such as the Tor Vergata model and the two-stream emission model are validated and implemented to develop new algorithms to better retrieve SM on the TP.

In short, progress has been made via the abovementioned efforts, which greatly promotes the in-depth application and development of L-band passive microwave RS technology in the TP. However, there are still many problems in the current research. For example, most work focuses on evaluating the accuracy of satellite-based SM products for a short-term period (e.g., less than 5 years), while the evaluation and improvement of the forward land emission model and SM retrieval algorithm are limited to the site or grid scale. There is still a lack of evaluating and improving both the land emission model and SM retrieval algorithms/products at the whole plateau scale, and the operational monitoring of unfrozen (liquid) water content in frozen ground is still missing. In view of the above research problems, in order to further enhance and expand the application of L-band passive microwave RS technology in the TP, the following research should be strengthened in the future.

Firstly, SMOS and SMAP satellite missions have provided long time series SM products for more than 12 and 7 years, respectively, while current work is mainly focused on evaluating the performance of these products for selected limited years. It is still unknown how accurate these products can capture the long-term trend of SM variations on the TP. Therefore, additional work is still needed to carry out long-term trend evaluation and analysis, whereby the long-term in situ SM dataset recently released by Zhang et al. [16] can be used as the ground reference for such assessment.

Secondly, to carry out the evaluation of the microwave emission model at the plateau scale. Specifically, to further validate and improve the parameterizations of the soil dielectric constant model, surface roughness and vegetation optical thickness are developed at the site or grid-scale for their applications to the whole plateau and to enhance the accuracy of microwave emission simulation at the plateau scale.

Thirdly, large amounts of research have been carried out to improve SM retrieval algorithms and products at the plateau scale based on the improved plateau-scale land emission model in combination with SM retrieval algorithms improved or newly developed at the site or grid scale. In addition, work has also been conducted to improve and develop SM products for a complete time series of consecutive years (including freezing periods) based on the L-band passive microwave RS observation via implementation of the four-phase dielectric mixing model that is applicable to both frozen and thawed soil conditions on the TP.

Finally, research has been conducted to further enhance the application of L-band satellite-based SM products on the TP, to assimilate satellite-based SM products or T_B^p observations to improve the simulation accuracy of plateau-scale water cycle and energy balance, and to evaluate and improve satellite-based precipitation products based on improved SM products. In addition, further work can be conducted to monitor drought changes and vegetation growth response to wet and dry transitions based on SM products, further expanding the application of L-band passive microwave RS products in the TP.

Author Contributions: Conceptualization, X.W. and J.W.; writing—original draft preparation, X.W.; writing—review and editing, J.W.; supervision, J.W.; funding acquisition, X.W. and J.W. All authors have read and agreed to the published version of the manuscript.

Funding: This research was funded by the National Natural Science Foundation of China (grant numbers 42030509, 41971308 and 41901317).

Data Availability Statement: Not applicable.

Conflicts of Interest: The authors declare no conflict of interest.

References

1. WMO; IOC; UNEP; ICSU. Implementation Plan for the Global Observing System for Climate in Support of the UNFCCC. GCOS-138, WMO-TD-1523. 2010. 180p. Available online: https://library.wmo.int/doc_num.php?explnum_id=3851 (accessed on 22 August 2022).
2. Koster, R.D.; Dirmeyer, P.A.; Guo, Z.; Bonan, G.; Chan, E.; Cox, P.; Gordon, C.T.; Kanae, S.; Kowalczyk, E.; Lawrence, D.; et al. Regions of Strong Coupling between Soil Moisture and Precipitation. *Science* **2004**, *305*, 1138–1140. [CrossRef] [PubMed]
3. Dorigo, W.; Himmelbauer, I.; Aberer, D.; Schremmer, L.; Petrakovic, I.; Zappa, L.; Preimesberger, W.; Xaver, A.; Annor, F.; Ardö, J.; et al. The International Soil Moisture Network: Serving Earth system science for over a decade. *Hydrol. Earth Syst. Sci.* **2021**, *25*, 5749–5804. [CrossRef]
4. Green, J.K.; Seneviratne, S.I.; Berg, A.M.; Findell, K.L.; Hagemann, S.; Lawrence, D.M.; Gentine, P. Large influence of soil moisture on long-term terrestrial carbon uptake. *Nature* **2019**, *565*, 476–479. [CrossRef]
5. Helbig, M.; Waddington, J.M.; Alekseychik, P.; Amiro, B.D.; Aurela, M.; Barr, A.G.; Black, T.A.; Blanken, P.D.; Carey, S.K.; Chen, J.; et al. Increasing contribution of peatlands to boreal evapotranspiration in a warming climate. *Nat. Clim. Chang.* **2020**, *10*, 555–560. [CrossRef]
6. Zheng, D.; Van der Velde, R.; Su, Z.; Wen, J.; Wang, X.; Booij, M.J.; Hoekstra, A.Y.; Lv, S.; Zhang, Y.; Ek, M.B. Impacts of Noah model physics on catchment-scale runoff simulations. *J. Geophys. Res. Atmos.* **2016**, *121*, 807–832. [CrossRef]
7. Zheng, D.; Van Der Velde, R.; Su, Z.; Wang, X.; Wen, J.; Booij, M.J.; Hoekstra, A.; Chen, Y. Augmentations to the Noah Model Physics for Application to the Yellow River Source Area. Part I: Soil Water Flow. *J. Hydrometeorol.* **2015**, *16*, 2659–2676. [CrossRef]
8. Zheng, D.; van der Velde, R.; Su, Z.; Wen, J.; Booij, M.J.; Hoekstra, A.Y.; Wang, X. Under-canopy turbulence and root water uptake of a Tibetan meadow ecosystem modeled by Noah-MP. *Water Resour. Res.* **2015**, *51*, 5735–5755. [CrossRef]
9. Pendergrass, A.G.; Meehl, G.A.; Pulwarty, R.; Hobbins, M.; Hoell, A.; AghaKouchak, A.; Bonfils, C.J.W.; Gallant, A.J.E.; Hoerling, M.; Hoffmann, D.; et al. Flash droughts present a new challenge for subseasonal-to-seasonal prediction. *Nat. Clim. Chang.* **2020**, *10*, 191–199. [CrossRef]
10. Brocca, L.; Tarpanelli, A.; Filippucci, P.; Dorigo, W.; Zaussinger, F.; Gruber, A.; Fernández-Prieto, D. How much water is used for irrigation? A new approach exploiting coarse resolution satellite soil moisture products. *Int. J. Appl. Earth Obs. Geoinf.* **2018**, *73*, 752–766. [CrossRef]
11. Rigden, A.J.; Mueller, N.D.; Holbrook, N.M.; Pillai, N.; Huybers, P. Combined influence of soil moisture and atmospheric evaporative demand is important for accurately predicting US maize yields. *Nat. Food* **2020**, *1*, 127–133. [CrossRef]
12. Zhang, K.; Li, X.; Zheng, D.; Zhang, L.; Zhu, G. Estimation of Global Irrigation Water Use by the Integration of Multiple Satellite Observations. *Water Resour. Res.* **2022**, *58*, e2021WR030031. [CrossRef]
13. Wu, G.; Duan, A.; Liu, Y.; Mao, J.; Ren, R.; Bao, Q.; He, B.; Liu, B.; Hu, W. Tibetan Plateau climate dynamics: Recent research progress and outlook. *Natl. Sci. Rev.* **2014**, *2*, 100–116. [CrossRef]
14. Zheng, D.; van der Velde, R.; Su, Z.; Wen, J.; Wang, X.; Yang, K. Impact of soil freeze-thaw mechanism on the runoff dynamics of two Tibetan rivers. *J. Hydrol.* **2018**, *563*, 382–394. [CrossRef]
15. Zheng, D.; Van Der Velde, R.; Su, Z.; Wen, J.; Wang, X.; Yang, K. Evaluation of Noah Frozen Soil Parameterization for Application to a Tibetan Meadow Ecosystem. *J. Hydrometeorol.* **2017**, *18*, 1749–1763. [CrossRef]
16. Zhang, P.; Zheng, D.; van der Velde, R.; Wen, J.; Zeng, Y.; Wang, X.; Wang, Z.; Chen, J.; Su, Z. Status of the Tibetan Plateau observatory (Tibet-Obs) and a 10-year (2009–2019) surface soil moisture dataset. *Earth Syst. Sci. Data* **2021**, *13*, 3075–3102. [CrossRef]

17. Su, Z.; Wen, J.; Dente, L.; van der Velde, R.; Wang, L.; Ma, Y.; Yang, K.; Hu, Z. The Tibetan Plateau observatory of plateau scale soil moisture and soil temperature (Tibet-Obs) for quantifying uncertainties in coarse resolution satellite and model products. *Hydrol. Earth Syst. Sci.* **2011**, *15*, 2303–2316. [CrossRef]
18. Yang, K.; Qin, J.; Zhao, L.; Chen, Y.; Tang, W.; Han, M.; Lazhu; Chen, Z.; Lv, N.; Ding, B.; et al. A Multiscale Soil Moisture and Freeze–Thaw Monitoring Network on the Third Pole. *Bull. Am. Meteorol. Soc.* **2013**, *94*, 1907–1916. [CrossRef]
19. Babaeian, E.; Sadeghi, M.; Jones, S.B.; Montzka, C.; Vereecken, H.; Tuller, M. Ground, Proximal, and Satellite Remote Sensing of Soil Moisture. *Rev. Geophys.* **2019**, *57*, 530–616. [CrossRef]
20. Wigneron, J.-P.; Jackson, T.; O'Neill, P.; De Lannoy, G.; de Rosnay, P.; Walker, J.; Ferrazzoli, P.; Mironov, V.; Bircher, S.; Grant, J.; et al. Modelling the passive microwave signature from land surfaces: A review of recent results and application to the L-band SMOS & SMAP soil moisture retrieval algorithms. *Remote Sens. Environ.* **2017**, *192*, 238–262. [CrossRef]
21. Kerr, Y.H.; Waldteufel, P.; Wigneron, J.P.; Martinuzzi, J.; Font, J.; Berger, M. Soil moisture retrieval from space: The Soil Moisture and Ocean Salinity (SMOS) mission. *IEEE Trans. Geosci. Remote Sens.* **2001**, *39*, 1729–1735. [CrossRef]
22. Lagerloef, G.; Colomb, F.R.; Le Vine, D.; Wentz, F.; Yueh, S.; Ruf, C.; Lilly, J.; Gunn, J.; Chao, Y.; Decharon, A.; et al. The Aquarius/SAC-D Mission: Designed to Meet the Salinity Remote-Sensing Challenge. *Oceanography* **2008**, *21*, 68–81. [CrossRef]
23. Entekhabi, D.; Njoku, E.G.; O'Neill, P.E.; Kellogg, K.H.; Crow, W.T.; Edelstein, W.N.; Entin, J.K.; Goodman, S.D.; Jackson, T.J.; Johnson, J.; et al. The Soil Moisture Active Passive (SMAP) Mission. *Proc. IEEE* **2010**, *98*, 704–716. [CrossRef]
24. Shi, J.; Dong, X.; Zhao, T.; Du, J.; Jiang, L.; Du, Y.; Liu, H.; Wang, Z.; Ji, D.; Xiong, C. WCOM: The science scenario and objectives of a global water cycle observation mission. In Proceedings of the IEEE International Geoscience and Remote Sensing Symposium, Quebec City, QC, Canada, 13–18 July 2014; pp. 3646–3649. [CrossRef]
25. Schwank, M.; Wigneron, J.-P.; Lopez-Baeza, E.; Volksch, I.; Matzler, C.; Kerr, Y.H. L-Band Radiative Properties of Vine Vegetation at the MELBEX III SMOS Cal/Val Site. *IEEE Trans. Geosci. Remote Sens.* **2012**, *50*, 1587–1601. [CrossRef]
26. Montzka, C.; Bogena, H.R.; Weihermuller, L.; Jonard, F.; Bouzinac, C.; Kainulainen, J.; Balling, J.E.; Loew, A.; Dall'Amico, J.T.; Rouhe, E.; et al. Brightness Temperature and Soil Moisture Validation at Different Scales During the SMOS Validation Campaign in the Rur and Erft Catchments, Germany. *IEEE Trans. Geosci. Remote Sens.* **2012**, *51*, 1728–1743. [CrossRef]
27. Panciera, R.; Walker, J.P.; Jackson, T.J.; Gray, D.A.; Tanase, M.A.; Ryu, D.; Monerris, A.; Yardley, H.; Rudiger, C.; Wu, X.; et al. The Soil Moisture Active Passive Experiments (SMAPEx): Toward Soil Moisture Retrieval from the SMAP Mission. *IEEE Trans. Geosci. Remote Sens.* **2013**, *52*, 490–507. [CrossRef]
28. Colliander, A.; Cosh, M.H.; Misra, S.; Jackson, T.J.; Crow, W.; Chan, S.; Bindlish, R.; Chae, C.; Collins, C.H.; Yueh, S.H. Validation and scaling of soil moisture in a semi-arid environment: SMAP validation experiment 2015 (SMAPVEX15). *Remote Sens. Environ.* **2017**, *196*, 101–112. [CrossRef]
29. Li, X.; Cheng, G.; Liu, S.; Xiao, Q.; Ma, M.; Jin, R.; Che, T.; Liu, Q.; Wang, W.; Qi, Y.; et al. Heihe Watershed Allied Telemetry Experimental Research (HiWATER): Scientific Objectives and Experimental Design. *Bull. Am. Meteorol. Soc.* **2013**, *94*, 1145–1160. [CrossRef]
30. Zhao, T.; Shi, J.; Lv, L.; Xu, H.; Chen, D.; Cui, Q.; Jackson, T.J.; Yan, G.; Jia, L.; Chen, L.; et al. Soil moisture experiment in the Luan River supporting new satellite mission opportunities. *Remote Sens. Environ.* **2020**, *240*, 111680. [CrossRef]
31. Colliander, A.; Jackson, T.J.; Bindlish, R.; Chan, S.; Das, N.; Kim, S.B.; Cosh, M.H.; Dunbar, R.S.; Dang, L.; Pashaian, L.; et al. Validation of SMAP surface soil moisture products with core validation sites. *Remote Sens. Environ.* **2017**, *191*, 215–231. [CrossRef]
32. Zhao, T.; Shi, J.; Bindlish, R.; Jackson, T.; Kerr, Y.; Cui, Q.; Li, Y.; Che, T. Refinement of SMOS multi-angular brightness temperature and its analysis over reference targets. *IEEE J. Sel. Top. Appl. Earth Obs. Remote Sens.* **2015**, *8*, 589–603. [CrossRef]
33. Zheng, D.; Wang, X.; Van Der Velde, R.; Zeng, Y.; Wen, J.; Wang, Z.; Schwank, M.; Ferrazzoli, P.; Su, Z. L-Band Microwave Emission of Soil Freeze–Thaw Process in the Third Pole Environment. *IEEE Trans. Geosci. Remote Sens.* **2017**, *55*, 5324–5338. [CrossRef]
34. Zheng, D.; Li, X.; Wang, X.; Wang, Z.; Wen, J.; van der Velde, R.; Schwank, M.; Su, Z. Sampling depth of L-band radiometer measurements of soil moisture and freeze-thaw dynamics on the Tibetan Plateau. *Remote Sens. Environ.* **2019**, *226*, 16–25. [CrossRef]
35. Zheng, D.; Li, X.; Zhao, T.; Wen, J.; van der Velde, R.; Schwank, M.; Wang, X.; Wang, Z.; Su, Z. Impact of Soil Permittivity and Temperature Profile on L-Band Microwave Emission of Frozen Soil. *IEEE Trans. Geosci. Remote Sens.* **2020**, *59*, 4080–4093. [CrossRef]
36. Dente, L.; Su, Z.; Wen, J. Validation of SMOS Soil Moisture Products over the Maqu and Twente Regions. *Sensors* **2012**, *12*, 9965–9986. [CrossRef]
37. Chen, Y.; Yang, K.; Qin, J.; Cui, Q.; Lu, H.; La, Z.; Han, M.; Tang, W. Evaluation of SMAP, SMOS, and AMSR2 soil moisture retrievals against observations from two networks on the Tibetan Plateau. *J. Geophys. Res. Atmos.* **2017**, *122*, 5780–5792. [CrossRef]
38. Zheng, D.; Wang, X.; van der Velde, R.; Ferrazzoli, P.; Wen, J.; Wang, Z.; Schwank, M.; Colliander, A.; Bindlish, R.; Su, Z. Impact of surface roughness, vegetation opacity and soil permittivity on L-band microwave emission and soil moisture retrieval in the third pole environment. *Remote Sens. Environ.* **2018**, *209*, 633–647. [CrossRef]
39. Zheng, D.; Wang, X.; van der Velde, R.; Schwank, M.; Ferrazzoli, P.; Wen, J.; Wang, Z.; Colliander, A.; Bindlish, R.; Su, Z. Assessment of Soil Moisture SMAP Retrievals and ELBARA-III Measurements in a Tibetan Meadow Ecosystem. *IEEE Geosci. Remote Sens. Lett.* **2019**, *16*, 1407–1411. [CrossRef]

40. Li, X.; Liu, S.; Xiao, Q.; Ma, M.; Jin, R.; Che, T.; Wang, W.; Hu, X.; Xu, Z.; Wen, J.; et al. A multiscale dataset for understanding complex eco-hydrological processes in a heterogeneous oasis system. *Sci. Data* **2017**, *4*, 170083. [CrossRef]
41. Zheng, D.; Li, X.; Wen, J.; Hofste, J.G.; van der Velde, R.; Wang, X.; Wang, Z.; Bai, X.; Schwank, M.; Su, Z. Active and Passive Microwave Signatures of Diurnal Soil Freeze-Thaw Transitions on the Tibetan Plateau. *IEEE Trans. Geosci. Remote Sens.* **2021**, *60*, 4301814. [CrossRef]
42. Kerr, Y.H.; Waldteufel, P.; Richaume, P.; Wigneron, J.P.; Ferrazzoli, P.; Mahmoodi, A.; Al Bitar, A.; Cabot, F.; Gruhier, C.; Juglea, S.E.; et al. The SMOS Soil Moisture Retrieval Algorithm. *IEEE Trans. Geosci. Remote Sens.* **2012**, *50*, 1384–1403. [CrossRef]
43. Bindlish, R.; Jackson, T.; Cosh, M.; Zhao, T.; O'Neill, P. Global Soil Moisture from the Aquarius/SAC-D Satellite: Description and Initial Assessment. *IEEE Geosci. Remote Sens. Lett.* **2015**, *12*, 923–927. [CrossRef]
44. O'Neill, P.; Chan, S.; Njoku, E.; Jackson, T.; Bindlish, R. Algorithm Theoretical Basis Document (ATBD): Level 2 & 3 Soil Moisture (Passive) Data Products [J/OL]. 2015. Initial Release, v.3, 1 October. Available online: http://smap.jpl.nasa.gov/science/dataproducts/ATBD/ (accessed on 22 August 2022).
45. Wu, X. Implementation of Two-Stream Emission Model for L-Band Retrievals on the Tibetan Plateau. *Remote Sens.* **2022**, *14*, 494. [CrossRef]
46. Mo, T.; Choudhury, B.J.; Schmugge, T.J.; Wang, J.R.; Jackson, T.J. A model for microwave emission from vegetation-covered fields. *J. Geophys. Res. Earth Surf.* **1982**, *87*, 11229–11237. [CrossRef]
47. Jackson, T.; Schmugge, T. Vegetation effects on the microwave emission of soils. *Remote Sens. Environ.* **1991**, *36*, 203–212. [CrossRef]
48. Wang, J.R.; Choudhury, B.J. Remote sensing of soil moisture content, over bare field at 1.4 GHz frequency. *J. Geophys. Res. Earth Surf.* **1981**, *86*, 5277–5282. [CrossRef]
49. Wigneron, J.-P.; Laguerre, L.; Kerr, Y. A simple parameterization of the L-band microwave emission from rough agricultural soils. *IEEE Trans. Geosci. Remote Sens.* **2001**, *39*, 1697–1707. [CrossRef]
50. Dobson, M.C.; Ulaby, F.T.; Hallikainen, M.T.; El-Rayes, M.A. Microwave Dielectric Behavior of Wet Soil-Part II: Dielectric Mixing Models. *IEEE Trans. Geosci. Remote Sens.* **1985**, *23*, 35–46. [CrossRef]
51. Wang, J.R.; Schmugge, T.J. An Empirical Model for the Complex Dielectric Permittivity of Soils as a Function of Water Content. *IEEE Trans. Geosci. Remote Sens.* **1980**, *GE-18*, 288–295. [CrossRef]
52. Mironov, V.L.; Kosolapova, L.G.; Fomin, S.V. Physically and Mineralogically Based Spectroscopic Dielectric Model for Moist Soils. *IEEE Trans. Geosci. Remote Sens.* **2009**, *47*, 2059–2070. [CrossRef]
53. Choudhury, B.J.; Schmugge, T.J.; Mo, T. A parameterization of effective soil temperature for microwave emission. *J. Geophys. Res. Earth Surf.* **1982**, *87*, 1301–1304. [CrossRef]
54. Zheng, D.; Van Der Velde, R.; Wen, J.; Wang, X.; Ferrazzoli, P.; Schwank, M.; Colliander, A.; Bindlish, R.; Su, Z. Assessment of the SMAP Soil Emission Model and Soil Moisture Retrieval Algorithms for a Tibetan Desert Ecosystem. *IEEE Trans. Geosci. Remote Sens.* **2018**, *56*, 3786–3799. [CrossRef]
55. Wigneron, J.P.; Chanzy, A.; Kerr, Y.; Shi, J.C.; Cano, A.; Rosnay, P.D.; Escorihuela, M.J.; Mironov, V.; Demontoux, F.; Grant, J. Improved Parameterization of the Soil Emission in L-MEB. *IEEE Trans. Geosci. Remote Sens.* **2011**, *49*, 1177–1189. [CrossRef]
56. Schwank, M.; Naderpour, R.; Mätzler, C. "Tau-Omega"- and Two-Stream Emission Models Used for Passive L-Band Retrievals: Application to Close-Range Measurements over a Forest. *Remote Sens.* **2018**, *10*, 1868. [CrossRef]
57. Wu, X.; Zheng, D. Surface Roughness Effect on L-Band Multiangular Brightness Temperature Modeling and Soil Liquid Water Retrieval of Frozen Soil. *IEEE Geosci. Remote Sens. Lett.* **2021**, *18*, 1615–1619. [CrossRef]
58. Wang, Q.; van der Velde, R.; Su, Z. Use of a discrete electromagnetic model for simulating Aquarius L-band active/passive observations and soil moisture retrieval. *Remote Sens. Environ.* **2018**, *205*, 434–452. [CrossRef]
59. Bai, X.; Zeng, J.; Chen, K.-S.; Li, Z.; Zeng, Y.; Wen, J.; Wang, X.; Dong, X.; Su, Z. Parameter Optimization of a Discrete Scattering Model by Integration of Global Sensitivity Analysis Using SMAP Active and Passive Observations. *IEEE Trans. Geosci. Remote Sens.* **2018**, *57*, 1084–1099. [CrossRef]
60. Schwank, M.; Stahli, M.; Wydler, H.; Leuenberger, J.; Matzler, C.; Fluhler, H. Microwave L-band emission of freezing soil. *IEEE Trans. Geosci. Remote Sens.* **2004**, *42*, 1252–1261. [CrossRef]
61. Wilheit, T.T. Radiative Transfer in a Plane Stratified Dielectric. *IEEE Trans. Geosci. Electron.* **1978**, *16*, 138–143. [CrossRef]
62. Chen, K.S.; Wu, T.D.; Tsang, L.; Li, Q.; Shi, J.; Fung, A.K. Emission of rough surfaces calculated by the integral equation method with comparison to three-dimensional moment method simulation. *IEEE Trans. Geosci. Remote Sens.* **2003**, *41*, 90–101. [CrossRef]
63. Wu, X. Implementation of Wilheit Model for Predicting L-Band Microwave Emission in the Third Pole Environment. *IEEE Geosci. Remote Sens. Lett.* **2021**, *19*, 4500505. [CrossRef]
64. Shi, J.; Chen, K.S.; Li, Q.; Jackson, T.J.; O'Neill, P.E.; Tsang, L. A parameterized surface reflectivity model and estimation of bare surface soil moisture with L-band radiometer. *IEEE Trans. Geosci. Remote Sens.* **2002**, *40*, 2674–2686.
65. Jin, R.; Li, X.; Yan, B.; Li, X.; Luo, W.; Ma, M.; Guo, J.; Kang, J.; Zhu, Z.; Zhao, S. A Nested Ecohydrological Wireless Sensor Network for Capturing the Surface Heterogeneity in the Midstream Areas of the Heihe River Basin, China. *IEEE Geosci. Remote Sens. Lett.* **2014**, *11*, 2015–2019. [CrossRef]
66. Kang, J.; Jin, R.; Li, X.; Zhang, Y. Mapping High Spatiotemporal-Resolution Soil Moisture by Upscaling Sparse Ground-Based Observations Using a Bayesian Linear Regression Method for Comparison with Microwave Remotely Sensed Soil Moisture Products. *Remote Sens.* **2021**, *13*, 228. [CrossRef]

67. Zhang, P.; Zheng, D.; van der Velde, R.; Wen, J.; Ma, Y.; Zeng, Y.; Wang, X.; Wang, Z.; Chen, J.; Su, Z. A dataset of 10-year regional-scale soil moisture and soil temperature measurements at multiple depths on the Tibetan Plateau. *Earth Syst. Sci. Data Discuss.* **2022**. [CrossRef]
68. Dente, L.; Vekerdy, Z.; Wen, J.; Su, Z. Maqu network for validation of satellite-derived soil moisture products. *Int. J. Appl. Earth Obs. Geoinf. ITC J.* **2012**, *17*, 55–65. [CrossRef]
69. Zhao, L.; Yang, K.; Qin, J.; Chen, Y.; Tang, W.; Lu, H.; Yang, Z.-L. The scale-dependence of SMOS soil moisture accuracy and its improvement through land data assimilation in the central Tibetan Plateau. *Remote Sens. Environ.* **2014**, *152*, 345–355. [CrossRef]
70. Zeng, J.; Li, Z.; Chen, Q.; Bi, H.; Qiu, J.; Zou, P. Evaluation of remotely sensed and reanalysis soil moisture products over the Tibetan Plateau using in-situ observations. *Remote Sens. Environ.* **2015**, *163*, 91–110. [CrossRef]
71. Liu, J.; Chai, L.; Lu, Z.; Liu, S.; Qu, Y.; Geng, D.; Song, Y.; Guan, Y.; Guo, Z.; Wang, J.; et al. Evaluation of SMAP, SMOS-IC, FY3B, JAXA, and LPRM Soil Moisture Products over the Qinghai-Tibet Plateau and Its Surrounding Areas. *Remote Sens.* **2019**, *11*, 792. [CrossRef]
72. Liu, J.; Chai, L.; Dong, J.; Zheng, D.; Wigneron, J.-P.; Liu, S.; Zhou, J.; Xu, T.; Yang, S.; Song, Y.; et al. Uncertainty analysis of eleven multisource soil moisture products in the third pole environment based on the three-corned hat method. *Remote Sens. Environ.* **2021**, *255*, 112225. [CrossRef]
73. Li, D.; Zhao, T.; Shi, J.; Bindlish, R.; Jackson, T.J.; Peng, B.; An, M.; Han, B. First Evaluation of Aquarius Soil Moisture Products Using *In Situ* Observations and GLDAS Model Simulations. *IEEE J. Sel. Top. Appl. Earth Obs. Remote Sens.* **2015**, *8*, 5511–5525. [CrossRef]
74. Zeng, J.; Shi, P.; Chen, K.-S.; Ma, H.; Bi, H.; Cui, C. Assessment and Error Analysis of Satellite Soil Moisture Products Over the Third Pole. *IEEE Trans. Geosci. Remote Sens.* **2021**, *60*, 4405418. [CrossRef]
75. Li, C.; Lu, H.; Yang, K.; Han, M.; Wright, J.S.; Chen, Y.; Yu, L.; Xu, S.; Huang, X.; Gong, W. The Evaluation of SMAP Enhanced Soil Moisture Products Using High-Resolution Model Simulations and In-Situ Observations on the Tibetan Plateau. *Remote Sens.* **2018**, *10*, 535. [CrossRef]
76. Ma, C.; Li, X.; Wei, L.; Wang, W. Multi-Scale Validation of SMAP Soil Moisture Products over Cold and Arid Regions in Northwestern China Using Distributed Ground Observation Data. *Remote Sens.* **2017**, *9*, 327. [CrossRef]
77. Owe, M.; de Jeu, R.; Walker, J. A methodology for surface soil moisture and vegetation optical depth retrieval using the microwave polarization difference index. *IEEE Trans. Geosci. Remote Sens.* **2001**, *39*, 1643–1654. [CrossRef]
78. Zhang, L.; Zhao, T.; Jiang, L.; Zhao, S. Estimate of phase transition water content in Freeze–Thaw process using microwave radi-ometer. *IEEE Trans. Geosci. Remote Sens.* **2010**, *48*, 4248–4255. [CrossRef]
79. Mironov, V.L.; Kosolapova, L.G.; Lukin, Y.I.; Karavaysky, A.Y.; Molostov, I.P. Temperature- and texture-dependent dielectric model for frozen and thawed mineral soils at a frequency of 1.4 GHz. *Remote Sens. Environ.* **2017**, *200*, 240–249. [CrossRef]
80. Wang, Q.; van der Velde, R.; Ferrazzoli, P.; Chen, X.; Bai, X.; Su, Z. Mapping soil moisture across the Tibetan Plateau plains using Aquarius active and passive L-band microwave observations. *Int. J. Appl. Earth Obs. Geoinf. ITC J.* **2019**, *77*, 108–118. [CrossRef]
81. Zeng, J.; Chen, K.S.; Cui, C.; Bai, X. A Physically Based Soil Moisture Index from Passive Microwave Brightness Temperatures for Soil Moisture Variation Monitoring. *IEEE Trans. Geosci. Remote Sens.* **2020**, *58*, 2782–2795. [CrossRef]

Review

Advances in the Quality of Global Soil Moisture Products: A Review

Yangxiaoyue Liu [1,2,*] and Yaping Yang [1,2]

1 State Key Laboratory of Resources and Environmental Information Systems, Institute of Geographic Sciences and Natural Resources Research, Chinese Academy of Sciences, Beijing 100101, China
2 Jiangsu Center for Collaborative Innovation in Geographical Information Resource Development and Application, Nanjing 210023, China
* Correspondence: lyxy@lreis.ac.cn

Abstract: Soil moisture is a crucial component of land–atmosphere interaction systems. It has a decisive effect on evapotranspiration and photosynthesis, which then notably impacts the land surface water cycle, energy transfer, and material exchange. Thus, soil moisture is usually treated as an indispensable parameter in studies that focus on drought monitoring, climate change, hydrology, and ecology. After consistent efforts for approximately half a century, great advances in soil moisture retrieval from in situ measurements, remote sensing, and reanalysis approaches have been achieved. The quality of soil moisture estimates, including spatial coverage, temporal span, spatial resolution, time resolution, time latency, and data precision, has been remarkably and steadily improved. This review outlines the recently developed techniques and algorithms used to estimate and improve the quality of soil moisture estimates. Moreover, the characteristics of each estimation approach and the main application fields of soil moisture are summarized. The future prospects of soil moisture estimation trends are highlighted to address research directions in the context of increasingly comprehensive application requirements.

Keywords: soil moisture; estimation method advances; applications; prospects

Citation: Liu, Y.; Yang, Y. Advances in the Quality of Global Soil Moisture Products: A Review. *Remote Sens.* **2022**, *14*, 3741. https://doi.org/10.3390/rs14153741

Academic Editor: George P. Petropoulos

Received: 5 May 2022
Accepted: 1 August 2022
Published: 4 August 2022

1. Introduction

Soil moisture (SM), the moisture content in the soil, is a crucial component in the hydrological cycle; it links atmospheric precipitation and underground water and is also an important parameter of energy exchange between the land surface and the atmosphere [1–4]. Consequently, SM is recognized as an essential element in studies aimed at analyzing and understanding Earth system processes, such as climate change and ecological evolution. Specifically, the available water content, which is essential for vegetation growth, is one of the most important components of soil and has crucial guiding significance for agricultural production. Currently, both ground and spaceborne sensors are used to derive the original SM information [2,5,6]. Numerous technologies, such as statistical models, data fusion, machine learning, and assimilation approaches, are widely used to improve SM quality [7–10]. Additionally, SM datasets with high spatial-temporal resolution are valuable for boosting agricultural production in terms of drought and flood monitoring, crop growth analysis, and yield estimation.

Significant efforts have been devoted to SM acquisition and estimation techniques during the past decades, and numerous global-scale SM estimates have been generated and are available for scientific studies [11–13]. To fulfill the increasingly comprehensive requirements for SM estimates, their quality, including spatial coverage, temporal span, spatial resolution, temporal resolution, time latency, and data precision, is notably improved through advanced methods. However, there is still a long way to go so as to further enhance the spatiotemporal integrity, accuracy, and stability of estimated SM. Therefore, it is necessary to rigorously summarize these data acquisition methods, progress in advanced techniques, and point out future challenges for SM retrieval.

The remainder of this paper is organized as follows. Section 1 introduces the meaning of improving SM products and the two main original SM data acquisition methods. Section 2 provides a comprehensive and systematic review of the methods for improving the quality of both ground- and satellite-observed SM products. The principles, advantages, and limitations of these methods are presented. Section 3 presents the application fields of SM products. Section 4 presents future prospects for advancing global SM products, and Section 5 concludes the article.

Currently, there are two main data acquisition methods:

(1) Point-scale original data acquisition: in situ measurements

Considering the scientific significance and application value of SM, the Soviet Union and Mongolia have started to record ground SM using monitoring sensors to retrieve national soil water content through networks since the 1950s [14–16]. In situ measurements can conveniently monitor SM at precise sites, depths, and hourly or sub-hourly frequencies. Both the sensors and networks are easily accessible and affordable. However, as various institutes have different research objectives, each SM network has its own station density, observation frequency, monitoring depth, sensor type, spatial coverage, and temporal period. SM can be expressed as a gravimetric unit (g/cm^3), volumetric unit (m^3/m^3), or a function of the field capacity according to usage habits [17]. Every SM network has its own method of sharing data, usually through a website in its own language. Therefore, it is difficult for researchers to derive SM records from different observation networks.

Facing these difficult problems, the International Soil Moisture Network (ISMN, https://ismn.geo.tuwien.ac.at/en/, accessed on 31 July 2022) is devoted to performing as a centralized data hosting facility for global in situ SM measurements [5,18,19]. This platform is initiated to collect global SM from operational networks and validation campaigns, standardize the techniques and protocols and make them available to users. Currently (June 2022), 73 networks and more than 2800 stations are located in Europe, North America, South America, Asia, Africa, Australia, and Oceania, which are collected by the ISMN and available to the public. In addition to SM, ISMN also integrates and provides SM-related meteorological variables, such as soil temperature and precipitation, which serve as critical supplementary references for the comprehensive analysis of soil water evolution characteristics. Currently, the ISMN is an increasingly popular data source for studies focused on SM validation worldwide [14,20–26]. With continuous network expansion and data updates, the ISMN has become an energetic and well-acknowledged global-scale SM ground observation database. Additionally, the National Soil Moisture Network has been established in the contiguous United States. There are 24 networks, and the SM data are retrieved in a timely manner with a one-day latency (http://nationalsoilmoisture.com/, accessed on 31 July 2022).

However, despite the increasingly standardized and abundant in situ measurements, it is still difficult for point-scale data to represent large-area SM conditions. Limited time and space coverage greatly restrict the application of in situ measurements in large-scale, long-term scientific studies and explorations. As a result, in situ measurements usually serve as a crucial reference for the evaluation of multi-scale SM estimates.

(2) Large-scale data acquisition: spaceborne remote-sensing technology

There is an urgent demand for access to near-real-time soil moisture data on a global scale. Since the 1970s, spaceborne remote sensing technology has gradually become a promising approach for obtaining global-scale continuous time-series surface SM data. The abundance of satellite-retrieved soil moisture data provides an unprecedented opportunity to conduct related analyses and applications.

A number of remotely sensed data, including optical, thermal infrared, and microwave bands, were employed to retrieve SM estimates [27]. In terms of optical and thermal infrared remote sensing data, soil surface spectral reflectance characteristics, soil surface emissivity, and surface temperature are mainly used to estimate SM [28]. However, retrieval models are mostly established on the basis of empirical relationships between SM and land surface condition indexes, that is, vegetation condition index [29], normalized difference vegetation index (NDVI) [30], temperature vegetation drought index (TVDI) [31], and soil wetness index [32], which can hardly satisfy large-scale and multi-climate zone applications. In addition, both optical and thermal remote sensing are vulnerable and sensitive to cloudy and rainy weather, dense vegetation coverage, and aerosol optical depth. Optical remote sensing can only measure reflection and emission from the land surface at a depth of 1 mm. For hydrological and agricultural analyses, SM data could be far more meaningful at a depth of several centimeters than at a mere 1 mm.

In comparison, microwave signals are impervious to rainy and cloudy weather, and their penetration depth can reach 0–5 cm, showing prominent advantages in SM retrieval. Microwave remote sensing technology can be divided into active and passive microwaves based on the working modes of different sensors. Active microwave sensors transmit signals to the detection targets and receive backscattered signals after the interaction between the signals and targets, whereas passive microwave sensors receive signals reflected and emitted from the underlying surface [33–35]. Currently, both active and passive microwave signals are employed to derive land surface soil water content. As shown in Table 1, a large number of spaceborne microwave SM products have been retrieved and published in the past half-century. Through their application in various hydrology-related scientific explorations, they efficiently boosted the understanding of spatial-temporal evolution characteristics of SM and the mechanism by which SM influences climate change across the globe. In addition to the listed global SM products, there are also studies and programs focused on SM deriving in a certain vegetation cover or climate zone to acquire regional SM with high accuracy [9,36,37].

Table 1. A brief summary of single spaceborne microwave-retrieved SM products and their basic properties.

Sensor Working Mode	Satellite Program	Time Range	Ascending/ Descending Time Node	Temporal Resolution	Spatial Resolution	Sensor	Sensor Band	Penetration Depth	Retrieving Algorithm	Publisher	Detailed Information
Active	European Remote-Sensing Satellite-1 (ERS-1)	1991.7.17–2000.3.10	22:15 (A), 10:30 (D)	Daily	50 × 50 km^2	Synthetic Aperture Radar (SAR)	C band (5.3 GHz)	<2 cm	The Integral Equation Model, the Semiempirical Change Detection Approach	The European Space Agency (ESA)	[38,39]
	European Remote-Sensing Satellite-2 (ERS-2)	1995.4.21–2011.9.5	22:30 (A), 10:30 (D)	Daily	25 × 25 km^2	SAR	C band (5.3 GHz)	<2 cm	The Backscattering Model, the Semiempirical Change Detection Approach	ESA	[33,40]
	Environmental Satellite (ENVISAT)	2002.3.1–2012.4.8	22:00 (A), 10:00 (D)	35 days	1 × 1 km^2	SAR	C band (5.3 GHz)	<2 cm	The Semiempirical Change Detection Approach	ESA	[41]
	Advanced Scatterometer on MetOp-A (ASCAT MetOp-A)	2006.10.19 ongoing	21:30 (A), 09:30 (D)	Daily	25 × 25 km^2, 50 × 50 km^2	SAR	C band (5.3 GHz)	<2 cm	The Semiempirical Change Detection Approach	ESA	[42,43]
	Advanced Scatterometer on MetOp-B (ASCAT MetOp-B)	2012.9.17 ongoing	21:30 (A), 09:30 (D)	Daily	25 × 25 km^2, 50 × 50 km^2	SAR	C band (5.3 GHz)	<2 cm	The Semiempirical Change Detection Approach	ESA	[20,44]
	Cyclone Global Navigation Satellite System (CYGNSS)	2017.3.18–2020.8.16	-	Every 6 h/daily	0.3° × 0.37°	Bistatic Radar	L band (1.4 GHz)	0–5 cm	The linear relationship between SMAP soil moisture and CYGNSS surface reflectivity observations	The National Aeronautics and Space Administration (NASA)	[45]
	Terra-Sar	2007.6 ongoing	18:00 (A), 06:00 (D)	Daily	2 × 2 m^2	SAR	X band (9.5 GHz)	<2 cm	The water-cloud model and self-organizing neural networks	ESA	[46–48]
	Sentinel-1	2014.4.3 ongoing	18:00 (A), 06:00 (D)	Daily	1 × 1 km^2	SAR	C band (5.404 GHz)	<2 cm	The change detection algorithm	ESA	[49,50]

Table 1. *Cont.*

Sensor Working Mode	Satellite Program	Time Range	Ascending/ Descending Time Node	Temporal Resolution	Spatial Resolution	Sensor	Sensor Band	Penetration Depth	Retrieving Algorithm	Publisher	Detailed Information
Passive	Scanning Multichannel Microwave Radiometer (SMMR)	1979.10–1987.8	12:00 (A), 24:00 (D)	Daily	150 × 150 km^2	Radiometer	C band (6.6 GHz), X band (10.7 GHz), K band (18 GHz)	<2 cm	The Land Parameter Retrieval Model	The National Snow and Ice Data Center (NSIDC)	[51,52]
	Special Sensor Microwave Imager (SSM/I)	1987 ongoing	F08 18:12 (A), 06:12 (D) F11 17:10 (A), 05:10 (D) F13 17:35 (A), 05:35 (D) F14 20:21 (A), 08:21 (D) F15 21:31 (A), 09:31 (D) F16 20:13 (A), 08:13 (D)	Daily	69 × 43 km^2	Radiometer	K band (19.4 GHz), Ka band (37.0 GHz)	<1.5 cm	The Land Parameter Retrieval Model	NSIDC	[51,53]
	Tropical Rainfall Measuring Mission Microwave Imager (TRMM TMI)	1997.12.7–2015.4.8	changes 24 h of local time in 46-day procession	Daily	59 × 36 km^2	Radiometer	X band (10.65 GHz), Ka band (37.0 GHz)	<2 cm	The Land Parameter Retrieval Model	the Goddard Earth Sciences Data and Information Services Center	[51,54]
	Advanced Microwave Scanning Radiometer for the Earth observing system (AMSR-E)	2002.6.1–2011.10.4	01:30 (A), 13:30 (D)	Daily	76 × 44 km^2	Radiometer	C band (6.9 GHz), X band (10.7 GHz)	<2 cm	The Land Parameter Retrieval Model, the Japanese Aerospace eXploration Agency algorithm	Earth Observation Research Center of Japan Aerospace Exploration Agency	[55,56]
	Advanced Microwave Scanning Radiometer 2 (AMSR-2)	2012.8.10 ongoing	01:30 (A), 13:30 (D)	Daily	35 × 62 km^2	Radiometer	C band (6.9 GHz), X band (10.7 GHz)	<2 cm	The Land Parameter Retrieval Model, the Japanese Aerospace eXploration Agency algorithm	The Earth Observation Research Center of Japan Aerospace Exploration Agency	[57,58]

Table 1. *Cont.*

Sensor Working Mode	Satellite Program	Time Range	Ascending/ Descending Time Node	Temporal Resolution	Spatial Resolution	Sensor	Sensor Band	Penetration Depth	Retrieving Algorithm	Publisher	Detailed Information
	Windsat/Coriolos	2003.2.13 ongoing	18:10 (A), 06:10 (D)	Daily	25×35 km^2	Radiometer	C band (6.9 GHz)	<2 cm	The Land Parameter Retrieval Model	the Goddard Earth Sciences Data and Information Services Center	[59,60]
	Soil Moisture and Ocean Salinity (SMOS)	2009.11.2 ongoing	06:00 (A), 18:00 (D)	Daily	25×25 km^2	Radiometer	L band (1.4 GHz)	<5 cm	The L-band Microwave Emission of the Biosphere model	ESA	[61]
	FengYun-3B (FY-3B)	2011.7.12–2019.8.19	13:40 (A), 01:40 (D)	Daily	25×25 km^2	the Microwave Radiation Imager	X band (10.65 GHz)	<2 cm	The Land Parameter Retrieval Model	China Meteorological Administration (CMA)	[62,63]
	FengYun-3C (FY-3C)	2014.5.29 ongoing	22:00 (A), 10:00 (D)	Daily	25×25 km^2	the Microwave Radiation Imager	X band (10.65 GHz)	<2 cm	The Land Parameter Retrieval Model	CMA	[64,65]
Passive (the active microwave scatterometer failed irreparably in July 2015.)	Soil Moisture Active Passive (SMAP)	2015.1.31 ongoing	18:00 (A), 06:00 (D)	Daily	36×36 km^2,	Radiometer	L band (1.4 GHz)	~5 cm	The H-pol Single Channel Algorithm, the V-pol Single Channel Algorithm, the Dual Channel Algorithm, Microwave Polarization Ratio Algorithm, and the Extended Dual Channel Algorithm	NASA	[66,67]

Specifically, active microwave-derived data have high spatial and low temporal resolution, although they are susceptible to surface roughness and vegetation cover. Comparatively, passive microwave-derived data often have high temporal resolution and low spatial resolution and can behave insensitively to surface roughness and vegetation cover. Additionally, both active and passive microwaves suffer from radio-frequency interference (RFI) [68,69]. Direct broadcast and communication satellites cause considerable RFI above the microwave band, which can be a critical reason for outliers and gap regions in satellite-retrieved SM products [70,71]. Basically, all single spaceborne microwave SM retrievals have large gap regions induced by RFI, dense vegetation coverage, veil of ice, and the relative motion between satellite revolution and Earth rotation [62], seriously impeding their spatiotemporal integrity.

Despite the enormous number of multi-source SM products mentioned above, scientific explorations and experiments pursuing high quality are ongoing. Attempts have mainly focused on improving the completeness, spatial representativeness, spatial resolution, and accuracy of currently accessible SM retrievals. Therefore, this review aims to provide an auxiliary reference for readers to understand the history and emerging trends of global SM retrieval methods.

2. Models to Improve the Quality of SM Products

2.1. Statistical Model

A statistical model can be established based on the significant statistical or empirical relationship between SM and land surface elements (such as surface temperature, vegetation index, evapotranspiration (ET), and albedo). These convenient and simple statistical models have been widely employed since inception and are mainly used for regional SM gap-filling and downscaling in terms of different research emphases [36,72–75]. Because of the variable coupling relationship along with various underlying surface hydrothermal features, the statistical model always has inter-regional applicability limitations. Furthermore, it is difficult to ensure the robustness and accuracy of statistical model-derived large-scale results.

2.1.1. Triangular (Tri)-Based Method

The Tri-based method can provide nonlinear solutions for SM estimation. Among the various statistical models, the Tri-based method is a classic method that estimates SM based on its close coupling relationship with land surface temperature (LST) and vegetation conditions [76–78]. Sandholt et al. [79] proposed a triangular feature space constructed using the LST and NDVI. The wet edge is composed of the lowest LST under different vegetation conditions, which indicates the maximum humidity. The dry edge, which indicates the minimum surface ET, is formed by the scatter of the highest LST under different NDVI values. As shown in Figure 1, if vegetation cover in a certain region ranges from bare soil to dense coverage and SM ranges from extreme drought to extreme humidity, the NDVI-LST scatter diagram is triangular in shape. A drought index, referred to as the TVDI, was defined and tightly linked to SM [80]. Then, a method was suggested to simulate SM using the combination of LST and NDVI based on the triangular feature space of TVDI. The Tri method equations are as follows.

$$Soil\ Moisture = a_{ij} \sum_{i=0}^{4} LST^{*i} \sum_{j=0}^{4} NDVI^{*j} \tag{1}$$

where a_{ij} is the correlation coefficient of every term in the polynomial, which is calculated using multiple regression. LST^* is calculated as follows:

$$LST^* = \frac{LST - LST_{min}}{LST_{max}(NDVI) - LST_{min}} \tag{2}$$

where $LST_{max}(NDVI)$, LST_{min} are the maximum and minimum values of the LST dataset calculated from NDVI, respectively. $NDVI^*$ is calculated as follows:

$$NDVI^* = \frac{NDVI - NDVI_{min}}{NDVI_{max} - NDVI_{min}} \tag{3}$$

where $NDVI_{max}$, $NDVI_{min}$ are the maximum and minimum values, respectively, of the NDVI dataset.

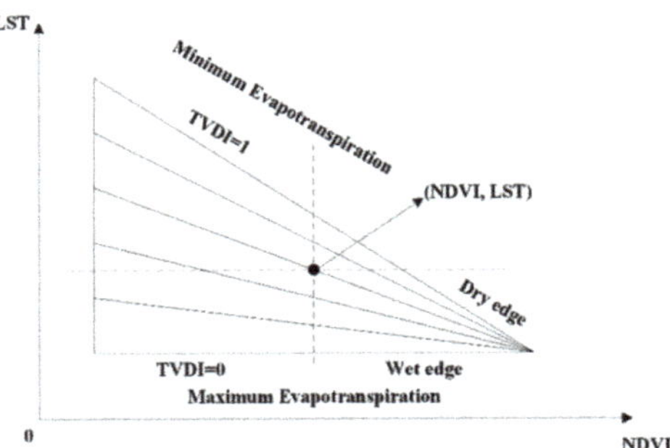

Figure 1. Triangular feature space of TVDI (figure reprinted from [72]).

Zhao et al. [36] systematically tested the performances of different vegetation indexes in the Tri model through a case study at the northeastern part of the Tibetan Plateau. The results demonstrated the advantage of NDVI in constructing the Tri model. The SM estimated by the NDVI-based model showed higher accuracy than those estimated by models constructed from the enhanced vegetation index (EVI) and soil-adjusted vegetation index (SAVI).

Many studies have attempted to estimate SM using the Tri method. The LST and NDVI datasets were acquired from high-resolution, remotely sensed products, and the established model could be effectively employed to improve the coarse-resolution SM [72,81–83]. Additionally, the Tri model neither requires any ancillary atmospheric data nor is it sensitive to atmospheric parameters. In general, this method is appropriate for flat regions with moderate vegetation coverage because NDVI is easily saturated in densely vegetated areas such as forests. This solution tends to exhibit better performance in regions with a single climate type and minimal artificial interference. Additionally, sufficient pixels are necessary to construct the "universal" triangular feature space. Sufficient pixels are also crucial for the accurate identification of wet and dry edges.

Apart from the classic vegetation and temperature combination, there are new approaches to parameterizing the Tri model. Shafian et al. [84] used thermal data and ground cover from Landsat imagery to establish the feature space to retrieve a perpendicular soil moisture index, which reduced the expense and complexity of the SM estimation. Sun [85] proposed a two-stage trapezoid to construct a feature space. This approach was established based on the theory that the vegetated surface temperature should vary after the bare soil surface temperature, as vegetation can absorb water from a deep soil layer to maintain transpiration. In addition, this two-stage method explicitly expresses the evolution of the feature space from a triangular to trapezoidal form.

2.1.2. Disaggregation Based on Physical and Theoretical Scale Change (DISPATCH) Algorithm

DISPATCH is another well-known and widely used algorithm capable of improving the spatial resolution of surface SM [86–90]. This approach was developed based on the tight interaction between surface SM and LST during the ET process. The DISPATCH method equation is a first-order Taylor series expansion and is expressed as follows [91]:

$$SM_D = SM_O + (\delta SEE/\delta SM)_O^{-1}(SEE_D - SEE_O) \tag{4}$$

where SM_D is the downscaled high-pixel resolution SM; SM_O is the original low-spatial resolution SM; SEE_D and SEE_O are the high- and low-resolution soil evaporative efficiency (SEE), respectively. $(\delta SEE/\delta SM)_O^{-1}$ is the inverse of the partial derivative of low-resolution SEE(SM). SEE is calculated as

$$SEE = ST_{max} - ST/ST_{max} - ST_{min} \tag{5}$$

where ST is the surface soil temperature. ST_{max} and ST_{min} correspond to the SM under extremely dry (SEE = 0) and extremely humid (SEE = 1) conditions, respectively. All ST were derived from the linear decomposition of LST into soil and vegetation using the following equation:

$$ST = LST - P_{veg}T_{veg}/1 - P_{veg} \tag{6}$$

where P_{veg} is the vegetation coverage percent, and T_{veg} is the vegetation temperature.

Merlin et al. [92] first proposed this algorithm and successively disaggregated the SMOS from 40 to 1 km with favorable accuracy. Then, they conducted a case study using DISPATCH to downscale SMOS SM in southeastern Australia [93]. This study found that the quality of the disaggregated product was good in summer and poor in winter. In addition, the coupling level in semi-arid areas was evidently stronger than that in temperate zones, and both vegetation coverage and vegetation water stress could influence ST retrieval. Hence, it is suggested that DISPATCH could perform better in low-vegetated semi-arid areas than in densely vegetated temperate regions. To enhance the disaggregation accuracy, Merlin et al. [94] designed a yearly SEE self-calibration model that could effectively make the DISPATCH algorithm more robust. This study proved the competence of DISPATCH in multi-scale SM downscaling through an evaluation study at 3 km and 100 m resolution in Spain. To extend the applicability of the DISPATCH approach, Ojha et al. [91] used TVDI instead of SEE in their model to include more densely vegetated areas. The results showed that the adoption of TVDI obviously increased the coverage percentage of the case study region, and the downscaled SM from the EVI-derived model displayed a higher correlation against in situ measurements than the one from the NDVI-derived model over vegetated areas. Apart from disaggregation, DISPATCH can also be utilized for coarse-resolution SM product evaluation [95].

2.2. Data Fusion

The data fusion method integrates multi-source remotely sensed data to produce SM estimations with higher accuracy, completeness, and reliability than the single satellite information source-retrieved ones. Through the fusion of multi-band, sensor working mode, and transit time remote sensing information, the quality of SM, including data accuracy, spatial coverage rate, temporal scope, and day-scale representativeness, can be efficiently improved. The Essential Climate Variable Soil Moisture (ECV SM), Soil Moisture Operational Product System (SMOPS), and Soil Moisture Active Passive (SMAP) are three well-known multiple microwave information-fused SM products. Because of their high performance in depicting soil water content conditions, they have received considerable attention since their inception.

(1) ECV SM

The ESA launched the ECV program, also known as the Climate Change Initiative, to monitor global climate evolution tendencies in 2010, and SM was simultaneously recognized as an ECV at the same time. The ECV SM, with global coverage, 0.25° pixel size, and daytime scale temporal resolution, was derived from the fusion of numerous satellite-based microwave products [96]. There are 13 versions available to the public to date, each updated with new sensors and an extended time series (https://esa-soilmoisture-cci.org/, accessed on 31 July 2022). Currently, the latest one is v07.1, which spans over 40 years from 1 November 1978 to 31 December 2021, combining information from 4 active and 12 passive microwave sensors. The ECV SM provides three SM estimations, which are derived from active sensors (ERS 1, ERS 2, Advanced Scatterometer on MetOp-A (ASCAT MetOp-A), and Advanced Scatterometer on MetOp-B (ASCAT MetOp-B)), passive sensors (Scanning Multichannel Microwave Radiometer (SMMR), Special Sensor Microwave Imager (SSM/I), TRMM TMI, Windsat/Coriolos, AMSR-E, AMSR-2, SMOS, FY-3B, FY-3C, FY-3D, SMAP, and GPM GMI), and their combinations.

The merging scheme of the ECV SM is described as follows: First, all the sensor retrievals are unified to a 0.25° grid and daily time stamps (00:00 UTC) through a hamming-window method and a nearest neighbor search. Then, the active estimation is retrieved using the TU Wien Water Retrieval Package, which is a change detection method to derive SM, as well as the official method to retrieve ASCAT L2 SM products [97]. Passive estimation is generated through the land parameter retrieval model, which is a forward model based on the radiative transfer model and has its own advantage of good frequency compatibility and a vegetation optical depth analytical solution [98]. The Global Land Data Assimilation System (GLDAS) Noah 2.1 was used for the active-passive combined estimation by offering a consistent climatology. The combined SM was finally derived through GLDAS Noah-based scaling, error characterization, and merging of each microwave sensor product. For more details about the merging algorithms of ECV SM and their evolutionary history, readers are referred to [99].

A number of studies have comprehensively and systematically evaluated the performance of ECV SM and almost consistently concluded that: (1) ECV SM expresses a good fitting degree to both ground observations and reanalysis products [100–102]. (2) The accuracy and robustness of ECV SM are steadily enhanced when the version is updated [99,103]. (3) Combined products are superior to the corresponding active and passive products [103,104]. (4) The spatiotemporal integrity and accuracy of the combined ECV SM display similar or better performances than each single microwave sensor retrieval [22,24].

(2) SMOPS

Although the ECV SM reveals a favorable capability in depicting land surface soil humidity conditions, the prevalent gap regions still hinder its spatial coverage integrity. The NOAA initiated the SMOPS program in 2012, which is dedicated to creating a global seamless SM product from accessible microwave satellite observations [105]. The first version of SMOPS-blended SMOS, ASCAT MetOp-A, and Windsat/Coriolos generated a 6 h and daily SM simultaneously. In 2016, the upgraded version 2 product with an extended time series introduced ASCAT MetOp-B and AMSR-2 into the system. Windsat/Coriolos were excluded. Both SMOPS V1.0 and V2.0 were generated using the single-channel retrieval algorithm, which could convert the brightness temperature of a single channel to emissivity [106]. The SM estimation can then be derived through the Fresnel equation by calculating the dielectric constant and dielectric mixing model. SMOPS V3.0, which contained 6 h and daily (00:00 UTC) SM products with a 0.25° grid, was developed in 2016, and SMAP was added to the blending system [107]. Moreover, a near real-time level-1 brightness temperature other than the officially released products was employed to satisfy the latency requirements.

SMOPS provides an almost seamless SM across the globe with high spatial coverage, which is a notable advantage compared to most satellite-based SM products. Small gap areas are mainly distributed in frozen (i.e., ice, snow) or dense vegetation-covered regions. Numerous studies have objectively assessed the quality of SMOPS and indicated that: (1) compared to the individual satellite-retrieved SM products, SMOPS exhibits much higher data availability; (2) the accuracy of SMOPS is continuously improved along with updated versions; (3) ECV shows higher accuracy, whereas SMOPS has superior spatial coverage [102,105,108].

(3) SMAP

Considering the merits of the L-band and fusion of active and passive microwave signals, NASA launched the SMAP program in 2010, utilizing L-band radar and radiometer instruments onboard the same spacecraft to detect surface SM conditions [66,67]. One of the main tasks of SMAP is to acquire an active and passive blended product to advance SM mapping by combining its strengths. Radar signals can achieve high pixel resolution; however, they are vulnerable to surface roughness and vegetation, which could significantly influence signal accuracy. In contrast, radiometer signals usually have coarse resolution, but they can be sensitive to SM and insensitive to surface roughness and vegetation. Therefore, the combined SMAP SM was expected to be capable of accurately expressing the surface soil water level with a relatively intermediate resolution. The brightness temperature disaggregation and time-series methods were used in the combination process. First, a linear relationship was established between variations in brightness temperature and radar backscatter using time-series approaches. This relationship was then employed to disaggregate brightness temperature. Finally, SM can be derived from the disaggregated brightness temperature and the corresponding retrieval algorithms. The 9 km combined SM has been validated by many scholars, and they found that it performed well in terms of fitting degree in the forested region [109]. However, on 7 July 2015, the radar failed irreparably after 3 months of operation. Although the time series of the combined SMAP SM product was only 86 days, it acted as a valuable precedent for SM merging using SMAP retrievals.

Many attempts have been made to renew the mission of generating a high-resolution SMAP SM product, and the signal from C-band SAR onboard Sentinel-1A/1B has been found to be an adequate substitute for the irreparable SMAP radar signal. By merging with Sentinel-1A/1B, a high-spatial-resolution SM product at 3 and 1 km has been generated. Meanwhile, the swath width of Sentinel-1A/1B is approximately 250 km, whereas that of SMAP can reach 1000 km. Because of this large difference, the overlap spatial coverage of SMAP and Sentinel is remarkably reduced, which then reduces the revisit interval from the original 3 days to 12 days. During the fusion process, the resampled 1 km Sentinel-1A/1B backscatter and the 9 km SMAP passive enhanced brightness temperature were input together as original data. The 1 km brightness temperature was obtained using the snapshot retrieval approach [110] on the overlapped area. Then, the high-resolution SM can be retrieved using the tau-omega model [111], together with the brightness temperature and ancillary datasets. For more details about the merging approaches of the SMAP/Sentinel SM product, readers can refer to [112]. Both 1 and 3 km resolution SMAP/Sentinel SM products have been validated against hundreds of in situ measurements, including dense and sparse networks across the globe. These encouraging results suggest that SMAP/Sentinel SM estimations could considerably match ground observations, demonstrating their capability to express soil water content with good accuracy and high resolution [112,113].

2.3. Assimilation and Reanalysis

The assimilation approach could effectively overcome the spatial scope and representativeness limitation of ground observations, overcome the depth limitation of spaceborne microwave-derived data, and achieve complete multi-depth coverage SM with definite physical meaning. It is efficient for the integration and improvement of SM from multiple independent sources [114]. Hence, spatial-temporal continuous SM profile information can be efficiently derived by assimilation systems [115,116]. The assimilation algorithm is an important part of the entire process that connects the observed and predicted data to optimize the estimation values. Commonly used SM assimilation methods include step-by-step correction [117], optimal interpolation [118], variational constraints [119], Kalman filters [120], and particle filters [121,122]. Recent studies note that deriving algorithms of filtering (i.e., ensemble Kalman filter) [123,124] and variational constraints (i.e., four-dimensional variational) [119,125] express favorable performance in estimating model parameters. As the central part of the assimilation process, the land surface model (LSM) simulates the physical processes occurring between the ground and atmosphere in the exchange of matter and energy. Many LSMs, such as Noah [126], the Community Land Model (CLM) [127], the Simple Biosphere Model [128], and the Boreal Ecosystem

Productivity Simulator [129], are frequently employed in the assimilation of land surface parameters (including SM).

Table 2 shows that many LSM-based SM estimations are released for various hydro-meteorological applications. It is worth noticing that the spatial extent of many LSM-based retrievals merely covers the specific nation or region the development organizations belong to, which remarkably restricts their scopes of application. In comparison, GLDAS, being one of the few global-scale assimilation systems, is well acknowledged as an eminent land surface modeling framework to produce optimal fields of land surface states and fluxes in near-real time across the world [130–133].

The SM profile information can also be retrieved from reanalysis approaches. The reanalysis process takes all available observations (i.e., ground- and spaceborne-based datasets) to calibrate the results from model running, whereas the assimilation process refers specifically to adding observation data for correction when the physical model is running. Many reanalysis retrievals have been released to simulate the global SM profile information (Table 2). ERA5 has attracted extensive attention since its advent as a fifth-generation reanalysis product of ECMWF. ERA5 is capable of generating higher spatial resolution (9 km) and temporal resolution (1 h for every atmospheric variable) retrievals than other reanalysis systems. In addition, it uses more satellite-based observations that are available to optimize the output results. Previous studies have revealed that the ERA5 SM exhibits higher skills than the other reanalysis products and a significant improvement over its predecessor [134], which may imply a promising application prospect for the ERA5 SM.

Table 2. A brief summary of assimilated and reanalyzed SM products and their basic properties.

Type	Program	LSM	Assimilation Algorithm	Spatial Extent	Spatial Resolution	Time Range	Temporal Resolution	Publisher	Detailed Information
Assimilation	Global Land Data Assimilation System (GLDAS)	Mosaic, CLM, Noah	Ensemble Kalman filter, extended Kalman filter, optimal interpolation	global	0.25° × 0.25°, 1° × 1°	1948.1.1 ongoing	3 h, 1 day, 1 month	NASA GSFC	[130]
	North American Land Data Assimilation System (NLDAS)	Mosaic, CLM, Noah	Ensemble Kalman filter, extended Kalman filter, optimal interpolation	67°W–125°W, 25°N–53°N	0.125° × 0.125°	1979.1.1 ongoing	1 h, 1 month	NASA GSFC	[135, 136]
	European Land Data Assimilation System (ELDAS)	Lokal Modell, ISBA and TERRA, TESSEL	Four-dimensional variational, Kalman filter, optimal interpolation	15°W–38°E, 35°N–72°N	0.2° × 0.2°, 1° × 1°	1999.10–2000.12	3 h, 1 day	The European Centre for Medium-Range Weather Forecasts (ECMWF)	[137, 138]
	CMA Land Data Assimilation System (CLDAS)	The Common Land Model, CLM, Noah	Three-dimensional variational, optimal interpolation	60°E–160°E, 0–65°N	0.0625° × 0.0625°	2012.1.1 ongoing	3 h, 1 day	CMA	[139, 140]
	Satellite Application Facility on Support to Operational Hydrology and Water Management (H SAF)	The Hydrology Tiled ECMWF Scheme for Surface Exchanges over Land	Four-dimensional variational	Global	1 km × 1 km; 12.5 km × 12.5 km; 25 km × 25 km;	2005 ongoing	1 day	European Organization for the Exploitation of Meteorological Satellites (EUMETSAT)	[26,141]
	The National Centers for Environmental Prediction/the National Center for Atmospheric Research (NCEP/NCAR)	The T62/28-level NCEP global operational spectral model	Three-dimensional variational, four-dimensional variational, optimal interpolation, SSI	Global	2.5° × 2.5°	1948.1.1 ongoing	6 h, 1 day, 1 month	The NOAA Earth System Research Laboratory Physical Sciences Laboratory	[142, 143]

Table 2. *Cont.*

Type	Program	LSM	Assimilation Algorithm	Spatial Extent	Spatial Resolution	Time Range	Temporal Resolution	Publisher	Detailed Information
Reanalysis	NCEP Climate Forecast System Reanalysis (CFSR)	NCEP Coupled Climate Forecast System Dynamical Model, the Seasonal Forecast Model	Three-dimensional variational, GSI	Global	0.5° × 0.5°, 2.5° × 2.5°	1979.1.1–2011.3.31	1 h, 6 h, 1 month	The NOAA National Centers for Environmental Information	[144]
	ECMWF Reanalysis v5 (ERA5)	Land-surface model (HTESSEL), ocean wave model	Four-dimensional variational	Global	9×9 km^2, 30×30 km^2	1950.1 ongoing	1 h, 1 day, 1 month	ECMWF	[145, 146]
	Modern Era Retrospective-Analysis for Research and Applications (MERRA)	The GEOS-5 atmospheric general circulation model	Three-dimensional variational, Gridpoint Statistical Interpolation (GSI)	Global	1/2° × 2/3°, 1.25° × 1.25°, 1° × 1.25°	1979–2016.2	1 h, 3 h, 6 h	NASA GSFC	[147]
	the Japan Meteorological Agency (JMA)	MRI/NPD unified non-hydrostatic model	Four-dimensional variational	Global	10×10 km^2	1958–2013	6 h, 1 day	The Japan Meteorological Agency	[148]
	CMA Reanalysis (CRA)	Noah	EnKF, three-dimensional variational	Global	~34×34 km^2	1979–2018	6 h	CMA	[149, 150]

2.4. Machine Learning

Recently, machine learning techniques have demonstrated great potential for simulating patterns and gaining insights into Earth's systems from scientific data. Machine-learning-based approaches exhibit notable competence in the simulation of nonlinear complex mapping relationships, such as SM. Machine learning algorithms are currently employed in SM estimation studies [151,152]. In terms of the different scale transition processes, the simulation can be divided into gap filling, downscaling, and upscaling (Figure 2). Gap filling means no scale transition during the entire simulation process, and the output estimations are dedicated to filling the gaps in the original SM products to improve spatial completeness. Great efforts have been made to downscale fields to acquire high pixel resolution SM estimations, which could depict regional SM spatial heterogeneity in detail and then be applied in the agricultural sector at the field scale. Comparatively, upscaling is usually dedicated to transferring point-scale in situ measurements to pixel-scale estimations, retrieving spatially continuous and representative SM products. Table 3 introduces the application of machine-learning methods to improve the performance of SM products. Meanwhile, an increasing number of published papers clearly state that machine-learning-based SM research is becoming a hot topic at present.

2.4.1. Traditional Machine Learning

Because of their greater ability in nonlinear and complex relationship simulations than traditional statistical regression methods, considerable attention has been devoted to using machine learning methodologies for enhancing SM products [7]. As shown in Table 3, several approaches, such as artificial neural networks (ANN), Bayesian, classification and regression trees (CART), extreme gradient boost (XGB), gradient boost decision trees (GBDT), K-nearest neighbor (KNN), random forest (RF), and support vector machine (SVM), are employed for both regional and global SM mapping [9,14,17,152–155]. Liu et al. [14] systematically compared the performance of six traditional machine learning approaches in surface SM downscaling from 0.25° to 1 km in four case study areas with different climates and land cover types. The results showed that the multi-regression tree-based RF achieved high performance with high goodness of fit and low regression bias, whereas the downscaled data from the ANN, CART, and SVM models occasionally showed abnormal values. Among the different case study regions, it was found that regions located in a single climate zone, with mild topographic variation and medium vegetation coverage tended to produce high-accuracy results. The contribution of each explanatory variable varied remarkably across the case study regions owing to their diverse complex hydrothermal and physical geographical conditions. On this basis, Liu et al. [154] further explored the capability of multiple regression tree-based machine learning algorithms to explicitly illuminate their characteristics in multi-scale surface SM disaggregation. Through inter-comparison among RF, GBDT, XGB, and CART, it was suggested that the best result was derived from GBDT in grasslands with a high correlation coefficient and low error, and both RF and XGB achieved favorable performances as well. Additionally, XGB was applied in multi-layer high-resolution SM estimation over the United States, and the downscaled SM favorably captured the temporal dynamics of in situ measurements with high accuracy [156]. The RF model was employed in a spatiotemporally continuous surface SM downscaling process at a field scale of 30 m resolution and displayed good performance in generating accurate SM estimations [9]. The GBDT algorithm was used for SM downscaling over the Tibetan Plateau and effectively improved the resolution of the SMAP SM from 36 to 1 km. High-resolution SM can preserve the accuracy of the original SMAP and express detailed spatial SM variability simultaneously [157]. Apart from the abovementioned studies, there is a host of research using multi-regression tree-derived machine learning methods to improve the resolution and spatial-temporal continuity of SM [65,158–161].

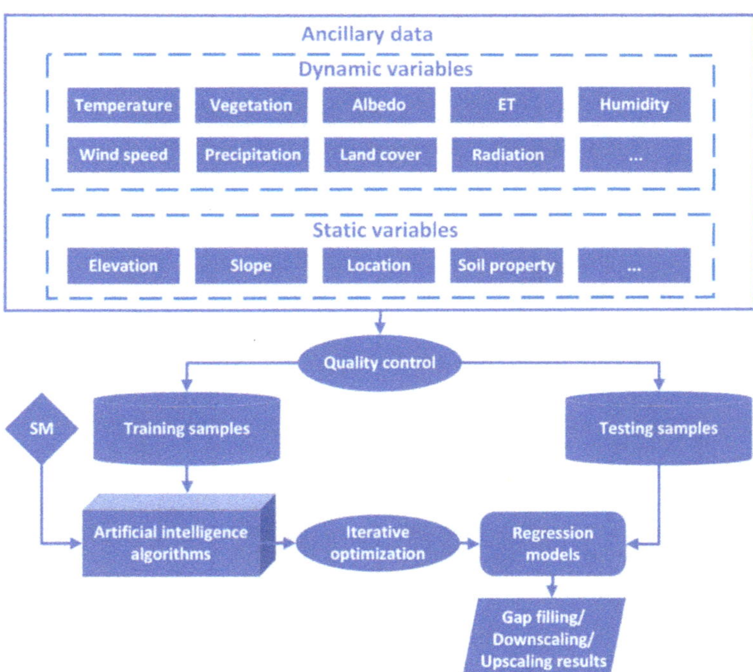

Figure 2. Flowchart of SM simulation using machine learning algorithms.

In general, great efforts have been made to clarify the performance of each member of the huge machine-learning family in simulating SM across various underlying surfaces. Among the numerous methodologies, multi-regression tree-derived approaches, such as RF, XGB, and GBDT, have revealed favorable capabilities in simulating and reconstructing SM products with good accuracy and fitting degree. Thus, this finding provides important guidance for the selection of machine learning methods in SM regression. Feature extraction, as a critical pre-processing step, could be very important in decreasing dimensionality and redundancy, increasing learning accuracy, and improving the understandability of results. However, for traditional machine-learning algorithms, the feature extraction and model training processes of classical machine-learning methods are two separate processes. The extracted features are used directly in subsequent calculations without any return adjustment, which results in error propagation. Under the joint action of climatic and human factors, the pattern of SM presents spatial-temporal distribution regularities. Classical machine learning methods only support the input of sample data in the form of discretization and rarely exploit the spatial-temporal dependencies of samples [162].

2.4.2. Deep Learning

In comparison, deep learning techniques are capable of constructing multi-layer neural networks by simulating the mechanism of the human brain, automatically extracting the spatial-temporal features of data, and then conducting spatial-temporal modeling and prediction based on deep understanding and mining [163–165]. Deep-learning methods can behave much better in learning high-dimensional features than classical machine-learning methods. A series of studies and applications have been carried out in the field of spatial data mining using deep learning methods, and relatively ideal results have been achieved in recent years [162]. Deep learning shows good potential for texture extraction and reconstruction. As presented in Table 3, many scholars have attempted to retrieve qualified SM estimations through deep learning algorithms, such as convolutional neural networks (CNN), gated recurrent units (GRU), long short-term memory (LSTM), deep feedforward neural networks (DFNN), and H2O models. Liu et al. [166] designed a novel LSTM-based multi-scale scheme for estimating surface SM by integrating remotely sensed data and in situ measurements over the United States. The model directly learned spatial-temporal patterns from in situ measurements, and the derived 9 km SM presented better accuracy than the 9 km products of the SMAP mission. This upscaling study revealed the significance of ground observations despite the availability of numerous satellite-retrieved products. Li et al. [167] tested the performance of CNN, LSTM, and ConvLSTM (a model integrating the merits of CNN and LSTM) in improving SMAP SM over China. The ERA5 SM information was transferred to SMAP to improve the prediction accuracy. The results illustrate that ConvLSTM outperformed CNN and LSTM in terms of a higher fitting degree and lower error. The transfer-based models exhibited better accuracy than the models without transfer learning, except in winter. ConvLSTM, combined with a physical model, was applied to estimate root-zone SM [168]. The GLDAS SM products were used as prediction data, and the spatiotemporal continuous root-zone SM derived from the physical model and in situ measurements were treated as target data. The estimated SM achieved high fitting coefficients compared with the original GLDAS SM, especially for the deep layers. Zhao et al. [169] investigated the capability of the deep belief network (DBN), improved DBN model, and residual network (ResNet) model in SM downscaling on the Tibetan Plateau. It was shown that the deep learning models had the advantage of fitting detailed SM texture patterns compared with RF. Compared to the DBN models, ResNet displayed an extraordinary ability to learn and simulate SM textures with high robustness.

The results and conclusions of these studies indicate that deep learning methods are suitable for SM simulations. Further, the well-designed deep learning model could outperform RF in SM estimation, suggesting the huge potential of deep learning methods in improving the quality of SM. The multiple deep learning algorithm-fused model usually behaved better than the single ones. In addition, because there are a number of algorithms inside the deep learning framework, more deep learning method-based explorations are necessary to determine comparatively eminent algorithms for SM estimation.

Table 3. A brief summary of machine learning algorithms utilized in enhancing the quality of SM products.

Type	Source	Algorithm	Target SM Data	Result	Conclusion
Gap filling	[170]	A two-layer machine learning-based framework	SMAP/Sentinel-1 SM product	3 km resolution SM estimations at four study regions (Arkansas, Arizona, Iowa, and Oklahoma) in a 3.5-year period between 1 April 2015 and 30 September 2018	The two-layer machine learning-based framework can reconstruct 3 km SM at gap regions with high fitting degree and low error
	[171]	Long Short-Term Memory (LSTM)	SMAP passive SM product	36 km resolution SM estimations in the continental United States from April 2015 to April 2017	The LSTM exhibit good spatial and temporal generalization capability in simulating SM
	[152]	RF	ECV active-passive combined SM product	Global gap-filled monthly ECV SM from January 2001 to December 2012	The gap-filled products achieve comparable performance as the original ECV
	[172]	Linear interpolation, cubic interpolation, SVM, and SVM combined with principal component analysis	ECV active-passive combined SM product	Gap-filled daily ECV SM in Southern Europe from 2003 to 2015	There are no substantial differences between the accuracy of the original and the SVM-reconstructed SM products
Downscaling	[153]	ANN, SVM, relevance vector machine, the generalized linear model	SMOS SM product	0.05° resolution SM estimations in southwestern England from February 2011 to January 2012	The ANN outperforms other algorithms in SM downscaling with higher accuracy
	[173]	RF	ECV active-passive combined SM product, CLDAS SM retrievals at 0–10 cm depth, in situ measurements	1 km resolution SM estimations at crop growth periods during 2015–2016 over Hebei Province, which is one of the major grain production regions of China	The RF model downscaled SM results display generally comparable and even better accuracy than the original SM products
	[14]	ANN, Bayesian, CART, KNN, RF, SVM	ECV active-passive combined SM product	1 km resolution SM estimations at four case study regions with quite different climate types and underlying surfaces across the globe	The RF model downscaled SM achieves excellent performance with a high correlation coefficient and a low regression error
	[154]	CART, GBDT, RF, XGBoost	SMAP passive and enhanced passive SM products	1 km resolution SM estimations in western Europe from January 2016 to December 2017	Multi-decision tree-derived RF, XGBoost, and GBDT could all achieve good performances in SM downscaling, and GBDT shows slight superiority to the other two methods
	[9]	RF	ECV, SMAP, and CLDAS SM products, in situ measurements	30 m resolution SM estimations during 1 March–31 October of years 2015, 2016, and 2017 in the Haihe Basin, which is one of the major grain production regions of China	The potential of the RF model is maximized to provide accurate and highly valuable SM estimations for hydrological research at the field scale
	[174]	H2O	AMSR-2 SM products, in situ measurements	4 km resolution SM estimations in the Korean Peninsula from 2014 to 2016	The H2O deep learning downscaled SM estimations indicate higher agreement with the in situ measurements than the original AMSR-2 and GLDAS SM products

Table 3. *Cont.*

Type	Source	Algorithm	Target SM Data	Result	Conclusion
	[175]	CNN, GRU	MODIS-based surface soil moisture data as a target variable obtained from the GLDAS 2.0 model	Forecasted SM in the Australian Murray Darling Basin on the 1st, 5th, 7th, 14th, 21st, and 30th day between 1 February 2003 and 31 March 2020	The CNN-GRU hybrid models are considerably superior in SM forecasting to standalone methods
Upscaling	[156]	XGBoost	Multi-layer in situ measurements (5, 10, 20, 50, and 100 cm depths), SMAP surface (0–5 cm), and root-zone (0–100 cm) SM products	1 km resolution SM retrievals at 5, 10, 20, 50, and 100 cm depths over the United States from 31 March 2015 to 29 February 2019	The multi-layer SM retrieyals well match the temporal dynamics of SM; the ubRMSE is less than 0.04 m^3/m^3 at most sites
	[176]	Bayesian	0–5 cm depth in situ measurements	100 km grid-box SM at 0–5 cm depth over the central Tibetan Plateau from 1 August 2010 to 20 September 2011	The upscaled SM revealed higher reliability and robustness compared to the point-scale data
	[177]	RF	In situ measurements from three networks, respectively	Gridded SM estimations with an approximate spatial resolution of 100 m at three networks located in North America	The RF model upscaled SM expresses high level matching degree against field samples and outperforms other common regression methods.
	[178]	DFNN	Top 10 cm in situ measurements at croplands	750 m resolution SM over the cropland of China on the 1st, 11th, and 21st day from May to October in 2012 to 2015	The deep learning model retrieved SM shows better accuracy than SMAP radar and GLDAS SM products

3. Applications

SM is a sensitive component of the Earth system that interacts with the atmosphere and Earth's surface at every moment. Although the in situ measured SM can precisely reflect the soil water content, the confined extent and point-scale value remarkably restrict its applicability. Moreover, the original remotely sensed SM can hardly provide high-resolution and spatial-temporal continuous SM records because of the inherent limitations of spaceborne microwave sensors. Comparatively, advanced SM products provide unprecedented opportunities for deriving datasets with improved spatial coverage, multi-depth information, high resolution, and extended time sequence from the 1950s to future scenarios. These multi-model improved SM products are broadly applied to advance the understanding of Earth system processes, which mainly include drought monitoring, climate change, hydrology, and ecology.

3.1. Drought Monitoring

Drought is usually induced by a deficiency of precipitation and excess ET, which jointly cause varying degrees of decline in SM. As drought can seriously affect crop growth and yield, agricultural departments have always attached great importance to real-time drought monitoring. Therefore, a wide variety of studies have explored the potential of SM for drought monitoring. First, for regions renowned for their advanced plant product industries, more ground stations could be arranged in cropland when establishing SM networks [18,179,180]. This arrangement style reflects the emphasis attached to cultivation-related drought monitoring by acquiring multi-depth SM recordings in real-time. Second, in regional- or national-scale drought forecasting studies, both in situ measurements and raster SM estimations are employed simultaneously to ensure data accuracy and spatial coverage [181–183]. Third, coarse-resolution SM products, retrieved from spaceborne sensors or LSMs, are mainly utilized for depicting large-scale (i.e., continental, global) drought characteristics [16,184]. In these studies, SM and other related auxiliary components, such as vegetation fraction, temperature, and precipitation, were used together in drought applications. These variables are co-converted to representative indices, such as the SM drought index [183], soil water deficit index [182], SM use efficiency [184], perpendicular drought index [16], modified perpendicular drought index [181], and enhanced combined drought index [185], to comprehensively indicate the duration, trend, intensity, and severity of drought conditions.

3.2. Climate Change

The Sixth Assessment Report of the Intergovernmental Panel on Climate Change was released in 2021 [186]. This unequivocally revealed a serious warning of unprecedented warming trends and increasingly frequent extreme weather events. Because every component inside the climate system constantly interacts with each other, the spatial and temporal patterns of SM are derived from the combined actions of all members. Consequently, SM products based on spaceborne sensors and LSMs have been widely used in climate-variability experiments and analyses. Dorigo et al. [187] evaluated the global trend in harmonized multi-satellite surface SM from 1988 to 2010 and found drying and wetting trends in different regions. Qiu et al. [188] compared the performance of satellite- and reanalysis-based SM products. The two types of products exhibit coincident patterns in non-irrigated areas. Moreover, the discrepancy was mainly induced by artificial interference such as irrigation and harvest. On the basis of ECV SM v4.2, Pan et al. [189] conducted seasonal and annual scale analysis, and the results revealed that "wet seasons get wetter, and dry seasons get dryer," proving the gradual extremity tendency. In addition to analyzing the evolutionary features of SM, integrated climate variability studies were carried out in terms of interactions and feedbacks between ET, temperature, precipitation, and SM [190–192].

3.3. Hydrology

SM plays an important role in the circulation of land–atmosphere hydrology and energy balance. It could "remember" exceptional signals from the land–atmosphere system and provide effective feedback to other components of the cycle, such as ET, precipitation, underground water, and runoff [193]. The Food and Agriculture Organization of the United Nations Irrigation and Drainage Paper No. 56 on crop Evapotranspiration listed SM availability as a key factor that could influence crop ET estimation [194]. Allam et al. [195] estimated evaporation over the upper Blue Nile Basin and used least-squares data assimilation methods to estimate soil water storage. SM datasets from the ECV, Climate Prediction Center, and Gravity Recovery and Climate Experiment terrestrial water storage were considered essential inputs during the assimilation procedure. The Global Land Evaporation Amsterdam Model v3 uses SM products retrieved from both spaceborne sensors (ECV and SMOS) and LSM (GLDAS Noah) to estimate terrestrial evaporation [196]. Previous studies have suggested a strong coupling between precipitation and SM [197,198]. By inverting the soil–water balance equation, an SM2RAIN algorithm was developed and used to estimate basin- and global-scale precipitation with satisfactory accuracy using in situ and satellite SM observations [199,200]. Swenson et al. [201] detected groundwater variability using in situ measurements in Oklahoma, U.S., and a time series of groundwater anomalies was successfully acquired after removing SM variability in the unsaturated zone. Additionally, remotely sensed SM has been proven capable of efficiently calibrating groundwater-land surface models [202]. Moreover, it is widely acknowledged that the spatial variability of SM and soil properties may have a dominant and complex impact on runoff in terms of changing storm size [203]. Therefore, multi-source SM products are widely utilized in advancing runoff models to help set the initial conditions and reduce prediction uncertainties [204,205].

3.4. Ecology

SM is a crucial regulator of the basic processes in terrestrial ecosystems. Its variability can remarkably impact the operational patterns of terrestrial ecosystems. SM can directly influence photosynthesis and the net primary productivity (NPP) of ecosystems by affecting the occurrence, intensity, and duration of vegetation water stress [96,206]. In addition, both nitrogen and carbon cycles are tightly linked to soil water movement [207]. Therefore, SM plays a significant role in ecosystem processes. Reich et al. [208] explicitly demonstrated the effect of SM on photosynthesis using in situ measurements. The results assumed that low SM may limit photosynthesis in boreal tree species during the growing season, despite warming temperatures. The impact of drought on NPP variability on a global scale was investigated, and a strong positive relationship between available moisture and NPP in arid and seasonally dry regions was demonstrated [209]. The SM balance was calculated using the Carnegie-Ames-Stanford approach and then converted to a water stress factor to express its impact on the NPP. In addition, dozens of global NPP estimation models have treated multi-depth SM (ranging from 0 to 2.5 m) as an important input parameter [210]. Li et al. [207] analyzed SM and other supplementary datasets from 1980 to 2015 in China's dryland derived from TerraClimate [211]. They found that water and soil conservation projects, such as reforestation, evidently increased the net primary production. However, SM continuously decreased, suggesting that the existing ecosystem was unlikely to be sustained. Satellite-derived SM together with related environmental drivers were employed to analyze the evaporation decline in the U.S. from 1961 to 2014, and a significant evaporation decrease of approximately 6% was detected [212].

4. Outlook

This study provides a brief introduction to the main types, deriving methodologies, quality-improving techniques, and applications of multi-source SM products. Generally, through development for more than half a century, great contributions and advancements have been made in SM acquisition and employment. However, to persistently enhance the performance and applicability of SM products, there is still a long way to go. Based on this review, we propose the following research priorities for future SM estimations.

4.1. Improved Spatial Coverage

Many studies employing SM as a key analysis object used seamless products to ensure complete coverage of the study area. Fortunately, assimilation- and reanalysis-based SM estimations have already overcome this problem in terms of the strength of numerous hydrological models. However, gap regions are prevalent for remotely sensed data. Owing to the limitation of microwave penetration, spaceborne sensors are unable to detect signals in frozen or dense vegetation ($\geq$5 kg/m^2)-covered regions. However, it is crucial to access spatial-temporal continuous SM over forests, which would enhance the understanding of the mechanisms by which forest structure affects soil water conditions. Forests have a significant impact on water movement in nature as well as the regulation of SM, precipitation, evaporation, runoff, and hydrological cycles. Unexpected RFI typically result in exceptional values. Moreover, the rotation difference between the satellite and the Earth could result in a strip-gap region. Hence, it is necessary to explore the capability of gap-filling methods (i.e., classical statistical algorithms and artificial approaches) and determine an adequate method to update the present products on the values of gap regions [72,171]. Data fusion is also an effective approach for improving spatial integrity by blending the quantities of qualified SM information. For example, the multi-source information-merged ECV and SMOPS SM products show an evidently higher coverage percentage than the single sensor-derived ones [102].

4.2. Higher Spatial Resolution

Compared to coarse-resolution SM products, fine-resolution SM products can be more appropriate for landscape scale, watershed scale, and field scale applications; for instance, hydrological simulation over the scale of drainage basins or SM spatial variability analysis on a field scale. Many studies have been conducted on SM downscaling using statistical models, data fusion, assimilation, and machine-learning algorithms. These works obtained good results by integrating high-resolution ancillary data collections from MODIS, Landsat, and Sentinel [11,14,113,176]. Moreover, machine learning approaches have notable advantages in terms of simplicity, efficiency, and competence. It was found that the multi-regression tree-based models could accurately reproduce SM with a downscaled resolution; however, these models did not consider spatial texture features. Comparatively, the advent of deep learning techniques provides an unparalleled opportunity for the simulation of spatially autocorrelated objects, such as SM. Therefore, it would be beneficial to develop a suitable model to estimate SM among the large deep-learning family [162]. In addition, high-resolution land surface observations from well-known optical sensors and SAR could serve as qualified explanatory variables for SM downscaling to hundreds or even dozens of meter grids [17,213].

4.3. Longer Time Span

It can be beneficial to analyze evolutionary trends over decades or even hundreds of years in climate change fields to capture the laws of climate origination and evolution. Thus, it is valuable that the time span of SM datasets can be continuously prolonged. Both satellite-based and assimilated SM products begin when the corresponding observation programs begin. For the sake of continuous acquisition of SM data, on the one hand, observations in existence should be maintained and ensured to work properly; on the other hand, new ground networks and satellites to provide continuous monitoring of SM are indispensable for extending time series. For instance, the National Satellite Meteorological Center of China launched the FY-3E satellite on 5th July 2021, which is dedicated to networking with FY-3C and FY-3D in orbit to observe SM and other meteorological parameters [214]. Additionally, forecasting SM with the help of future scenarios and hydrologic models could also provide access to acquire SM predictions, which may favor the investigation of future climate variations [190,215].

4.4. Higher Temporal Resolution

In addition to pursuing a high spatial resolution, improving the frequency would also be a key research priority for future SM products. Hourly monitoring data can be of great benefit in investigating subtle SM fluctuations induced by artificial irrigation, rainfall, and ET within a day, which is valuable for agricultural and land–atmosphere interaction applications [195,199,200,216]. At present, both in situ measurements and LSMs are capable of providing sub-hourly and sub-daily observations. Additionally, the SMAP publishes three-hourly surface and root zone SM estimates with ~2.5-day latency, which are derived from the assimilation of both ascending and descending brightness temperature data into the catchment LSM [217]. It is suggested that LSM is an effective and promising approach for generating high temporal resolution SM estimates. Furthermore, with an increasing number of satellites launched with different transit moments from each other, it would be promising to acquire observations more and more times per day across the globe [214].

4.5. Shorter Time Latency

It is imperative to access real-time or near-real-time SM recordings to conduct drought monitoring and early flood warning. Croplands also have high timeliness requirements for SM product availability to arrange irrigation or drainage without delay. In situ measurement data can be quickly collected through sensors and the internet. However, in terms of remotely sensed and assimilated products, there is always a latency of dozens of hours. For instance, the SMOPS data latency for 6-h products is 3 h and that for daily products is 6 h. The SMAP data latency for available data products is as follows: (1) Level 1 products, within 12 h of acquisition; (2) Level 2 products, within 24 h of acquisition; (3) Level 3 products, within 50 h of acquisition; and (4) Level 4 products, within 7 days for SM [129]. ERA5 is continuously updated with a latency of approximately 5 days [145]. Consequently, there is an urgent need to accelerate and optimize the processes of data transmission, algorithm operation, and data distribution, which should include, but not be limited to, the improvement of related equipment, techniques, and methodologies.

4.6. Developing Multi-Depth Products

Land surface and root-zone SM recordings are of equal importance for advancing the understanding of Earth's system processes. Furthermore, root-zone SM counts more than top-layer SM in vegetation growth. It is critical to develop multi-depth SM products to comprehensively master the soil wetness profile. In situ measurements can detect multi-depth SM using probes at different depths [18]. Assimilation- and reanalysis-based products can effectively describe soil water movement and then generate root-zone SM estimates to fulfill the requirements of considerably progressing hydrological and agricultural applications [145]. In addition, satellite-based programs have started to produce root-zone values through a data assimilation system. For instance, the SMAP project integrates its own observations with complementary information into an LSM and produces 3 h and 9 km surface (0–5 cm) and root-zone (0–100 cm) SM estimations through both spatial and temporal interpolations and extrapolations [66,218]. The ECV program also initiated a program to develop root-zone SM products using Noah-MP and ISBA LSMs, which are dedicated to linking vegetation phenology and biomass carbon allocation to moisture availability in the soil.

4.7. Higher Data Accuracy

Significant efforts have been devoted to reducing errors to continuously close the gap between SM estimations and real SM conditions. In a previous study, ground probes were periodically calibrated and maintained to ensure their operation under good conditions [18]. AMSR-2 retrieves SM using an X-band signal and applies a neighboring C-band to escape

RFI [58]. The SMAP program designed effective L-band SM detection sensors together with advanced anti-RFI devices and improved algorithms to detect and remove harmful interference in the L-band [68,219]. A series of developments in model physics, core dynamics, and data assimilation have been steadily achieved, which have contributed to significant improvements in SM consistency [145]. Despite this progress, there is still considerable room to pursue higher accuracy. Artificial intelligence-driven algorithms display great potential for simulating the SM model. Increasing SM datasets will become available as more ground networks and satellite programs are being planned. A significant benefit can be expected from combining these advanced technologies and datasets.

4.8. Better Model Performance and Interpretability

In recent decades, numerous models have been built and updated to estimate SM, and the overall quality of the corresponding products has been evidently enhanced. Traditional physical models are widely employed in spaceborne and assimilation systems to retrieve SM. These sophisticated and exquisite models are carefully designed and theoretically interpretable [145,146]. In comparison, artificial intelligence-driven approaches, especially the deep learning family, exhibit outstanding capabilities in SM regression and prediction [17,162]. In addition, they have the advantages of being highly efficient, simple, and convenient. However, their inner operational mechanisms are difficult to explain. Consequently, it could be favorable to develop hybrid models by combining physical and artificial intelligence methods, which would be able to exploit the strengths and discard the weaknesses of both methods. The hybrid model is expected to improve both model performance and interpretability.

5. Conclusions

Much attention has been paid to SM monitoring since ancient times. Before the existence of modern technology and equipment, subjective perceptions were prevalently employed to detect local SM conditions for proper irrigation arrangements. With the emergence of advanced probes, spaceborne sensors, and algorithms, spatial-temporal continuous SM records are becoming increasingly easily available. Because SM plays an important role in the land–atmosphere interaction system, vast amounts of multi-source SM datasets have been utilized in numerous studies on drought monitoring, climate change, ecology, and hydrology. However, the current status and characteristics of SM estimates should be clarified before they can be used in practical applications. The review of SM has generally been limited to certain retrieval algorithms, scale-conversion techniques, or applications. Therefore, there is an urgent need for a relatively comprehensive demonstration of advances in the quality of global SM products.

In this study, we introduce the primary retrieval methodologies of SM and the current approaches used to enhance the quality of SM products. Owing to the complex driving mechanism of its spatial-temporal distribution and evolution, great efforts have been made to advance retrieval methods. Numerous statistics, data fusion, assimilation, and machine learning-based approaches have been continuously designed and improved to enhance the reliability (including spatial-temporal completeness, resolution, and accuracy) of retrieved SM products. Although some of the established models are explainable, whereas others remain unexplainable in mechanism, they basically give renewed impetus to advancing the quality of SM estimations. In addition, a large quantity of SM-related original datasets and land–atmosphere parameters collected from different sensors, bands, and time nodes have been taken as ancillary references during the retrieval process to promote the reasonability of the response of SM to land–atmosphere variation.

Despite the steady progress in SM estimation models, there is still a large margin for improvement, such as pursuing higher spatial coverage, finer spatial resolution, longer time span, higher temporal resolution, shorter time latency, multi-depth products, higher data accuracy, and better model performance and interpretability. Moreover, it is critical to propose targeted solutions to mitigate the influences of various vegetation canopies and human activity interference, which could fundamentally improve the accuracy of spaceborne received signals and retrieved SM.

This review is expected to provide a reference for understanding the advances achieved in global SM estimation in terms of different approaches. Although many previous studies are referred to in this review, it could be difficult to include all publications on this topic. More complete research is necessary to contribute to the generalization of studies focused on SM in the future.

Author Contributions: Y.L.: conceptualization, methodology, writing—original draft preparation. Y.Y.: supervision, writing—reviewing and editing. All authors have read and agreed to the published version of the manuscript.

Funding: This research was jointly funded by the Strategic Priority Research Program of the Chinese Academy of Sciences (XDA28060400), the Second Tibetan Plateau Scientific Expedition and Research Program (2019QZKK09), the National Natural Science Foundation of China (42101475), the Special Program of Network Security and Informatization of Chinese Academy of Sciences (CAS-WX2021SF-0106-03), the Geographic Resources and Ecology Knowledge Service System of China Knowledge Center for Engineering Sciences and Technology (CKCEST-2015-1-4), the National Earth System Science Data Center (http://www.geodata.cn/, accessed on 31 July 2022), and the Guangzhou Science and Technology Plan Program (202102020676).

Data Availability Statement: Not applicable.

Acknowledgments: We appreciate the anonymous reviewers for their valuable comments and suggestions in improving this manuscript.

Conflicts of Interest: The authors declare no conflict of interest.

Acronyms

AMSR-2	the Advanced Microwave Scanning Radiometer 2
AMSR-E	the Advanced Microwave Scanning Radiometer for the Earth observing system
ANN	Artificial Neural Network
ASCAT MetOp-A	the Advanced Scatterometer on MetOp-A
ASCAT MetOp-B	the Advanced Scatterometer on MetOp-B
CART	Classification and Regression Trees
CFSR	the NCEP Climate Forecast System Reanalysis
CLDAS	the CMA Land Data Assimilation System
CMA	China Meteorological Administration
CRA	the CMA Reanalysis
CLM	the Community Land Model
CNN	Convolutional Neural Network
CYGNSS	Cyclone Global Navigation Satellite System
DEM	Digital Elevation Model
DBN	Deep Belief Network
DFNN	Deep Feedforward Neural Network
DISPATCH	Disaggregation based on physical and theoretical scale change
ECV SM	the Essential Climate Variable Soil Moisture
ECMWF	the European Centre for Medium-Range Weather Forecasts
ELDAS	the European Land Data Assimilation System
ENVISAT	the Environmental Satellite
ERA5	the ECMWF Reanalysis v5
ERS-1	the European Remote-Sensing Satellite-1
ERS-2	the European Remote-Sensing Satellite-2
ESA	the European Space Agency
ET	Evapotranspiration
EUMETSAT	European Organization for the Exploitation of Meteorological Satellites
EVI	enhanced vegetation index
FY-3B	FengYun-3B
FY-3C	FengYun-3C
GBDT	Gradient Boost Decision Tree
GLDAS	the Global Land Data Assimilation System
GPM GMI	the Global Precipitation Measurement Microwave Imager
GRU	Gated Recurrent Unit
GSFC	Goddard Space Flight Center

H SAF	Satellite Application Facility on Support to Operational Hydrology and Water Management
ISMN	the International Soil Moisture Network
JMA	the Japan Meteorological Agency
KNN	K Nearest Neighbor
LSM	the Land Surface Model
LST	land surface temperature
LSTM	Long Short Term Memory
MERRA	the Modern Era Retrospective-Analysis for Research and Applications
MODIS	the Moderate Resolution Imaging Spectroradiometer
NCEP/NCAR	the National Centers for Environmental Prediction/the National Center for Atmospheric Research
NDVI	Normalized Difference Vegetation Index
NLDAS	the North American Land Data Assimilation System
NPP	Net Primary Productivity
NSIDC	the National Snow and Ice Data Center
ResNet	Residual Network
RF	Random Forest
RFI	Radio Frequency Interference
SAR	Synthetic Aperture Radar
SAVI	Soil Adjusted Vegetation Index
SEE	the Soil Evaporative Efficiency
SM	Soil Moisture
SMMR	the Scanning Multichannel Microwave Radiometer
SMAP	the Soil Moisture Active Passive
SMOPS	the Soil Moisture Operational Product System
SMOS	the Soil Moisture and Ocean Salinity
SSM/I	the Special Sensor Microwave Imager
ST	Soil Temperature
SVM	Support Vector Machine
Tri	the Triangular-Based Method
TRMM TMI	the Tropical Rainfall Measuring Mission Microwave Imager
TVDI	Temperature Vegetation Drought Index
UTC	Coordinated Universal Time
XGB	Extreme Gradient Boost

References

1. Babaeian, E.; Sadeghi, M.; Jones, S.B.; Montzka, C.; Vereecken, H.; Tuller, M. Ground, proximal, and satellite remote sensing of soil moisture. *Rev. Geophys.* **2019**, *57*, 530–616. [CrossRef]
2. Li, Z.-L.; Leng, P.; Zhou, C.; Chen, K.-S.; Zhou, F.-C.; Shang, G.-F. Soil moisture retrieval from remote sensing measurements: Current knowledge and directions for the future. *Earth-Sci. Rev.* **2021**, *218*, 103673. [CrossRef]
3. Baatz, R.; Hendricks Franssen, H.; Euskirchen, E.; Sihi, D.; Dietze, M.; Ciavatta, S.; Fennel, K.; Beck, H.; De Lannoy, G.; Pauwels, V. Reanalysis in Earth system science: Toward terrestrial ecosystem reanalysis. *Rev. Geophys.* **2021**, *59*, e2020RG000715. [CrossRef]
4. Kornelsen, K.C.; Coulibaly, P. Advances in soil moisture retrieval from synthetic aperture radar and hydrological applications. *J. Hydrol.* **2013**, *476*, 460–489. [CrossRef]
5. Gruber, A.; Dorigo, W.A.; Zwieback, S.; Xaver, A.; Wagner, W. Characterizing Coarse-Scale Representativeness of in situ Soil Moisture Measurements from the International Soil Moisture Network. *Vadose Zone J.* **2013**, *12*, 522–525. [CrossRef]
6. Wang, L.; Qu, J.J. Satellite remote sensing applications for surface soil moisture monitoring: A review. *Front. Earth Sci. China* **2009**, *3*, 237–247. [CrossRef]
7. Ali, I.; Greifeneder, F.; Stamenkovic, J.; Neumann, M.; Notarnicola, C. Review of Machine Learning Approaches for Biomass and Soil Moisture Retrievals from Remote Sensing Data. *Remote Sens.* **2015**, *7*, 221–236. [CrossRef]
8. Li, Y.; Shu, H.; Mousa, B.; Jiao, Z. Novel Soil Moisture Estimates Combining the Ensemble Kalman Filter Data Assimilation and the Method of Breeding Growing Modes. *Remote Sens.* **2020**, *12*, 889. [CrossRef]
9. Abowarda, A.S.; Bai, L.; Zhang, C.; Long, D.; Li, X.; Huang, Q.; Sun, Z. Generating surface soil moisture at 30 m spatial resolution using both data fusion and machine learning toward better water resources management at the field scale. *Remote Sens. Environ.* **2021**, *255*, 112301. [CrossRef]
10. Llamas, R.; Guevara, M.; Rorabaugh, D.; Taufer, M.; Vargas, R. Spatial Gap-Filling of ESA CCI Satellite-Derived Soil Moisture Based on Geostatistical Techniques and Multiple Regression. *Remote Sens.* **2020**, *12*, 665. [CrossRef]
11. Sadeghi, M.; Babaeian, E.; Tuller, M.; Jones, S. The optical trapezoid model: A novel approach to remote sensing of soil moisture applied to Sentinel-2 and Landsat-8 observations. *Remote Sens. Environ.* **2017**, *198*, 52–68. [CrossRef]
12. Babaeian, E.; Sadeghi, M.; Franz, T.E.; Jones, S.; Tuller, M. Mapping soil moisture with the OPtical TRApezoid Model (OPTRAM) based on long-term MODIS observations. *Remote Sens. Environ.* **2018**, *211*, 425–440. [CrossRef]
13. Leng, P.; Song, X.; Duan, S.-B.; Li, Z.-L. A practical algorithm for estimating surface soil moisture using combined optical and thermal infrared data. *Int. J. Appl. Earth Obs. Geoinf.* **2016**, *52*, 338–348. [CrossRef]

14. Liu, Y.; Jing, W.; Wang, Q.; Xia, X. Generating high-resolution daily soil moisture by using spatial downscaling techniques: A comparison of six machine learning algorithms. *Adv. Water Resour.* **2020**, *141*, 103601. [CrossRef]
15. Walker, J.P.; Houser, P.R. A methodology for initializing soil moisture in a global climate model: Assimilation of near-surface soil moisture observations. *J. Geophys. Res. Atmos.* **2001**, *106*, 11761–11774. [CrossRef]
16. Sheffield, J.; Wood, E.F. Global Trends and Variability in Soil Moisture and Drought Characteristics, 1950–2000, from Observation-Driven Simulations of the Terrestrial Hydrologic Cycle. *J. Clim.* **2006**, *21*, 432–458. [CrossRef]
17. Peng, J.; Loew, A.; Merlin, O.; Verhoest, N.E.C. A review of spatial downscaling of satellite remotely sensed soil moisture. *Rev. Geophys.* **2017**, *55*, 341–366. [CrossRef]
18. Dorigo, W.A.; Wagner, W.; Hohensinn, R.; Hahn, S.; Paulik, C.; Xaver, A.; Gruber, A.; Drusch, M.; Mecklenburg, S.; Oevelen, P.V. The International Soil Moisture Network: A data hosting facility for global in situ soil moisture measurements. *Hydrol. Earth Syst. Sci.* **2011**, *15*, 1675–1698. [CrossRef]
19. Dorigo, W.A.; Xaver, A.; Vreugdenhil, M.; Gruber, A.; Hegyiová, A.; Sanchis-Dufau, A.D.; Zamojski, D.; Cordes, C.; Wagner, W.; Drusch, M. Global Automated Quality Control of In Situ Soil Moisture Data from the International Soil Moisture Network. *Vadose Zone J.* **2013**, *12*, 918–924. [CrossRef]
20. Paulik, C.; Dorigo, W.; Wagner, W.; Kidd, R. Validation of the ASCAT Soil Water Index using in situ data from the International Soil Moisture Network. *Int. J. Appl. Earth Obs. Geoinf.* **2014**, *30*, 1–8. [CrossRef]
21. Chen, Y.; Yang, K.; Qin, J.; Cui, Q.; Lu, H.; Zhu, L.; Han, M.; Tang, W. Evaluation of SMAP, SMOS, and AMSR2 soil moisture retrievals against observations from two networks on the Tibetan Plateau. *J. Geophys. Res. Atmos.* **2017**, *122*, 5780–5792. [CrossRef]
22. Ma, H.; Zeng, J.; Chen, N.; Zhang, X.; Cosh, M.H.; Wang, W. Satellite surface soil moisture from SMAP, SMOS, AMSR2 and ESA CCI: A comprehensive assessment using global ground-based observations. *Remote Sens. Environ.* **2019**, *231*, 111215. [CrossRef]
23. Griesfeller, A.; Lahoz, W.A.; De Jeu, R.A.M.; Dorigo, W.; Haugen, L.E.; Svendby, T.M.; Wagner, W. Evaluation of satellite soil moisture products over Norway using ground-based observations. *Int. J. Appl. Earth Obs. Geoinf.* **2016**, *45*, 155–164. [CrossRef]
24. Dorigo, W.A.; Gruber, A.; Jeu, R.A.M.D.; Wagner, W.; Stacke, T.; Loew, A.; Albergel, C.; Brocca, L.; Chung, D.; Parinussa, R.M. Evaluation of the ESA CCI soil moisture product using ground-based observations. *Remote Sens. Environ.* **2015**, *162*, 380–395. [CrossRef]
25. Beck, H.E.; Pan, M.; Miralles, D.G.; Reichle, R.H.; Dorigo, W.A.; Hahn, S.; Sheffield, J.; Karthikeyan, L.; Balsamo, G.; Parinussa, R.M. Evaluation of 18 satellite-and model-based soil moisture products using in situ measurements from 826 sensors. *Hydrol. Earth Syst. Sci.* **2021**, *25*, 17–40. [CrossRef]
26. Albergel, C.; De Rosnay, P.; Gruhier, C.; Muñoz-Sabater, J.; Hasenauer, S.; Isaksen, L.; Kerr, Y.; Wagner, W. Evaluation of remotely sensed and modelled soil moisture products using global ground-based in situ observations. *Remote Sens. Environ.* **2012**, *118*, 215–226. [CrossRef]
27. Chen, S.; Liu, Y.; Wen, Z. Satellite retrieval of soil moisture: An overview. *Adv. Earth Sci.* **2012**, *27*, 1192–1203.
28. Goward, S.N.; Xue, Y.; Czajkowski, K.P. Evaluating land surface moisture conditions from the remotely sensed temperature/vegetation index measurements: An exploration with the simplified simple biosphere model. *Remote Sens. Environ.* **2002**, *79*, 225–242. [CrossRef]
29. Quiring, S.M.; Ganesh, S. Evaluating the utility of the Vegetation Condition Index (VCI) for monitoring meteorological drought in Texas. *Agric. For. Meteorol.* **2010**, *150*, 330–339. [CrossRef]
30. Gu, Y.; Hunt, E.; Wardlow, B.; Basara, J.B.; Brown, J.F.; Verdin, J.P. Evaluation of MODIS NDVI and NDWI for vegetation drought monitoring using Oklahoma Mesonet soil moisture data. *Geophys. Res. Lett.* **2008**, *35*, 1092–1104. [CrossRef]
31. Patel, N.; Anapashsha, R.; Kumar, S.; Saha, S.; Dadhwal, V. Assessing potential of MODIS derived temperature/vegetation condition index (TVDI) to infer soil moisture status. *Int. J. Remote Sens.* **2009**, *30*, 23–39. [CrossRef]
32. Mallick, K.; Bhattacharya, B.K.; Patel, N. Estimating volumetric surface moisture content for cropped soils using a soil wetness index based on surface temperature and NDVI. *Agric. For. Meteorol.* **2009**, *149*, 1327–1342. [CrossRef]
33. Walker, J.P.; Houser, P.R.; Willgoose, G.R. Active microwave remote sensing for soil moisture measurement: A field evaluation using ERS-2. *Hydrol. Processes* **2004**, *18*, 1975–1997. [CrossRef]
34. Barrett, B.W.; Dwyer, E.; Whelan, P. Soil moisture retrieval from active spaceborne microwave observations: An evaluation of current techniques. *Remote Sens.* **2009**, *1*, 210–242. [CrossRef]
35. Wagner, W.; Dorigo, W.; de Jeu, R.; Fernandez, D.; Benveniste, J.; Haas, E.; Ertl, M. Fusion of Active and Passive Microwave Observations to Create AN Essential Climate Variable Data Record on Soil Moisture. *ISPRS Ann. Photogramm. Remote Sens. Spatial Inf. Sci.* **2012**, *I-7*, 315–321. [CrossRef]
36. Zhao, W.; Li, A.; Jin, H.; Zhang, Z.; Bian, J.; Yin, G. Performance evaluation of the triangle-based empirical soil moisture relationship models based on Landsat-5 TM data and in situ measurements. *IEEE Trans. Geosci. Remote Sens.* **2017**, *55*, 2632–2645. [CrossRef]
37. Carlson, T.N.; Perry, E.M.; Schmugge, T.J. Remote estimation of soil moisture availability and fractional vegetation cover for agricultural fields. *Agric. For. Meteorol.* **1990**, *52*, 45–69. [CrossRef]
38. Magagi, R.D.; Kerr, Y.H. Retrieval of soil moisture and vegetation characteristics by use of ERS-1 wind scatterometer over arid and semi-arid areas. *J. Hydrol.* **1997**, *188–189*, 361–384. [CrossRef]
39. Altese, E.; Bolognani, O.; Mancini, M.; Troch, P.A. Retrieving soil moisture over bare soil from ERS 1 synthetic aperture radar data: Sensitivity analysis based on a theoretical surface scattering model and field data. *Water Resour. Res.* **1996**, *32*, 653–661. [CrossRef]

40. Wang, C.; Qi, J.; Moran, S.; Marsett, R. Soil moisture estimation in a semiarid rangeland using ERS-2 and TM imagery. *Remote Sens. Environ.* **2004**, *90*, 178–189. [CrossRef]
41. Pathe, C.; Wagner, W.; Sabel, D.; Doubkova, M.; Basara, J.B. Using ENVISAT ASAR global mode data for surface soil moisture retrieval over Oklahoma, USA. *IEEE Trans. Geosci. Remote Sens.* **2009**, *47*, 468–480. [CrossRef]
42. Bartalis, Z.; Wagner, W.; Naeimi, V.; Hasenauer, S.; Scipal, K.; Bonekamp, H.; Figa, J.; Anderson, C. Initial soil moisture retrievals from the METOP-A Advanced Scatterometer (ASCAT). *Geophys. Res. Lett.* **2007**, *34*. [CrossRef]
43. Gruber, A.; Paloscia, S.; Santi, E.; Notarnicola, C.; Pasolli, L.; Smolander, T.; Pulliainen, J.; Mittelbach, H.; Dorigo, W.; Wagner, W. Performance inter-comparison of soil moisture retrieval models for the MetOp-A ASCAT instrument. In Proceedings of the 2014 IEEE Geoscience and Remote Sensing Symposium, Quebec City, QC, Canada, 13–18 July 2014; pp. 2455–2458.
44. Brocca, L.; Crow, W.T.; Ciabatta, L.; Massari, C.; De Rosnay, P.; Enenkel, M.; Hahn, S.; Amarnath, G.; Camici, S.; Tarpanelli, A. A review of the applications of ASCAT soil moisture products. *IEEE J. Sel. Top. Appl. Earth Obs. Remote Sens.* **2017**, *10*, 2285–2306. [CrossRef]
45. Chew, C.; Small, E. Description of the UCAR/CU soil moisture product. *Remote Sens.* **2020**, *12*, 1558. [CrossRef]
46. Aubert, M.; Baghdadi, N.; Zribi, M.; Douaoui, A.; Loumagne, C.; Baup, F.; El Hajj, M.; Garrigues, S. Analysis of TerraSAR-X data sensitivity to bare soil moisture, roughness, composition and soil crust. *Remote Sens. Environ.* **2011**, *115*, 1801–1810. [CrossRef]
47. Kseneman, M.; Gleich, D.; Potočnik, B. Soil-moisture estimation from TerraSAR-X data using neural networks. *Mach. Vis. Appl.* **2012**, *23*, 937–952. [CrossRef]
48. Baghdadi, N.; Aubert, M.; Zribi, M. Use of TerraSAR-X Data to Retrieve Soil Moisture Over Bare Soil Agricultural Fields. *IEEE Geosci. Remote Sens. Lett.* **2012**, *9*, 512–516. [CrossRef]
49. Balenzano, A.; Mattia, F.; Satalino, G.; Lovergine, F.P.; Palmisano, D.; Peng, J.; Marzahn, P.; Wegmüller, U.; Cartus, O.; Dąbrowska-Zielińska, K.; et al. Sentinel-1 soil moisture at 1 km resolution: A validation study. *Remote Sens. Environ.* **2021**, *263*, 112554. [CrossRef]
50. Balenzano, A.; Mattia, F.; Satalino, G.; Lovergine, F.P.; Palmisano, D.; Davidson, M.W.J. Dataset of Sentinel-1 surface soil moisture time series at 1 km resolution over Southern Italy. *Data Brief* **2021**, *38*, 107345. [CrossRef]
51. Owe, M.; de Jeu, R.; Holmes, T. Multisensor historical climatology of satellite-derived global land surface moisture. *J. Geophys. Res. Earth Surf.* **2008**, *113*. [CrossRef]
52. Reichle, R.H.; Koster, R.D.; Liu, P.; Mahanama, S.P.; Njoku, E.G.; Owe, M. Comparison and assimilation of global soil moisture retrievals from the Advanced Microwave Scanning Radiometer for the Earth Observing System (AMSR-E) and the Scanning Multichannel Microwave Radiometer (SMMR). *J. Geophys. Res. Atmos.* **2007**, *112*. [CrossRef]
53. Ridder, K.D. Surface soil moisture monitoring over Europe using Special Sensor Microwave/Imager (SSM/I) imagery. *J. Geophys. Res.* **2003**, *108*, 4422. [CrossRef]
54. Drusch, M.; Wood, E.F.; Gao, H. Observation operators for the direct assimilation of TRMM microwave imager retrieved soil moisture. *Geophys. Res. Lett.* **2005**, *32*. [CrossRef]
55. Njoku, E.G.; Jackson, T.J.; Lakshmi, V.; Chan, T.K.; Nghiem, S.V. Soil moisture retrieval from AMSR-E. *IEEE Trans. Geosci. Remote Sens.* **2003**, *41*, 215–229. [CrossRef]
56. Brocca, L.; Hasenauer, S.; Lacava, T.; Melone, F.; Moramarco, T.; Wagner, W.; Dorigo, W.; Matgen, P.; Martínez-Fernández, J.; Llorens, P. Soil moisture estimation through ASCAT and AMSR-E sensors: An intercomparison and validation study across Europe. *Remote Sens. Environ.* **2011**, *115*, 3390–3408. [CrossRef]
57. Bindlish, R.; Cosh, M.H.; Jackson, T.J.; Koike, T.; Fujii, H.; Chan, S.K.; Asanuma, J.; Berg, A.; Bosch, D.D.; Caldwell, T. GCOM-W AMSR2 Soil Moisture Product Validation Using Core Validation Sites. *IEEE J. Sel. Top. Appl. Earth Obs. Remote Sens.* **2018**, *11*, 209–219. [CrossRef]
58. Parinussa, R.M.; Holmes, T.R.H.; Wanders, N.; Dorigo, W.A.; De Jeu, R.A.M. A Preliminary Study toward Consistent Soil Moisture from AMSR2. *J. Hydrometeorol.* **2013**, *16*, 932–947. [CrossRef]
59. Li, L.; Gaiser, P.W.; Gao, B.C.; Bevilacqua, R.M.; Jackson, T.J.; Njoku, E.G.; Rudiger, C.; Calvet, J.C.; Bindlish, R. WindSat Global Soil Moisture Retrieval and Validation. *IEEE Trans. Geosci. Remote Sens.* **2010**, *48*, 2224–2241. [CrossRef]
60. Gaiser, P.W.; St Germain, K.M.; Twarog, E.M.; Poe, G.A.; Purdy, W.; Richardson, D.; Grossman, W.; Jones, W.L.; Spencer, D.; Golba, G. The WindSat spaceborne polarimetric microwave radiometer: Sensor description and early orbit performance. *IEEE Trans. Geosci. Remote Sens.* **2004**, *42*, 2347–2361. [CrossRef]
61. Al Bitar, A.; Leroux, D.; Kerr, Y.H.; Merlin, O.; Richaume, P.; Sahoo, A.; Wood, E.F. Evaluation of SMOS soil moisture products over continental US using the SCAN/SNOTEL network. *IEEE Trans. Geosci. Remote Sens.* **2012**, *50*, 1572–1586. [CrossRef]
62. Liu, Y.; Zhou, Y.; Lu, N.; Tang, R.; Liu, N.; Li, Y.; Yang, J.; Jing, W.; Zhou, C. Comprehensive assessment of Fengyun-3 satellites derived soil moisture with in-situ measurements across the globe. *J. Hydrol.* **2021**, *594*, 125949. [CrossRef]
63. Parinussa, R.M.; Wang, G.; Holmes, T.R.H.; Liu, Y.Y.; Dolman, A.J.; Jeu, R.A.M.D.; Jiang, T.; Zhang, P.; Shi, J. Global surface soil moisture from the Microwave Radiation Imager onboard the Fengyun-3B satellite. *Int. J. Remote Sens.* **2014**, *35*, 7007–7029. [CrossRef]
64. Wu, S.; Chen, J. Instrument Performance and Cross Calibration of FY-3C MWRI. In Proceedings of the 2016 IEEE International Geoscience and Remote Sensing Symposium (IGARSS), Beijing, China, 10–15 July 2016; pp. 388–391.
65. Zhang, S.; Weng, F.; Yao, W. A Multivariable Approach for Estimating Soil Moisture from Microwave Radiation Imager (MWRI). *J. Meteorol. Res.* **2020**, *34*, 732–747. [CrossRef]

66. Entekhabi, D.; Njoku, E.G.; O'Neill, P.E.; Kellogg, K.H.; Crow, W.T.; Edelstein, W.N.; Entin, J.K.; Goodman, S.D.; Jackson, T.J.; Johnson, J.; et al. The Soil Moisture Active Passive (SMAP) Mission. *Proc. IEEE* **2010**, *98*, 704–716. [CrossRef]
67. Chan, S.K.; Bindlish, R.; O'Neill, P.E.; Njoku, E.; Jackson, T.; Colliander, A.; Fan, C.; Burgin, M.; Dunbar, S.; Piepmeier, J. Assessment of the SMAP Passive Soil Moisture Product. *IEEE Trans. Geosci. Remote Sens.* **2016**, *54*, 4994–5007. [CrossRef]
68. Piepmeier, J.R.; Johnson, J.T.; Mohammed, P.N.; Bradley, D.; Ruf, C.; Aksoy, M.; Garcia, R.; Hudson, D.; Miles, L.; Wong, M. Radio-Frequency Interference Mitigation for the Soil Moisture Active Passive Microwave Radiometer. *IEEE Trans. Geosci. Remote Sens.* **2014**, *52*, 761–775. [CrossRef]
69. Lacava, T.; Faruolo, M.; Pergola, N.; Coviello, I.; Tramutoli, V. A comprehensive analysis of AMSRE C- and X-bands Radio Frequency Interferences. In Proceedings of the Microwave Radiometry and Remote Sensing of the Environment, Rome, Italy, 5–9 March 2012; pp. 1–4.
70. Draper, D.W. Radio Frequency Environment for Earth-Observing Passive Microwave Imagers. *IEEE J. Sel. Top. Appl. Earth Obs. Remote Sens.* **2018**, *11*, 1913–1922. [CrossRef]
71. Zou, X.; Zhao, J.; Weng, F.; Qin, Z. Detection of Radio-Frequency Interference Signal Over Land From FY-3B Microwave Radiation Imager (MWRI). *Adv. Meteorol. Sci. Technol.* **2013**, *50*, 4994–5003. [CrossRef]
72. Liu, Y.; Yao, L.; Jing, W.; Di, L.; Yang, J.; Li, Y. Comparison of two satellite-based soil moisture reconstruction algorithms: A case study in the state of Oklahoma, USA. *J. Hydrol.* **2020**, *590*, 125406. [CrossRef]
73. Mohseni, F.; Mokhtarzade, M. A new soil moisture index driven from an adapted long-term temperature-vegetation scatter plot using MODIS data. *J. Hydrol.* **2020**, *581*, 124420. [CrossRef]
74. Yang, J.; Zhang, D. Soil moisture estimation with a remotely sensed dry edge determination based on the land surface temperature-vegetation index method. *J. Appl. Remote Sens.* **2019**, *13*, 024511. [CrossRef]
75. Peng, J.; Loew, A.; Zhang, S.; Wang, J.; Niesel, J. Spatial downscaling of satellite soil moisture data using a vegetation temperature condition index. *IEEE Trans. Geosci. Remote Sens.* **2016**, *54*, 558–566. [CrossRef]
76. Holzman, M.; Rivas, R.; Piccolo, M. Geoinformation. Estimating soil moisture and the relationship with crop yield using surface temperature and vegetation index. *Int. J. Appl. Earth Obs. Geoinf.* **2014**, *28*, 181–192.
77. Chen, J.; Wang, C.; Jiang, H.; Mao, L.; Yu, Z. Estimating soil moisture using Temperature–Vegetation Dryness Index (TVDI) in the Huang-huai-hai (HHH) plain. *Int. J. Remote Sens.* **2011**, *32*, 1165–1177. [CrossRef]
78. Yuan, L.; Li, L.; Zhang, T.; Chen, L.; Zhao, J.; Hu, S.; Cheng, L.; Liu, W. Soil moisture estimation for the Chinese Loess Plateau using MODIS-derived ATI and TVDI. *Remote Sens.* **2020**, *12*, 3040. [CrossRef]
79. Sandholt, I.; Rasmussen, K.; Andersen, J. A simple interpretation of the surface temperature/vegetation index space for assessment of surface moisture status. *Remote Sens. Environ.* **2002**, *79*, 213–224. [CrossRef]
80. Carlson, T. An Overview of the Triangle Method for Estimating Surface Evapotranspiration and Soil Moisture from Satellite Imagery. *Sensors* **2007**, *7*, 1612–1629. [CrossRef]
81. Rahmati, M.; Oskouei, M.M.; Neyshabouri, M.R.; Walker, J.P.; Fakherifard, A.; Ahmadi, A.; Mousavi, S.B. Soil moisture derivation using triangle method in the lighvan watershed, north western Iran. *J. Soil Sci. Plant Nutr.* **2015**, *15*, 167–178. [CrossRef]
82. Tang, R.; Li, Z.-L.; Tang, B. An application of the Ts–VI triangle method with enhanced edges determination for evapotranspiration estimation from MODIS data in arid and semi-arid regions: Implementation and validation. *Remote Sens. Environ.* **2010**, *114*, 540–551. [CrossRef]
83. Zhang, F.; Zhang, L.W.; Shi, J.J.; Huang, J.F. Soil Moisture Monitoring Based on Land Surface Temperature-Vegetation Index Space Derived from MODIS Data. *Pedosphere* **2014**, *24*, 450–460. [CrossRef]
84. Shafian, S.; Maas, S.J. Index of soil moisture using raw Landsat image digital count data in Texas high plains. *Remote Sens.* **2015**, *7*, 2352–2372. [CrossRef]
85. Sun, H. Two-Stage Trapezoid: A New Interpretation of the Land Surface Temperature and Fractional Vegetation Coverage Space. *IEEE J. Sel. Top. Appl. Earth Obs. Remote Sens.* **2016**, *9*, 336–346. [CrossRef]
86. Merlin, O.; Malbéteau, Y.; Notfi, Y.; Bacon, S.; Er-Raki, S.; Khabba, S.; Jarlan, L. Performance metrics for soil moisture downscaling methods: Application to DISPATCH data in central Morocco. *Remote Sens.* **2015**, *7*, 3783–3807. [CrossRef]
87. Djamai, N.; Magagi, R.; Goïta, K.; Merlin, O.; Kerr, Y.; Roy, A. A combination of DISPATCH downscaling algorithm with CLASS land surface scheme for soil moisture estimation at fine scale during cloudy days. *Remote Sens. Environ.* **2016**, *184*, 1–14. [CrossRef]
88. Fontanet, M.; Fernàndez-Garcia, D.; Ferrer, F. The value of satellite remote sensing soil moisture data and the DISPATCH algorithm in irrigation fields. *Hydrol. Earth Syst. Sci.* **2018**, *22*, 5889–5900. [CrossRef]
89. Dumedah, G.; Walker, J.P.; Merlin, O. Root-zone soil moisture estimation from assimilation of downscaled Soil Moisture and Ocean Salinity data. *Adv. Water Resour.* **2015**, *84*, 14–22. [CrossRef]
90. Malbéteau, Y.; Merlin, O.; Balsamo, G.; Er-Raki, S.; Khabba, S.; Walker, J.; Jarlan, L. Toward a surface soil moisture product at high spatiotemporal resolution: Temporally interpolated, spatially disaggregated SMOS data. *J. Hydrometeorol.* **2018**, *19*, 183–200. [CrossRef]
91. Ojha, N.; Merlin, O.; Suere, C.; Escorihuela, M.J. Extending the Spatio-Temporal Applicability of DISPATCH Soil Moisture Downscaling Algorithm: A Study Case Using SMAP, MODIS and Sentinel-3 Data. *Front. Environ. Sci.* **2021**, *9*, 555216. [CrossRef]
92. Merlin, O.; Chehbouni, A.G.; Kerr, Y.H.; Njoku, E.G.; Entekhabi, D. A combined modeling and multispectral/multiresolution remote sensing approach for disaggregation of surface soil moisture: Application to SMOS configuration. *IEEE Trans. Geosci. Remote Sens.* **2005**, *43*, 2036–2050. [CrossRef]

93. Merlin, O.; Rudiger, C.; Bitar, A.A.; Richaume, P.; Walker, J.P.; Kerr, Y.H. Disaggregation of SMOS Soil Moisture in Southeastern Australia. *IEEE Trans. Geosci. Remote Sens.* **2012**, *50*, 1556–1571. [CrossRef]
94. Merlin, O.; Escorihuela, M.J.; Mayoral, M.A.; Hagolle, O.; Al Bitar, A.; Kerr, Y. Self-calibrated evaporation-based disaggregation of SMOS soil moisture: An evaluation study at 3km and 100m resolution in Catalunya, Spain. *Remote Sens. Environ.* **2013**, *130*, 25–38. [CrossRef]
95. Malbéteau, Y.; Merlin, O.; Molero, B.; Rüdiger, C.; Bacon, S. DisPATCh as a tool to evaluate coarse-scale remotely sensed soil moisture using localized in situ measurements: Application to SMOS and AMSR-E data in Southeastern Australia. *Int. J. Appl. Earth Obs. Geoinf.* **2016**, *45*, 221–234. [CrossRef]
96. Dorigo, W.; Wagner, W.; Albergel, C.; Albrecht, F.; Balsamo, G.; Brocca, L.; Chung, D.; Ertl, M.; Forkel, M.; Gruber, A. ESA CCI Soil Moisture for improved Earth system understanding: State-of-the art and future directions. *Remote Sens. Environ.* **2017**, *203*, 185–215. [CrossRef]
97. Naeimi, V.; Scipal, K.; Bartalis, Z.; Hasenauer, S.; Wagner, W. An improved soil moisture retrieval algorithm for ERS and METOP scatterometer observations. *IEEE Trans. Geosci. Remote Sens.* **2009**, *47*, 1999–2013. [CrossRef]
98. De Jeu, R.A.M.; Holmes, T.R.H.; Panciera, R.; Walker, J.P. Parameterization of the Land Parameter Retrieval Model for L-Band Observations Using the NAFE'05 Data Set. *IEEE Geosci. Remote Sens. Lett.* **2009**, *6*, 630–634. [CrossRef]
99. Gruber, A.; Scanlon, T.; van der Schalie, R.; Wagner, W.; Dorigo, W. Evolution of the ESA CCI Soil Moisture climate data records and their underlying merging methodology. *Earth Syst. Sci. Data* **2019**, *11*, 717–739. [CrossRef]
100. Wang, S.; Mo, X.; Liu, S.; Lin, Z.; Hu, S. Validation and trend analysis of ECV soil moisture data on cropland in North China Plain during 1981–2010. *Int. J. Appl. Earth Obs. Geoinf.* **2016**, *48*, 110–121. [CrossRef]
101. McNally, A.; Shukla, S.; Arsenault, K.R.; Wang, S.; Peters-Lidard, C.D.; Verdin, J.P. Evaluating ESA CCI soil moisture in East Africa. *Int. J. Appl. Earth Obs. Geoinf.* **2016**, *48*, 96–109. [CrossRef]
102. Wang, Y.; Leng, P.; Peng, J.; Marzahn, P.; Ludwig, R. Global assessments of two blended microwave soil moisture products CCI and SMOPS with in-situ measurements and reanalysis data. *Int. J. Appl. Earth Obs. Geoinf.* **2021**, *94*, 102234. [CrossRef]
103. González-Zamora, Á.; Sánchez, N.; Pablos, M.; Martínez-Fernández, J. CCI soil moisture assessment with SMOS soil moisture and in situ data under different environmental conditions and spatial scales in Spain. *Remote Sens. Environ.* **2018**, *225*, 469–482. [CrossRef]
104. Liu, Y.; Yang, Y.; Yue, X. Evaluation of Satellite-Based Soil Moisture Products over Four Different Continental In-Situ Measurements. *Remote Sens.* **2018**, *10*, 1161. [CrossRef]
105. Liu, J.; Zhan, X.; Hain, C.; Yin, J.; Fang, L.; Li, Z.; Zhao, L. NOAA Soil Moisture Operational Product System (SMOPS) and Its Validations. In Proceedings of the 2016 IEEE International Geoscience and Remote Sensing Symposium (IGARSS), Beijing, China, 10–15 July 2016; pp. 3477–3480.
106. Jackson, T.J., III. Measuring surface soil moisture using passive microwave remote sensing. *Hydrol. Processes* **1993**, *7*, 139–152. [CrossRef]
107. Yin, J.; Zhan, X.; Liu, J. NOAA Satellite Soil Moisture Operational Product System (SMOPS) Version 3.0 Generates Higher Accuracy Blended Satellite Soil Moisture. *Remote Sens.* **2020**, *12*, 2861. [CrossRef]
108. Yin, J.; Zhan, X.; Liu, J.; Schull, M. An intercomparison of Noah model skills with benefits of assimilating SMOPS blended and individual soil moisture retrievals. *Water Resour. Res.* **2019**, *55*, 2572–2592. [CrossRef]
109. Pan, M.; Cai, X.; Chaney, N.; Entekhabi, D.; Wood, E.F. An Initial Assessment of SMAP Soil Moisture Retrievals Using High Resolution Model Simulations and In-situ Observations: SMAP Comparisons. *Geophys. Res. Lett.* **2016**, *43*, 9662–9668. [CrossRef]
110. Jagdhuber, T.; Entekhabi, D.; Das, N.N.; Link, M.; Baur, M.; Akbar, R.; Montzka, C.; Kim, S.; Yueh, S.; Baris, I. Physics-based modeling of active-passive microwave covariations for geophysical retrievals. In Proceedings of the IGARSS 2018–2018 IEEE International Geoscience and Remote Sensing Symposium, Valencia, Spain, 22–27 July 2018; pp. 250–253.
111. Mo, T.; Choudhury, B.; Schmugge, T.; Wang, J.R.; Jackson, T. A model for microwave emission from vegetation-covered fields. *J. Geophys. Res. Ocean.* **1982**, *87*, 11229–11237. [CrossRef]
112. Das, N.N.; Entekhabi, D.; Dunbar, R.S.; Chaubell, M.J.; Colliander, A.; Yueh, S.; Jagdhuber, T.; Chen, F.; Crow, W.; O'Neill, P.E. The SMAP and Copernicus Sentinel 1A/B microwave active-passive high resolution surface soil moisture product. *Remote Sens. Environ.* **2019**, *233*, 111380. [CrossRef]
113. Kim, H.; Lee, S.; Cosh, M.H.; Lakshmi, V.; Kwon, Y.; McCarty, G.W. Assessment and Combination of SMAP and Sentinel-1A/B-Derived Soil Moisture Estimates With Land Surface Model Outputs in the Mid-Atlantic Coastal Plain, USA. *IEEE Trans. Geosci. Remote Sens.* **2020**, *59*, 991–1011. [CrossRef]
114. Draper, C.; Reichle, R.; De Lannoy, G.; Liu, Q. Assimilation of passive and active microwave soil moisture retrievals. *Geophys. Res. Lett.* **2012**, *39*. [CrossRef]
115. Bi, H.; Ma, J.; Zheng, W.; Zeng, J. Comparison of soil moisture in GLDAS model simulations and in situ observations over the Tibetan Plateau. *J. Geophys. Res. Atmos.* **2016**, *121*, 2658–2678. [CrossRef]
116. Kumar, S.V.; Reichle, R.H.; Koster, R.D.; Crow, W.T.; Peters-Lidard, C.D. Role of subsurface physics in the assimilation of surface soil moisture observations. *J. Hydrometeorol.* **2009**, *10*, 1534–1547. [CrossRef]
117. Weisse, A.; Michel, C.; Aubert, D.; Loumagne, C. *Assimilation of Soil Moisture in a Hydrological Model for Flood Forecasting*; IAHS Publication: Maastricht, The Netherlands, 2001; pp. 249–256.

118. Bouttier, F.; Mahfouf, J.; Noilhan, J. Sequential assimilation of soil moisture from atmospheric low-level parameters. Part I: Sensitivity and calibration studies. *J. Appl. Meteorol. Climatol.* **1993**, *32*, 1335–1351. [CrossRef]
119. Reichle, R.H.; Entekhabi, D.; McLaughlin, D.B. Downscaling of radio brightness measurements for soil moisture estimation: A four-dimensional variational data assimilation approach. *Water Resour. Res.* **2001**, *37*, 2353–2364. [CrossRef]
120. De Rosnay, P.; Drusch, M.; Vasiljevic, D.; Balsamo, G.; Albergel, C.; Isaksen, L. A simplified extended Kalman filter for the global operational soil moisture analysis at ECMWF. *Q. J. R. Meteorol. Soc.* **2013**, *139*, 1199–1213. [CrossRef]
121. Lan, X.; Guo, Z.; Tian, Y.; Lei, X.; Wang, J. Review in soil moisture remote sensing estimation based on data assimilation. *Adv. Earth Sci.* **2015**, *30*, 668–679.
122. Montzka, C.; Moradkhani, H.; Weihermüller, L.; Franssen, H.-J.H.; Canty, M.; Vereecken, H. Hydraulic parameter estimation by remotely-sensed top soil moisture observations with the particle filter. *J. Hydrol.* **2011**, *399*, 410–421. [CrossRef]
123. Huang, C.; Li, X.; Lu, L.; Gu, J. Experiments of one-dimensional soil moisture assimilation system based on ensemble Kalman filter. *Remote Sens. Environ.* **2008**, *112*, 888–900. [CrossRef]
124. Brandhorst, N.; Erdal, D.; Neuweiler, I. Soil moisture prediction with the ensemble Kalman filter: Handling uncertainty of soil hydraulic parameters. *Adv. Water Resour.* **2017**, *110*, 360–370. [CrossRef]
125. Balsamo, G.; Bouyssel, F.; Noilhan, J. A simplified bi-dimensional variational analysis of soil moisture from screen-level observations in a mesoscale numerical weather-prediction model. *Q. J. R. Meteorol. Soc. A J. Atmos. Sci. Appl. Meteorol. Phys. Oceanogr.* **2004**, *130*, 895–915. [CrossRef]
126. Srivastava, P.K.; Han, D.; Ricoramirez, M.A.; Oneill, P.E.; Islam, T.; Gupta, M.; Dai, Q. Performance evaluation of WRF-Noah Land surface model estimated soil moisture for hydrological application: Synergistic evaluation using SMOS retrieved soil moisture. *J. Hydrol.* **2015**, *529*, 200–212. [CrossRef]
127. Decker, M.; Zeng, X. Impact of Modified Richards Equation on Global Soil Moisture Simulation in the Community Land Model (CLM3.5). *J. Adv. Model. Earth Syst.* **2009**, *1*. [CrossRef]
128. Liston, G.; Sud, Y.; Walker, G. *Design of a Global Soil Moisture Initialization Procedure for the Simple Biosphere Model*; Technical Memorandum; NASA: Washington, DC, USA, 1993.
129. He, L.; Chen, J.M.; Mostovoy, G.; Gonsamo, A. Soil Moisture Active Passive Improves Global Soil Moisture Simulation in a Land Surface Scheme and Reveals Strong Irrigation Signals Over Farmlands. *Geophys. Res. Lett.* **2021**, *48*, e2021GL092658. [CrossRef]
130. Rodell, M.; Houser, P.; Jambor, U.; Gottschalck, J.; Mitchell, K.; Meng, C.-J.; Arsenault, K.; Cosgrove, B.; Radakovich, J.; Bosilovich, M. The global land data assimilation system. *Bull. Am. Meteorol. Soc.* **2004**, *85*, 381–394. [CrossRef]
131. Spennemann, P.C.; Rivera, J.A.; Saulo, A.C.; Penalba, O.C. A comparison of GLDAS soil moisture anomalies against standardized precipitation index and multisatellite estimations over South America. *J. Hydrometeorol.* **2015**, *16*, 158–171. [CrossRef]
132. Wu, Z.; Feng, H.; He, H.; Zhou, J.; Zhang, Y. Evaluation of Soil Moisture Climatology and Anomaly Components Derived From ERA5-Land and GLDAS-2.1 in China. *Water Resour. Manag.* **2021**, *35*, 629–643. [CrossRef]
133. Ji, L.; Senay, G.B.; Verdin, J.P. Evaluation of the Global Land Data Assimilation System (GLDAS) air temperature data products. *J. Hydrometeorol.* **2015**, *16*, 2463–2480. [CrossRef]
134. Li, M.; Wu, P.; Ma, Z. A comprehensive evaluation of soil moisture and soil temperature from third-generation atmospheric and land reanalysis data sets. *Int. J. Climatol.* **2020**, *40*, 5744–5766. [CrossRef]
135. Mitchell, K.E.; Lohmann, D.; Houser, P.R.; Wood, E.F.; Schaake, J.C.; Robock, A.; Cosgrove, B.A.; Sheffield, J.; Duan, Q.; Luo, L. The multi-institution North American Land Data Assimilation System (NLDAS): Utilizing multiple GCIP products and partners in a continental distributed hydrological modeling system. *J. Geophys. Res. Atmos.* **2004**, *109*. [CrossRef]
136. Cosgrove, B.A.; Lohmann, D.; Mitchell, K.E.; Houser, P.R.; Wood, E.F.; Schaake, J.C.; Robock, A.; Marshall, C.; Sheffield, J.; Duan, Q. Real-time and retrospective forcing in the North American Land Data Assimilation System (NLDAS) project. *J. Geophys. Res. Atmos.* **2003**, *108*, 8842. [CrossRef]
137. Van den Hurk, B. Overview of the European Land Data Assimilation System (ELDAS) Project. *AGU Fall Meet. Abstr.* **2002**, *2002*, H62D-0886.
138. Jacobs, C.; Moors, E.; Ter Maat, H.; Teuling, A.; Balsamo, G.; Bergaoui, K.; Ettema, J.; Lange, M.; Van Den Hurk, B.; Viterbo, P. Evaluation of European Land Data Assimilation System (ELDAS) products using in situ observations. *Tellus A Dyn. Meteorol. Oceanogr.* **2008**, *60*, 1023–1037. [CrossRef]
139. Shi, C.; Xie, Z.; Qian, H.; Liang, M.; Yang, X. China land soil moisture EnKF data assimilation based on satellite remote sensing data. *Sci. China Earth Sci.* **2011**, *54*, 1430–1440. [CrossRef]
140. Shi, C.; Jiang, L.; Zhang, T.; Xu, B.; Han, S. Status and plans of CMA land data assimilation system (CLDAS) project. In Proceedings of the EGU General Assembly Conference, Vienna, Austria, 27 April–2 May 2014; p. 5671.
141. Hasenauer, S.; Wagner, W.; Scipal, K.; Naeimi, V.; Bartalis, Z. *Implementation of Near Real-Time Soil Moisture Products in the SAF Network Based on MetOp ASCAT Data*; Citeseer: Princeton, NJ, USA, 2006.
142. Kistler, R.; Kalnay, E.; Collins, W.; Saha, S.; White, G.; Woollen, J.; Chelliah, M.; Ebisuzaki, W.; Kanamitsu, M.; Kousky, V. The NCEP–NCAR 50-year reanalysis: Monthly means CD-ROM and documentation. *Bull. Am. Meteorol. Soc.* **2001**, *82*, 247–268. [CrossRef]
143. Kalnay, E.; Kanamitsu, M.; Kistler, R.; Collins, W.; Deaven, D.; Gandin, L.; Iredell, M.; Saha, S.; White, G.; Woollen, J. The NCEP/NCAR 40-year reanalysis project. *Bull. Am. Meteorol. Soc.* **1996**, *77*, 437–472. [CrossRef]

144. Saha, S.; Nadiga, S.; Thiaw, C.; Wang, J.; Wang, W.; Zhang, Q.; Van den Dool, H.; Pan, H.-L.; Moorthi, S.; Behringer, D. The NCEP climate forecast system. *J. Clim.* **2006**, *19*, 3483–3517. [CrossRef]
145. Hersbach, H.; Bell, B.; Berrisford, P.; Hirahara, S.; Horányi, A.; Muñoz-Sabater, J.; Nicolas, J.; Peubey, C.; Radu, R.; Schepers, D. The ERA5 global reanalysis. *Q. J. R. Meteorol. Soc.* **2020**, *146*, 1999–2049. [CrossRef]
146. Hoffmann, L.; Günther, G.; Li, D.; Stein, O.; Wu, X.; Griessbach, S.; Heng, Y.; Konopka, P.; Müller, R.; Vogel, B. From ERA-Interim to ERA5: The considerable impact of ECMWF's next-generation reanalysis on Lagrangian transport simulations. *Atmos. Chem. Phys.* **2019**, *19*, 3097–3124. [CrossRef]
147. Rienecker, M.M.; Suarez, M.J.; Gelaro, R.; Todling, R.; Bacmeister, J.; Liu, E.; Bosilovich, M.G.; Schubert, S.D.; Takacs, L.; Kim, G.-K. MERRA: NASA's modern-era retrospective analysis for research and applications. *J. Clim.* **2011**, *24*, 3624–3648. [CrossRef]
148. Saito, K.; Fujita, T.; Yamada, Y.; Ishida, J.-i.; Kumagai, Y.; Aranami, K.; Ohmori, S.; Nagasawa, R.; Kumagai, S.; Muroi, C. The operational JMA nonhydrostatic mesoscale model. *Mon. Weather. Rev.* **2006**, *134*, 1266–1298. [CrossRef]
149. Liu, Z.; Shi, C.; Zhou, Z.; Jiang, L.; Liang, X.; Zhang, T.; Liao, J.; Liu, J.; Wang, M.; Yao, S. CMA Global Reanalysis (CRA-40): Status and Plans. In Proceedings of the 5th International Conference on Reanalysis, Rome, Italy, 13–17 November 2017; pp. 13–17.
150. Liang, X.; Jiang, L.; Pan, Y.; Shi, C.; Liu, Z.; Zhou, Z. A 10-Yr Global Land Surface Reanalysis Interim Dataset (CRA-Interim/Land): Implementation and Preliminary Evaluation. *J. Meteorol. Res.* **2020**, *34*, 101–116. [CrossRef]
151. Lary, D.J.; Alavi, A.H.; Gandomi, A.H.; Walker, A.L. Machine learning in geosciences and remote sensing. *Geosci. Front.* **2016**, *7*, 3–10. [CrossRef]
152. Jing, W.; Zhang, P.; Zhao, X. Reconstructing Monthly ECV Global Soil Moisture with an Improved Spatial Resolution. *Water Resour. Manag.* **2018**, *32*, 2523–2537. [CrossRef]
153. Srivastava, P.K.; Han, D.; Ramirez, M.R.; Islam, T. Machine Learning Techniques for Downscaling SMOS Satellite Soil Moisture Using MODIS Land Surface Temperature for Hydrological Application. *Water Resour. Manag.* **2013**, *27*, 3127–3144. [CrossRef]
154. Liu, Y.; Xia, X.; Yao, L.; Jing, W.; Zhou, C.; Huang, W.; Li, Y.; Yang, J. Downscaling Satellite Retrieved Soil Moisture Using Regression Tree-based Machine Learning Algorithms Over Southwest France. *Earth Space Sci.* **2020**, *7*, e2020EA001267. [CrossRef]
155. Sabaghy, S.; Walker, J.P.; Renzullo, L.J.; Jackson, T.J. Spatially enhanced passive microwave derived soil moisture: Capabilities and opportunities. *Remote Sens. Environ.* **2018**, *209*, 551–580. [CrossRef]
156. Karthikeyan, L.; Mishra, A.K. Multi-layer high-resolution soil moisture estimation using machine learning over the United States. *Remote Sens. Environ.* **2021**, *266*, 112706. [CrossRef]
157. Wei, Z.; Meng, Y.; Zhang, W.; Peng, J.; Meng, L. Downscaling SMAP soil moisture estimation with gradient boosting decision tree regression over the Tibetan Plateau. *Remote Sens. Environ.* **2019**, *225*, 30–44. [CrossRef]
158. Abbaszadeh, P.; Moradkhani, H.; Zhan, X. Downscaling SMAP radiometer soil moisture over the CONUS using an ensemble learning method. *Water Resour. Res.* **2019**, *55*, 324–344. [CrossRef]
159. Jia, Y.; Jin, S.; Savi, P.; Gao, Y.; Tang, J.; Chen, Y.; Li, W. GNSS-R soil moisture retrieval based on a XGboost machine learning aided method: Performance and validation. *Remote Sens.* **2019**, *11*, 1655. [CrossRef]
160. Liu, Y.; Yang, Y.; Jing, W.; Yue, X. Comparison of Different Machine Learning Approaches for Monthly Satellite-Based Soil Moisture Downscaling over Northeast China. *Remote Sens.* **2017**, *10*, 31. [CrossRef]
161. Zhao, W.; Sánchez, N.; Lu, H.; Li, A. A spatial downscaling approach for the SMAP passive surface soil moisture product using random forest regression. *J. Hydrol.* **2018**, *563*, 1009–1024. [CrossRef]
162. Reichstein, M.; Camps-Valls, G.; Stevens, B.; Jung, M.; Denzler, J.; Carvalhais, N. Deep learning and process understanding for data-driven Earth system science. *Nature* **2019**, *566*, 10. [CrossRef] [PubMed]
163. Deng, L.; Yu, D. *Deep Learning: Methods and Applications*; Now Publishers: Norwell, MA, USA, 2014; Volume 7, pp. 197–387.
164. Lecun, Y.; Bengio, Y.; Hinton, G. Deep learning. *Nature* **2015**, *521*, 436–444. [CrossRef] [PubMed]
165. Kamilaris, A.; Prenafeta-Boldú, F.X. Deep learning in agriculture: A survey. *Comput. Electron. Agric.* **2018**, *147*, 70–90. [CrossRef]
166. Liu, J.; Rahmani, F.; Lawson, K.; Shen, C. A Multiscale Deep Learning Model for Soil Moisture Integrating Satellite and In Situ Data. *Geophys. Res. Lett.* **2022**, *49*, e2021GL096847. [CrossRef]
167. Li, Q.; Wang, Z.; Shangguan, W.; Li, L.; Yao, Y.; Yu, F. Improved daily SMAP satellite soil moisture prediction over China using deep learning model with transfer learning. *J. Hydrol.* **2021**, *600*, 126698. [CrossRef]
168. Aa, Y.; Wang, G.; Hu, P.; Lai, X.; Xue, B.; Fang, Q. Root-zone soil moisture estimation based on remote sensing data and deep learning. *Environ. Res.* **2022**, *212*, 113278. [CrossRef]
169. Zhao, H.; Li, J.; Yuan, Q.; Lin, L.; Yue, L.; Xu, H. Downscaling of soil moisture products using deep learning: Comparison and analysis on Tibetan Plateau. *J. Hydrol.* **2022**, *607*, 127570. [CrossRef]
170. Mao, H.; Kathuria, D.; Duffield, N.; Mohanty, B.P. Gap Filling of High-Resolution Soil Moisture for SMAP/Sentinel-1: A Two-layer Machine Learning-based Framework. *Water Resour. Res.* **2019**, *55*, 6986–7009. [CrossRef]
171. Fang, K.; Shen, C.; Kifer, D.; Yang, X. Prolongation of SMAP to spatiotemporally seamless coverage of continental US using a deep learning neural network. *Geophys. Res. Lett.* **2017**, *44*, 11030–11039. [CrossRef]
172. Almendra-Martín, L.; Martínez-Fernández, J.; Piles, M.; González-Zamora, Á. Comparison of gap-filling techniques applied to the CCI soil moisture database in Southern Europe. *Remote Sens. Environ.* **2021**, *258*, 112377. [CrossRef]
173. Long, D.; Bai, L.; Yan, L.; Zhang, C.; Yang, W.; Lei, H.; Quan, J.; Meng, X.; Shi, C. Generation of spatially complete and daily continuous surface soil moisture of high spatial resolution. *Remote Sens. Environ.* **2019**, *233*, 111364. [CrossRef]

174. Lee, C.S.; Sohn, E.; Park, J.D.; Jang, J.-D. Estimation of soil moisture using deep learning based on satellite data: A case study of South Korea. *GIScience Remote Sens.* **2019**, *56*, 43–67. [CrossRef]
175. Ahmed, A.; Deo, R.C.; Raj, N.; Ghahramani, A.; Feng, Q.; Yin, Z.; Yang, L. Deep Learning Forecasts of Soil Moisture: Convolutional Neural Network and Gated Recurrent Unit Models Coupled with Satellite-Derived MODIS, Observations and Synoptic-Scale Climate Index Data. *Remote Sens.* **2021**, *13*, 554. [CrossRef]
176. Qin, J.; Yang, K.; Lu, N.; Chen, Y.; Zhao, L.; Han, M. Spatial upscaling of in-situ soil moisture measurements based on MODIS-derived apparent thermal inertia. *Remote Sens. Environ.* **2013**, *138*, 1–9. [CrossRef]
177. Clewley, D.; Whitcomb, J.B.; Akbar, R.; Silva, A.R.; Berg, A.; Adams, J.R.; Caldwell, T.; Entekhabi, D.; Moghaddam, M. A method for upscaling in situ soil moisture measurements to satellite footprint scale using random forests. *IEEE J. Sel. Top. Appl. Earth Obs. Remote Sens.* **2017**, *10*, 2663–2673. [CrossRef]
178. Zhang, D.; Zhang, W.; Huang, W.; Hong, Z.; Meng, L. Upscaling of surface soil moisture using a deep learning model with VIIRS RDR. *ISPRS Int. J. Geo-Inf.* **2017**, *6*, 130. [CrossRef]
179. Rossing, W.; Zander, P.; Josien, E.; Groot, J.; Meyer, B.; Knierim, A. Integrative modelling approaches for analysis of impact of multifunctional agriculture: A review for France, Germany and The Netherlands. *Agric. Ecosyst. Environ.* **2007**, *120*, 41–57. [CrossRef]
180. Van der Veer Martens, B.; Illston, B.G.; Fiebrich, C.A. The Oklahoma Mesonet: A Pilot Study of Environmental Sensor Data Citations. *Data Sci. J.* **2017**, *16*, 47. [CrossRef]
181. Ghulam, A.; Qin, Q.; Teyip, T.; Li, Z.-L. Modified perpendicular drought index (MPDI): A real-time drought monitoring method. *ISPRS J. Photogramm. Remote Sens.* **2007**, *62*, 150–164. [CrossRef]
182. Liu, D.; Mishra, A.K.; Yu, Z.; Yang, C.; Konapala, G.; Vu, T. Performance of SMAP, AMSR-E and LAI for weekly agricultural drought forecasting over continental United States. *J. Hydrol.* **2017**, *553*, 88–104. [CrossRef]
183. Park, S.; Im, J.; Park, S.; Rhee, J. Drought monitoring using high resolution soil moisture through multi-sensor satellite data fusion over the Korean peninsula. *Agric. For. Meteorol.* **2017**, *237–238*, 257–269. [CrossRef]
184. Do, N.; Kang, S. Assessing drought vulnerability using soil moisture-based water use efficiency measurements obtained from multi-sensor satellite data in Northeast Asia dryland regions. *J. Arid. Environ.* **2014**, *105*, 22–32. [CrossRef]
185. Enenkel, M.; Steiner, C.; Mistelbauer, T.; Dorigo, W.; Wagner, W.; See, L.; Atzberger, C.; Schneider, S.; Rogenhofer, E. A Combined Satellite-Derived Drought Indicator to Support Humanitarian Aid Organizations. *Remote Sens.* **2016**, *8*, 340. [CrossRef]
186. Pedersen, J.S.T.; Santos, F.D.; van Vuuren, D.; Gupta, J.; Coelho, R.E.; Aparício, B.A.; Swart, R. An assessment of the performance of scenarios against historical global emissions for IPCC reports. *Glob. Environ. Chang.* **2021**, *66*, 102199. [CrossRef]
187. Dorigo, W.; de Jeu, R.; Chung, D.; Parinussa, R.; Liu, Y.; Wagner, W.; Fernández-Prieto, D. Evaluating global trends (1988–2010) in harmonized multi-satellite surface soil moisture. *Geophys. Res. Lett.* **2012**, *39*. [CrossRef]
188. Qiu, J.; Gao, Q.; Wang, S.; Su, Z. Comparison of temporal trends from multiple soil moisture data sets and precipitation: The implication of irrigation on regional soil moisture trend. *Int. J. Appl. Earth Obs. Geoinf.* **2016**, *48*, 17–27. [CrossRef]
189. Pan, N.; Wang, S.; Liu, Y.; Zhao, W.; Fu, B. Global Surface Soil Moisture Dynamics in 1979–2016 Observed from ESA CCI SM Dataset. *Water* **2019**, *11*, 883. [CrossRef]
190. Seneviratne, S.I.; Corti, T.; Davin, E.L.; Hirschi, M.; Jaeger, E.B.; Lehner, I.; Orlowsky, B.; Teuling, A.J. Investigating soil moisture–climate interactions in a changing climate: A review. *Earth-Sci. Rev.* **2010**, *99*, 125–161. [CrossRef]
191. Rodriguez-Iturbe, I.; D'odorico, P.; Porporato, A.; Ridolfi, L. On the spatial and temporal links between vegetation, climate, and soil moisture. *Water Resour. Res.* **1999**, *35*, 3709–3722. [CrossRef]
192. Pastor, J.; Post, W. Influence of climate, soil moisture, and succession on forest carbon and nitrogen cycles. *Biogeochemistry* **1986**, *2*, 3–27. [CrossRef]
193. Li, M.; Wu, P.; Sexton, D.M.; Ma, Z. Potential shifts in climate zones under a future global warming scenario using soil moisture classification. *Clim. Dyn.* **2021**, *56*, 2071–2092. [CrossRef]
194. Pereira, L.S.; Allen, R.G.; Smith, M.; Raes, D. Crop evapotranspiration estimation with FAO56: Past and future. *Agric. Water Manag.* **2015**, *147*, 4–20. [CrossRef]
195. Allam, M.M.; Jain Figueroa, A.; McLaughlin, D.B.; Eltahir, E.A. Estimation of evaporation over the upper blue nile basin by combining observations from satellites and river flow gauges. *Water Resour. Res.* **2016**, *52*, 644–659. [CrossRef]
196. Martens, B.; Miralles, D.G.; Lievens, H.; Van Der Schalie, R.; De Jeu, R.A.; Fernández-Prieto, D.; Beck, H.E.; Dorigo, W.A.; Verhoest, N.E. GLEAM v3: Satellite-based land evaporation and root-zone soil moisture. *Geosci. Model Dev.* **2017**, *10*, 1903–1925. [CrossRef]
197. Koster, R.D.; Dirmeyer, P.A.; Zhichang, G.; Gordon, B.; Edmond, C.; Peter, C.; Gordon, C.T.; Shinjiro, K.; Eva, K.; David, L. Regions of strong coupling between soil moisture and precipitation. *Science* **2004**, *305*, 1138–1140. [CrossRef]
198. Koster, R.D.; Suarez, M.J.; Higgins, R.W.; Van den Dool, H.M. Observational evidence that soil moisture variations affect precipitation. *Geophys. Res. Lett.* **2003**, *30*. [CrossRef]
199. Brocca, L.; Moramarco, T.; Melone, F.; Wagner, W. A new method for rainfall estimation through soil moisture observations. *Geophys. Res. Lett.* **2013**, *40*, 853–858. [CrossRef]
200. Brocca, L.; Ciabatta, L.; Massari, C.; Moramarco, T.; Hahn, S.; Hasenauer, S.; Kidd, R.; Dorigo, W.; Wagner, W.; Levizzani, V. Soil as a natural rain gauge: Estimating global rainfall from satellite soil moisture data. *J. Geophys. Res. Atmos.* **2014**, *119*, 5128–5141. [CrossRef]

201. Swenson, S.; Famiglietti, J.; Basara, J.; Wahr, J. Estimating profile soil moisture and groundwater variations using GRACE and Oklahoma Mesonet soil moisture data. *Water Resour. Res.* **2008**, *44*. [CrossRef]
202. Sutanudjaja, E.; Van Beek, L.; De Jong, S.; Van Geer, F.; Bierkens, M. Calibrating a large-extent high-resolution coupled groundwater-land surface model using soil moisture and discharge data. *Water Resour. Res.* **2014**, *50*, 687–705. [CrossRef]
203. Merz, B.; Plate, E.J. An analysis of the effects of spatial variability of soil and soil moisture on runoff. *Water Resour. Res.* **1997**, *33*, 2909–2922. [CrossRef]
204. Brocca, L.; Melone, F.; Moramarco, T.; Wagner, W.; Naeimi, V.; Bartalis, Z.; Hasenauer, S. Improving runoff prediction through the assimilation of the ASCAT soil moisture product. *Hydrol. Earth Syst. Sci.* **2010**, *14*, 1881–1893. [CrossRef]
205. Tramblay, Y.; Bouvier, C.; Martin, C.; Didon-Lescot, J.-F.; Todorovik, D.; Domergue, J.-M. Assessment of initial soil moisture conditions for event-based rainfall–runoff modelling. *J. Hydrol.* **2010**, *387*, 176–187. [CrossRef]
206. Reichstein, M.; Bahn, M.; Ciais, P.; Frank, D.; Mahecha, M.D.; Seneviratne, S.I.; Zscheischler, J.; Beer, C.; Buchmann, N.; Frank, D.C. Climate extremes and the carbon cycle. *Nature* **2013**, *500*, 287–295. [CrossRef]
207. Li, C.; Fu, B.; Wang, S.; Stringer, L.C.; Wang, Y.; Li, Z.; Liu, Y.; Zhou, W. Drivers and impacts of changes in China's drylands. *Nat. Rev. Earth Environ.* **2021**, *2*, 858–873. [CrossRef]
208. Reich, P.B.; Sendall, K.M.; Stefanski, A.; Rich, R.L.; Hobbie, S.E.; Montgomery, R.A. Effects of climate warming on photosynthesis in boreal tree species depend on soil moisture. *Nature* **2018**, *562*, 263–267. [CrossRef]
209. Chen, T.; Werf, G.; Jeu, R.d.; Wang, G.; Dolman, A. A global analysis of the impact of drought on net primary productivity. *Hydrol. Earth Syst. Sci.* **2013**, *17*, 3885–3894. [CrossRef]
210. Churkina, G.; Running, S.W.; Schloss, A.L.; the participants of the Potsdam NPP Model Intercomparison. Comparing global models of terrestrial net primary productivity (NPP): The importance of water availability. *Glob. Chang. Biol.* **1999**, *5*, 46–55. [CrossRef]
211. Abatzoglou, J.T.; Dobrowski, S.Z.; Parks, S.A.; Hegewisch, K.C. TerraClimate, a high-resolution global dataset of monthly climate and climatic water balance from 1958–2015. *Sci. Data* **2018**, *5*, 1–12. [CrossRef]
212. Rigden, A.J.; Salvucci, G.D. Stomatal response to humidity and CO 2 implicated in recent decline in US evaporation. *Glob. Change Biol.* **2017**, *23*, 1140–1151. [CrossRef] [PubMed]
213. Xu, C.; Qu, J.J.; Hao, X.; Cosh, M.H.; Prueger, J.H.; Zhu, Z.; Gutenberg, L. Downscaling of Surface Soil Moisture Retrieval by Combining MODIS/Landsat and In Situ Measurements. *Remote Sens.* **2018**, *10*, 210. [CrossRef]
214. Zhang, P.; Hu, X.; Lu, Q.; Zhu, A.; Lin, M.; Sun, L.; Chen, L.; Xu, N. *FY-3E: The First Operational Meteorological Satellite Mission in an Early Morning Orbit*; Springer: Berlin/Heidelberg, Germany, 2021.
215. Falloon, P.; Jones, C.D.; Ades, M.; Paul, K.J.G.B.C. Direct soil moisture controls of future global soil carbon changes: An important source of uncertainty. *Glob. Biogeochem. Cycles* **2011**, *25*. [CrossRef]
216. Mladenova, I.E.; Bolten, J.D.; Crow, W.T.; Sazib, N.; Cosh, M.H.; Tucker, C.J.; Reynolds, C. Evaluating the operational application of SMAP for global agricultural drought monitoring. *IEEE J. Sel. Top. Appl. Earth Obs. Remote Sens.* **2019**, *12*, 3387–3397. [CrossRef]
217. Reichle, R.H.; Liu, Q.; Ardizzone, J.V.; Crow, W.T.; De Lannoy, G.J.; Dong, J.; Kimball, J.S.; Koster, R.D. The contributions of gauge-based precipitation and SMAP brightness temperature observations to the skill of the SMAP Level-4 soil moisture product. *J. Hydrometeorol.* **2021**, *22*, 405–424. [CrossRef]
218. Reichle, R.H.; De Lannoy, G.J.M.; Liu, Q.; Ardizzone, J.V.; Colliander, A.; Conaty, A.; Crow, W.; Jackson, T.J.; Jones, L.A.; Kimball, J.S.; et al. Assessment of the SMAP Level-4 Surface and Root-Zone Soil Moisture Product Using In Situ Measurements. *J. Hydrometeorol.* **2017**, *18*, 2621–2645. [CrossRef]
219. O'Neill, P.; Entekhabi, D.; Njoku, E.; Kellogg, K. The NASA Soil Moisture Active Passive (SMAP) Mission: Overview. In Proceedings of the Geoscience and Remote Sensing Symposium, Honolulu, HI, USA, 25–30 July 2010; pp. 704–716.

MDPI
St. Alban-Anlage 66
4052 Basel
Switzerland
www.mdpi.com

Remote Sensing Editorial Office
E-mail: remotesensing@mdpi.com
www.mdpi.com/journal/remotesensing

www.ingramcontent.com/pod-product-compliance
Lightning Source LLC
LaVergne TN
LVHW070104170726
843515LV00007B/1606

* 9 7 8 3 0 3 6 5 9 0 9 4 3 *